# The Marine Shells of the West Coast of North America

## VOL. II

### PART III

BY

IDA SHEPARD OLDROYD

Curator of the Geological Museum

STANFORD UNIVERSITY PRESS
STANFORD, CALIFORNIA

Stanford University Press
Stanford, California
Printed in the United States of America
ISBN 0-8047-0987-4
LC 26-20866
Original edition 1927
Reissued in 1978
Last figure below indicates year of this printing:
87 86 85 84 83 82 81 80 79 78

# PREFACE TO PART III

Volume II contains the Marine Gastropods and is in three parts, of which this is the third and last part.

Many of the plates here used are from the *Proceedings* and *Bulletins* of the United States National Museum. In some cases species for comparison from Lower California and south of there are included on the plates, but are not described in the text.

The United States National Museum had photographs made of some of the unfigured species to use in the plates. The American Museum of Natural History, New York City, also had photographs made of some of the type figures. Many of the figures in *Bulletin 112* and in some of the *Proceedings* have been refigured for use in making up the plates.

The line or lines preceding the description of a shell in the text give the reference to the original description and figure.

Acknowledgments are given in the Preface to Part I.

I. S. Oldroyd

June 15, 1927

# CONTENTS OF VOLUME II, PART III

# RANGE OF SPECIES

## CIRCUMBOREAL SPECIES

Amicula vestitus
Hanleya hanleyi
Lepidochitona sharpei
Lepidopleurus cancellatus
Margarites cinereus
Margarites helicinus

Margarites marginatus
Margarites olivaceus
Margarites sordidus
Margarites umbilicalis
Natica affinis
Natica aleutica

Polinices gronlandicus
Polinices nanus
Solariella obscura
Solariella varicosa
Velutina laevigata
Velutina zonata

## RANGE TO GULF OF CALIFORNIA AND LOWER CALIFORNIA

Acanthochiton avicula
Acmaea asmi
Acmaea depicta
Acmaea, v. funiculata
Acmaea incessa
Acmaea limatula
Acmaea paleacea
Acmaea, v. pelta
Acmaea persona
Acmaea rosacea
Acmaea scabra
Acmaea spectrum
Acmaea triangularis
Astraea undosa
Calliostoma eximium
Calliostoma gemmulatum
Callistochiton decoratus
Callistochiton palmulatus
Callistochiton,
v. punctocostatus
Calyptraea contorta
Chaetopleura beanii
Chaetopleura gemmea
Chaetopleura parallela
Chaetopleura prasinata
Chaetopleura thamnopora
Cidarina ceratophora
Cidarina cidaris
Cidarina equatorialis
Crepidula exuviata
Crepidula, v. fimbriata
Crepidula nummaria
Cyclostrema baldridgae
Diadora aspera

Diadora, v. densiclathrata
Diadora murina
Fissurella, v. crucifera
Haliotis, v. bonita
Haliotis californiensis
Haliotis corrugata
Haliotis cracherodii
Haliotis fulgens
Haliotis, v. holzneri
Haliotis rufescens
Ischnochiton acrior
Ischnochiton, v. acutior
Ischnochiton aureotinctus
Ischnochiton clathratus
Ischnochiton corrugatus
Ischnochiton fallax
Ischnochiton magdalenensis
Ischnochiton sarcosus
Ischnochiton scabricostatus
Ischnochiton serratus
Lamellaria rhombica
Lepidochitona dentiens
Lepidochitona hartwegii
Lepidochitona, v. nuttalli
Lepidopleurus farallonis
Lepidopleurus rugatus
Leptothyra bacula
Leptothyra lurida
Liotia acuticostata
Liotia cookeana
Lottia gigantea
Lucapinella callomarginata
Megatebennus bimaculatus

Megathura crenulata
Mopalia ciliata
Mopalia, v. hindsii
Mopalia, v. lignosa
Mopalia muscosa
Natica salimba
Norrisia norrisii
Nuttallana fluxa
Oldroydia percrassa
Pallochiton lanuginosus
Phasianella compta
Phasianella pulloides
Phasianella, v. punctulata
Phasianella typica
Placiphorella velata
Placiphorella stimpsoni
Polinices recluzianus
Puncturella cucullata
Scissilabra dalli
Sinum californicum
Sinum debile
Tegula aureotincta
Tegula funebralis
Tegula impressa
Tegula ligulata
Tegula regina
Tegula rugosa
Tegula, v. tincta
Tegula, v. umbilicata
Teinostoma invallata
Teinostoma supravallata
Turcica caffea
Vitrinella oldroydi

## RANGE TO JAPAN

| | | |
|---|---|---|
| Chlamydochiton amiculatus | Lepidochitona marmorea | Natica clausa |
| Cryptochiton stelleri | Lepidochitona submarmorea | Natica janthostoma |
| Haliotis kamtschatkana | Margarites sordidus | Solariella peramabilis |

## RANGE TO PHILIPPINES

Lepidopleurus belknapi

# LIST OF FAMILIES IN PART III

Volume II, Part III contains the following families:

| | | |
|---|---|---|
| Cerithiidae | Synceratidae | Turbinidae |
| Trichotropidae | Capulidae | Liotiidae |
| Caecidae | Hipponicidae | Trochidae |
| Vermetidae | Calyptraeidae | Vitrinellidae |
| Turritellidae | Crepidulidae | Scissurellidae |
| Littorinidae | Naticidae | Haliotidae |
| Lacunidae | Vanicoroidae | Fissurellidae |
| Fossaridae | Lamellariidae | Lepidopleuridae |
| Litiopidae | Velutinidae | Lepidochitonidae |
| Rissoidae | Lepetidae | Ischnochitonidae |
| Rissoinidae | Acmaeidae | Chitonidae |
| Anaplocamidae | Cocculinidae | Cryptochitonidae |
| Truncatellidae | Phasianellidae | Mopaliidae |

# CLASS GASTROPODA *(Concluded)*

## Family CERITHIIDAE

### Genus **DIASTOMA** Deshayes, 1861

Shell turreted, whorls with numerous transverse ribs, and with a few intermediate varices. Inner margin of the aperture partially detached from the previous whorl; the aperture itself is strongly contracted posteriorly. (Tryon, *Structural and Systematic Conchology.*)

TYPE. ?

DISTRIBUTION. California, Pleistocene, of Santa Barbara, California.

### Diastoma oldroydae Bartsch, 1911

### Plate 73, fig. 2

*Proceedings of the United States National Museum,* **39**:583; fig. 3.

Shell broadly conic, light brown. Nuclear whorls decollated, postnuclear whorls slightly rounded, shouldered at the summit, marked by weak, axial ribs, of which sixteen occur upon the fourth and fifth, eighteen upon the sixth and penultimate turn. Intercostal spaces about twice as wide as the ribs. In addition to the axial ribs the whorls are crossed by four slender, poorly developed, spiral cords, the intersection of which, with the ribs, renders them feebly nodulose. On the last whorl a slender, additional thread appears between the cord at the summit and the one adjacent to it. The spaces inclosed between the ribs and the spiral cords are rectangular, depressed areas, having their long axes parallel with their spiral sculpture. Periph. ry of the last whorl marked by a shallow sulcus. Base moderately long, well-rounded, marked by six equal and equally spaced, spiral cords, the axial sculpture being represented by incremental lines only. Sutures channeled. Aperture irregular, large, decidedly channeled and somewhat twisted anteriorly; posterior angle obtuse; outer lip moderately thin, showing the external sculpture within by transmitted light; columella very strongly curved and revolute; parietal wall covered with a thick callus which renders the peritreme complete. A strong, white varix is present about a quarter turn behind the lip. Length, 3.8; diameter, 1.6 mm. (Bartsch.)

TYPE in United States National Museum, No. 213024. Type locality, San Pedro, California.

RANGE. Known only from type locality.

[ 9

## Diastoma stearnsi Bartsch, 1911

### Plate 73, fig. 3

*Proceedings of the United States National Museum,* **39**:584; fig. 4.

Shell elongate-conic. Early whorls and columella light chestnut brown, the remaining creamy white. Nuclear whorls decollated, only the last two volutions remaining, which are well-rounded and smooth. Post-nuclear whorls strongly rounded, marked by low, well-rounded, vertical axial ribs which are about as wide as the intercostal spaces; of these ribs sixteen occur upon the first, and twenty-four upon the remaining whorls. In addition to these ribs, the whorls are marked by seven spiral cords, which are about as wide as the spaces that separate them and render the axial ribs nodulose. The cord at the summit is a little wider than the rest, likewise the space that separates it from the adjacent cord and this renders the summit of the whorls crenulated. The first subperipheral cord is apparent on all the whorls and forms the seventh one, between the sutures. The axial ribs usually encroach upon this only feebly, the nodulations therefore being less marked than on the spiral cords posterior to it. Sutures strongly constricted; periphery of the last whorl and base well-rounded, the later marked by six equal and equally spaced spiral cords which are a little less than the spaces that separate them. Aperture not quite complete in our specimen, broadly ovate, channeled anteriorly; posterior angle acute; outer lip thin, showing the external markings within; columella moderately strong, somewhat twisted, curved, decidedly reflected; parietal wall glazed with a thick callus. Length, 9; diameter, 3 mm. (Bartsch.)

TYPE in United States National Museum, No. 32212. Type locality, San Diego, California.

RANGE. Known only from type locality.

## Diastoma fastigiata Carpenter, 1864

### Plate 73, fig. 1

*Supplementary Report,* British Association for the Advancement of Science, 655. *Annals and Magazine of Natural History,* series 3, 283. *Proceedings of the United States National Museum,* **39**: 581; fig. 1, 1911.

B. testa parva, gracili, pallide rufo-cinerea, marginibus spirae vix excurvatis; anfr. nucl. iii., laevibus, tumidis, apice acuto; norm. ix., planatis, suturis alte impressis; anfr. primis iii., carinatis, postea costis radiantibus circ. xiii, obtusis, satis extantibus, ad suturas interruptis, interstitiis undatis, liris spiralibus iv., in spira se monstrantibus, costas undatim superantibus, quarum antica in testa jun. plerumque extat; anfr. ultimo parum contracto, basi elongata, liris spiralibus vi., contiguis ornata; apertura gibbosa; labro acuto, interdum varicoso, antice angulatim emarginato; labio tenui. Long., 25; long. spir., 19; lat., 09. poll. (Carpenter.)

Shell elongate-conic, yellowish white. Nuclear whorls two, well-rounded, smooth. Post-nuclear whorls flattened, much wider at the periphery than at the summit, overhanging, ornamented on the first seven whorls by four equal and equally spaced spiral cords, of which the posterior is at the summit and the anterior at some little distance above the suture; on the last whorl an additional slender cord appears between that on the summit and its neighbor. In addition to these spiral cords, the whorls are marked by strong, broad, low, axial ribs, of which twelve occur upon the first and second, fourteen upon the third and fourth, and sixteen upon each of the remaining whorls. The intersections of the spiral cords and axial ribs are nodulose. In addition to these cords and ribs, the entire surface of the shell is marked by numerous, very fine, spiral striations and slender lines of growth. Periphery of the last whorl marked by a sulcus which is as wide as the space between the cords on the spire and is crossed by the slender continuations of the axial ribs. Base well-rounded, marked by eight spiral cords. Aperture elongate, decidedly effuse at the junction of the outer and basal lips, channeled anteriorly and subchanneled at the posterior angle; outer lip thin, rendered somewhat sinuous by external sculpture; columella short and twisted; parietal wall covered with a thick callus, which extends over the edge of the columella and renders the peritreme complete. In the two specimens before me a strong varix is present a fourth of a turn behind the lip. Length, 4.7; diameter, 1.5 mm. (Bartsch.)

TYPE. Specimen in United States National Museum, No. 162561. Type locality, Lower Pleistocene beds at Santa Barbara, California.

RANGE. Known only from type locality.

## Genus **ALABINA** Dall, 1902

Shell conic, small, of numerous whorls; aperture oval, outer lip thin, columella short, moderately strong, strongly curved and slightly reflected at the base. Canal undeveloped. (Oldroyd.) Dr. Dall named the genus but did not describe it or cite a type, hence *Mesalia tenuisculpta* Carpenter, 1863, the first species mentioned under the new genus, is chosen as the type.

DISTRIBUTION. California to Lower California.

## Alabina calena Dall, 1919

*Proceedings of the United States National Museum,* **56**:345.

Shell small, with a very pale olivaceous periostracum, acute, slender, with about eight rounded whorls, the apex eroded; suture distinct, deep, marked by a small thread against which it is laid; spiral sculpture of (on the spire three) flattish cords with subequal channeled interspaces, on the last whorl four, the third at the periphery largest; axial sculpture on the upper spire of obscure radials which undulate the cords but do not show in the interspaces, and later are obsolete; there are also very fine silky regular incremental lines; the base is discoid, slightly flattish, imperforate, the aperture ovate, simple, the outer lip thin sharp, the inner lip white glazed; the lippar short, thick, and attenuated, the lip beyond it produced, patulous. Height, 10; diameter, 4 mm. (Dall.)

TYPE in United States National Museum, No. 271070. Type locality, U. S. Fish Commission Station 3195, off San Luis Obispo Bay in 252 fathoms.

RANGE. Known only from type locality.

## Alabina californica Dall and Bartsch, 1901

### Plate 95, fig. 7

*Nautilus,* **15**:58–59. *Proceedings of the United States National Museum,* **39**: Pl. 62, fig. 7.

Shell elongate-conic, yellowish white. Nuclear whorls two and one-tenth, moderately rounded, smooth. Post-nuclear whorls strongly rounded, somewhat inflated, appressed at the summit, marked by broad, low, strong, protractive, axial ribs, of which 14 occur upon the first to third, 16 upon the fourth, and 18 upon the penultimate whorl. In addition to the axial ribs, the whorls are marked on the spire by three feebly impressed, spiral lines which pass over the ribs as well as the broad, intercostal spaces. The middle one of these three lines is halfway between the sutures. The other two divide space anterior and posterior to it into equal halves. Sutures strongly constricted. Periphery and base of the last whorl well-rounded, smooth, excepting faint lines of growth. Aperture broadly ovate; posterior angle acute; outer lip thin; columella short, moderately strong, strongly curved and slightly reflected over the reinforcing base. Length, 5.3; diameter, 2.2 mm. (Dall and Bartsch.)

TYPE in United States National Museum, No. 162548. Type locality, Deadman Island, San Pedro.

RANGE. Known only from the type locality.

Described as *Bittium (Elachista) californicum.*

## Alabina phanea Bartsch, 1911

### Plate 95, fig. 6

*Proceedings of the United States National Museum,* 39:412; Pl. 62, fig. 6.

Shell elongate-conic, white, excepting the nuclear whorls which are yellowish brown. Nuclear whorls small; the first one and one-half well-rounded and smooth; the next two marked by two strong spiral keels, which divide the space between the sutures into three subequal parts. Post-nuclear whorls strongly, slopingly shouldered, ornamented with moderately strong, decidedly curved, axial ribs, of which 16 occur upon the first, 18 upon the second to fourth, 20 upon the fifth, 22 upon the sixth, and 18 upon the penultimate turn. In addition to these ribs, the whorls are marked by three broad, low, spiral cords, the weakest of which is at the summit, while the other two divide the space between the sutures into three equal portions. The intersections of the axial ribs and the spiral keels form strong tubercles which are truncated posteriorly and slope gently anteriorly. The two middle cords are much more strongly tuberculate than the one at the summit. The spaces inclosed between the axial ribs and the spiral cords are moderately impressed, squarish pits. Sutures strongly impressed. Periphery of the last whorl marked by a channel which is crossed by the continuations of the axial ribs that terminate at its anterior border. Base short, well-rounded, ornamented with four rather broad, weak spiral cords. Aperture subquadrate, channeled anteriorly; posterior angle decidedly obtuse; outer lip thin, showing the external sculpture within; columella decidedly oblique, strongly revolute and somewhat twisted; parietal wall covered with a moderately thick callus. Length, 3.6; diameter, 1.1 mm. (Bartsch.)

Type in United States National Museum, No. 198924. Type locality, San Diego, California.

Range. Known only from type locality.

## Alabina barbarensis Bartsch, 1911

### Plate 96, fig. 3

*Proceedings of the United States National Museum,* 39:411; Pl. 61, fig. 3.

Shell broadly conic, creamy white. (Nuclear whorls decollated.) Post-nuclear whorls flattened, appressed at the summit, marked by four slender, incised, spiral lines, which divide the space between the sutures into five equal, flat cords; axial sculpture consisting of lines of growth only. Sutures strongly impressed. Periphery of the last whorl angulated. Base well-rounded, marked by six spiral lines, which divide it into six cords, the posterior five of which are equal, the one about the umbilicus being wider than the rest. Aperture ovate, feebly channeled anteriorly. Posterior angle acute; outer lip thin at the edge; columella decidedly curved, oblique, strongly reflected over the reinforcing base; parietal wall covered with a thick callus. Length, 6.2; diameter, 2.8 mm. (Bartsch.)

TYPE in United States National Museum, No. 203676. Type locality, Postpliocene of Santa Barbara, California.

RANGE. Known only from the type locality.

## Alabina tenuisculpta Carpenter, 1864
### Plate 96, fig. 6

*Supplementary Report,* British Association for the Advancement of Science. *Proceedings of the United States National Museum,* **39**:415. *Proceedings of the California Academy of Sciences,* **3**:216.

M. t. tenui, regulariter-turrita, fusco-cinerea; anfr. nucl. laevibus, normalibus, apice acuto; norm. viii. rotundatis, suturis impressis; lirulis spiralibus haud extantibus, plus minusve distantibus, irregularibus cincta, quarum anfr. primis duae anticae majores; lirulis circa basim rotundatam obtusis, subregularibus; rugulis incrementi irregularibus, interdum decussantibus; apertura subrotundata, peritremati haud continua; labro acuto, postice flexuoso; labio tenui. Long., 0.28; long. spir., 0.18; lat., 0.10 poll. (Carpenter.)

Shell acicular, light chestnut brown. Nuclear whorls very small, two and one-half, increasing regularly in size, well-rounded. Post-nuclear whorls with a strong, very wide, sloping shoulder which extends over the posterior half of the whorls between the sutures; and is bounded at the summit by a slender spiral thread. The first five post-nuclear turns have a strong, median, spiral cord and a second as strong as the median, about halfway between these two; a slender spiral thread can be seen on the third to the sixth whorl. On the last three whorls the median cord is almost completely lost, while the one above the sutural line retains its strength. On these three whorls additional fine spiral lines make their appearance. The axial sculpture consists of fine lines of growth only. Sutures weakly channeled. Periphery and base of the last whorl well-rounded, marked by eight subequal and unequally spaced, spiral keels and fine lines of growth. The space immediately surrounding the umbilical area has no spiral sculpture. Aperture broadly ovate; posterior angle obtuse; outer lip thin; columella slender, decidedly curved and slightly reflected over the narrow umbilicus; parietal wall glazed with a thin callus. (Bartsch.)

TYPE in United States National Museum, No. 40933. San Diego, California.

RANGE. San Pedro, California, to Magdalena Bay, Lower California. Described as *Mesalia tenuisculpta.*

## Alabina tenuisculpta phalacra Bartsch, 1911
### Plate 96, fig. 5

*Proceedings of the United States National Museum,* **39**:417; Pl. 61, fig. 5.

Shell broadly elongate-conic, light chestnut brown. Nuclear whorls 2½, small, increasing regularly in size, smooth. Post-nuclear whorls slightly shouldered at the summit, marked by slender, axial ribs, of which 18 occur upon the first to fourth, 20 upon the fifth and sixth, and 24 upon the penultimate turn. Upon the early whorls the ribs are almost vertical; on the last two they are decidedly retractive. Here also they are less regular. In addition to the axial ribs, the whorls are marked by three obsolete spiral cords, of which one is median, another a little nearer to the median cord than the summit, and the third about halfway between the median and the suture. Periphery and base of the last whorl inflated, the latter narrowly umbilical, marked by five subequal subobsolete cords. The space immediately about the umbilical area is smooth. Aperture very broadly ovate; posterior angle obtuse; outer lip thin, showing the external sculpture within; columella slender, curved, and slightly revolute; parietal wall glazed with a thin callus. Length, 7.5; diameter, 3 mm. (Bartsch.)

TYPE in United States National Museum, No. 32205a. Type locality, San Diego, California.

RANGE. Known only from type locality.

## Alabina diegensis Bartsch, 1911
### Plate 96, fig. 4

*Proceedings of the United States National Museum,* **39**:416; Pl. 61, fig. 4.

Shell elongate-conic, chestnut brown, excepting the extreme apex and the last volution, which are paler. Nuclear whorls three, small, increasing regularly in size, well-rounded, without sculpture. The early post-nuclear whorls have a decidedly sloping shoulder which extends from the middle to the whorls, between the sutures to the summit. This shoulder is marked on the first whorl by a single cord that limits it anteriorly, on the second by an additional cord, which divides the shoulder in two equal halves, while on the third, two additional cords, a little less strong than the other two, divide the space between the summit and the first cord and the space between the next two cords into equal halves. The sculpture on the anterior half of the whorls between the suture consists of a single cord on the first and second, which is halfway between the median cord and the suture. On the third, an additional cord a little less strong appears between the two. This cord becomes equal in strength to the other two on the succeeding turns. The space between the suture and the first supra-sutural cord remains plain, barring exceedingly fine microscopic spiral striations, and is as wide as the space between this cord and the median cord. In addition to the spiral sculpture, the whorls are marked by very many irregular decidedly curved and regularly distributed axial riblets, which render their intersections with the spiral cords very weakly nodulose. The summits of the whorls are roundly shouldered and make the sutures appear constricted. Periphery and base of the last whorl somewhat inflated, well-rounded, the latter marked by seven equal and equally spaced, low, rounded, spiral cords and feeble axial threads. The space immediately surrounding the umbilical area is free from all spiral sculpture. Aperture very broadly ovate; outer lip thin, showing the external sculpture within; columella slender, strongly curved and slightly revolute; parietal wall glazed with a thin callus. Length, 7.5; diameter, 2.7 mm. (Bartsch.)

TYPE in United States National Museum, No. 195219. Type locality, San Pedro, California.

RANGE. San Pedro to San Diego, California.

## Alabina occidentalis Hemphill, 1894
### Plate 95, fig. 2

*Zoë*, **4**: 395. *Proceedings of the United States National Museum,* **39**; Pl. 62, fig. 2.

Shell elongate-conic, subdiaphanous. Nuclear whorls 1½, small, well rounded, smooth. Postnuclear whorls with a very strong, concavely sloping shoulder, which extends from the appressed summit over the posterior third of the whorls terminating there in a well-rounded keel. The shoulder is smooth, excepting fine lines of growth; the portion between the suture and the shoulder is well rounded, marked with lines of growth and 9 to 11 equal and equally spaced incised spiral lines. Sutures strongly constricted; periphery of the last whorl well rounded. Base moderately long, somewhat inflated, well rounded, narrowly umbilicated, marked by slender lines of growth and numerous fine spiral striations. Aperture very broadly ovate; posterior angle obtuse; outer lip thin, columella slender, slightly oblique, revolute. (Bartsch.)

TYPE. Specimen described and figured in United States National Museum, No. 127551. Type locality, Mud Flats, near San Diego, California.

RANGE. San Diego, California.

Described as *Eulimella occidentalis*.

## Alabina turrita Carpenter, 1866
### Plate 95, fig. 4

*Proceedings of the California Academy of Sciences,* **3**: 219.

St. t. minuta, albida, solidiore, laevi, turrita; anfr. nucl. iii. subnaticoideis, apice mamillato; norm. v. planatis, angustis, suturis parum impressis; basi subito rotundata, haud umbilicata; apertura subrotundata; labropostice parum contracto. Long., 0.06; long. spir., 0.04; lat., 0.02. (Carpenter.)

TYPE in United States National Museum, No. 15566. Type locality, San Pedro, California.

Shell small, elongate-conic, white. Nuclear whorls minute, apparently not differentiated from the remaining turns. Postnuclear whorls strongly flattened, somewhat overhanging, separated by a deeply channeled suture, apparently without sculpture. Periphery and base of last whorl well rounded. Aperture ovate, posterior angle acute, outer lip thin; columella short, moderately strong, slightly reflected. (Bartsch.)

## Genus **BITTIUM** (Leach) Gray, 1847

Shell elevated, with numerous granular whorls and irregular varices; anterior canal short, not recurved; inner lip simple, outer lip not reflected, usually with an exterior rib. Operculum four-whorled with central nucleus. (Tryon, *Manual of Conchology*.)

TYPE. *Cerithium reticulatum* Da Costa.

DISTRIBUTION. Europe, Canary Island, Azores, West Indies, New England to Florida, Sitka, Alaska, to Monterey, San Pedro, San Diego, Australia, Japan, China, New Zealand, Hawaiian Islands.

### Subgenus BITTIUM S.S.

#### Bittium johnstonae Bartsch, 1911

*Proceedings of the United States National Museum*, **40**:387; Pl. 53, fig. 6.

Shell very elongate-conic, chestnut brown with strong white varices at irregular intervals. (Nuclear whorls decollated.) Post-nuclear whorls well-rounded, the early turns ornamented with four spiral keels, all of which except the anterior one, which is at the summit, are of equal size and about as wide as the spaces that separate them; the one at the summit being much more slender. On the last three whorls the keel at the summit is divided into two slender threads, which occupy the same amount of space as the single keel does on the earlier turns. In addition to the spiral keels, the whorls are marked by strong, well-rounded, axial ribs which are about two-thirds as wide as the spaces that separate them. These ribs render the spiral keels tuberculate at their intersection. There are 12 ribs upon the third, 14 upon the fourth, 16 upon the fifth and sixth, and 18 upon the seventh turn. On the last two whorls, four axial ribs occur between each pair of varices. Varices very large and conspicuous. Sutures strongly constricted. Periphery of the last whorl well-rounded, marked by a slender spiral cord which is not tuberculate. Base rather prolonged, well-rounded, marked by four equal and equally spaced spiral cords, of which the last is immediately behind the columella. Entire surface of base marked by numerous, rather strong lines of growth. Aperture ovate, decidedly channeled anteriorly; outer lip thin at the edge, rendered sinuous by the external sculpture which is visible within; columella very short, curved, and covered on its entire surface by a strong callus, which also extends over the parietal wall. Length, 7.9; diameter, 2.6 mm. (Bartsch.)

TYPE in United States National Museum, No. 196208. Type locality, Lower California.

RANGE. Laguna Beach, California, to Lower California.

Subgenus STYLIDIUM Dall, 1907

**Bittium eschrichtii** Middendorff, 1849

Plate 79, fig. 4

*Beitrage zu einer Malacologia Rossica,* 2: 68; Pl. 11, fig. 1.

Testa turrita, apice acuto, corneo-cinerea calcarea, faucibus intense violaceis, apertura alba; anfractibus applanatis, sulcis longitudinalibus, linearibus, exaratis; suturis canaliculatis; basi sensim sensimque, nec angula, in convexitatem anfractus ultimi transeuntellabro tenui, vix crenulato; columella callosa. (Middendorff.)

Es kommt diese Art der *Turritella erosa* Couthouy, welche ich mit *T. polaris* Beck (Möller), *Index,* p. 10) für synonym halte, am nächsten, unterscheidet sich aber von dieser durch ihre höhere und deshalb mehr ovale Apertur, die viel mehr abgeschrägte Basis (welche bei *T. erosa* fast flach abfällt) die platteren Windungen dickere Schale u.s.w.

Die Windungen der *T. eschrichtii* sind ganz flach-konisch, und nicht konvex wie alle Abbildungen der *T. erosa* das von dieser Art augenscheinlicher nachweisen als ich es selbst in der Natur finde. Drei bis vier flachrückige, scharfkantige, erhaben Längsbänder, welche wenig breiter oder bis doppelt so breit sind als die sie untereinander trennenden Zwischenräume, umziehen jede Windung. Zuweilen verläuft in einigen dieser Zwischenräume, oder selbst in der rinnenartig vertieften Nath, ein erhabner feiner Streifen. Diese Längsbänder sind schärfer ausgeprägt als es bei *T. erosa* der Fall ist. Bei einzelen Exemplaren erscheinen die Zwischenräume der Längsbänder durch feine Anwachsstreifen wie genetzt. Von hinten betrachtet, beträgt die Höhe der letzten Windung fast 1/3 der Gesammtlänge; dagegen bei *T. erosa* nur 1/4, weil bei letzterer die Basis fast rechtwinklig sich abflacht (basi paene concava; Möll.) während sie bei unser Art sich allmählig verjüngt Deshalb ist auch die Apertur nicht rundlich wie bei *T. erosa* sondern oval, und die Masse der columella externa beiner Arten sind ganz verschieden. (Middendorf.)

Shell broadly elongate-conic, rather coarse, varying in color from white to chestnut brown. The nucleus consists of single, smooth white whorl, well-rounded. Post-nuclear whorls well-rounded, marked by four strong, somewhat flattened, spiral keels between the sutures, which are separated by deep, strong, spiral grooves about two-thirds as wide as the keels. In addition to this spiral sculpture, the whorls are marked by numerous fine, spiral striations and fine lines of growth. Periphery of the last whorl marked by a sulcus as wide as the sulci between the keels on the spire. Base well-rounded marked by eight equally spaced spiral cords, which grow successively weaker from the periphery to the umbilical area. In addition to these cords the base is marked by fine spiral lines and fine lines of growth. Aperture oval, somewhat effuse anteriorly; posterior angle obtuse; outer lip thin, rendered wavy by the external sculpture; columella short, very broad at base, somewhat twisted and reflected; parietal wall glazed with a thin callus. Length, 14; diameter, 5 mm. (Bartsch.)

TYPE in Museum, St. Petersburg. Type locality, Sitka, Alaska.

RANGE. Sitka, Alaska, to Puget Sound, Washington. Fossil, Fossil Pliocene Rock, Coos Bay, Oregon.

## Bittium eschrichtii icelum Bartsch, 1907

### Plate 77, fig. 3

*Proceedings of the United States National Museum,* **33**:178.

In *B. eschrichtii* only the early whorls show axial ribs. In the present form they are well developed on all the turns, weakening only on the last. The type has 9 whorls (the nucleus being lost), and measures: length, 15; diameter, 5.5 mm. (Bartsch.)

TYPE in United States National Museum, No. 15209a. Type locality, Neah Bay, Washington.

RANGE. Straits of Juan de Fuca to Monterey, California.

## Bittium eschrichtii montereyense Bartsch, 1907

### Plate 79, fig. 5

*Proceedings of the United States National Museum,* **33**:178.

This form is the southern race of *B. eschrichtii*. It differs from the typical form in being less strongly spirally keeled, much more smooth, more slender, and in every way more elegant than *eschrichtii*. The typical form varies in color from brown to white, and is very rarely spotted. In *montereyense* the variegated forms predominate; that is, the shells are whitish, mottled with rust brown. Length, 13.8; diameter, 5 mm. (Bartsch.)

TYPE in United States National Museum, No. 32221.

RANGE. Crescent City, California, to Cape San Lucas, Lower California. Pliocene, California.

### Bittium paganicum Dall, 1919

*Proceedings of the United States National Museum,* **56**:345.

Shell small, slender, dark reddish brown, more or less divided by paler spiral lines on the later whorls; smooth except for incremental lines, with more than six rather flattish whorls, the apex in every case eroded; suture distinct, not deep; base rounded, imperforate, the aperture ovate, the outer lip thin, sharp, somewhat arcuate, and produced anteriorly; inner lip erased; pillar lip slightly twisted, not thickened; operculum paucispiral. Height, 8; diameter at truncation, 1.5; at base, 3 mm. (Dall.)

TYPE in United States National Museum, No. 271078. Type locality, U.S. Fish Commission, Station 4508, nine miles off Point Pinos, Monterey Bay, California, in 356 fathoms.

RANGE. Known only from type locality.

### Subgenus SEMIBITTIUM Cossmann, 1896
### Bittium vancouverense Dall and Bartsch, 1910

Canada, Department of Mines, Geological Survey Branch, *Memoir No. 14-N,* 1910, 10; Pl. 1, fig. 8.

Shell elongate-conic, grayish white outside and dark, purplish brown within. Nuclear whorls at least two, apparently smooth, worn in all specimens. Post-nuclear whorls slightly rounded, ornamented with three strong, equal and equally spaced, nodulose, spiral keels, the first of which is a little below the summit. The spaces separating the spiral keels are of equal width. Immediately below the third keel is a strong, peripheral sulcus, which equals those between the spiral keels. In addition to the spiral sculpture, the whorls are marked by almost vertical, axial ribs, which are not quite as wide as the spiral keels. These render the keels nodulose at their intersections. Of these ribs, 12 occur upon the first, 14 upon the second and third, 16 upon the fourth, 18 upon the fifth, 24 upon the sixth, and 30 upon the penultimate turn. The spaces inclosed between the spiral keels and the axial ribs are well-impressed, rounded pits. All the tubercles are truncated on the posterior margin and slope gently anteriorly. Base of the last whorl moderately long, ornamented with seven spiral cords, of which the two immediately below the periphery are the strongest and broadest, while the two bounding the umbilical area are wider than those intervening. Sutures channeled. Aperture irregular, channeled anteriorly; posterior angle obtuse; outer lip thin, sinuous, showing the external sculpture within; columella stout, short, twisted and reflected; parietal wall glazed with a moderately thick callus. Length, 7.8; diameter, 2.7 mm. (Dall and Bartsch.)

TYPE in Dominion Geological Survey Collection. Type locality, Ucluelet, Barkley Sound, Vancouver Island, British Columbia.

RANGE. Known only from type locality.

## Bittium attenuatum Carpenter, 1864
### Plate 78, figs. 1, 2, 5

*Supplementary Report,* British Association for the Advancement of Science, 655. *Proceedings of the United States National Museum,* **40**: 393; Pl. 54, figs. 1, 2, 5.

B. t. valde gracili, attenuata; anfr. nucl. . . . . (detritis); normalibus 10 planatis, suturis haud impressis; t. juniore lirulis spiralibus 2 anticis conspicuis, aliis posticis parum conspicuis, supra costulas circiter 11, radiantes transeuntibus; t. adulta costulis et lirulis anticis obsoletis; lirulis 2. suturalibus; basi prolongata, striis circiter 6 ornata; apertura ovali; columella intorta, parum emarginata. Long., .4; long. spir., 31; lat., 11 poll. (Carpenter.)

Shell elongate-conic, varying in color from plain white to brown, variously banded or maculated. Nuclear whorls two, moderately rounded, smooth. Post-nuclear whorls slightly rounded, ornamented with weak spiral bands, which are best developed on the early whorls. These spiral bands are truncated anteriorly and slope gently to the posterior boundary. The early whorls have three spiral bands, of which the posterior, at the summit, is less strongly developed than the other two, which divide the remaining space between the sutures equally. On the middle whorls the posterior keel has a tendency to become divided, while on the later turns it becomes obsolete. In addition to the spiral sculpture, the whorls are marked by poorly developed, rounded, protractive, axial ribs which render their intersections with the spiral cords nodulose. Of these ribs 12 occur upon the third and 14 upon each of the remaining turns upon which ribs are discernible. The spaces inclosed between the ribs and spiral bands are shallow, impressed, rectangular pits. In addition to the axial ribs, the whorls are marked by numerous fine lines of growth. Periphery of the last whorl marked by a spiral band, which is separated from the first band above the periphery by a sulcus as wide as the sulci on the spire. Base short, well-rounded, marked by six spiral keels, of which the first below the periphery is much larger than the rest, which are subequal and subequally spaced. Aperture ovate, somewhat effuse anteriorly; posterior angle acute; outer lip thin, showing the external markings within; columella short, very broad at base, oblique, and reflected. One of the cotypes measures: length, 10.2; diameter, 3 mm. (Bartsch.)

TYPE in United States National Museum. Type locality, Monterey, California.

COTYPES in United States National Museum, No. 15584. Type locality, Monterey, California.

RANGE. Vancouver Island, British Columbia, to San Diego, California. Fossil, Lower San Pedro Series, Deadman Island, California.

### Bittium attenuatum boreale Bartsch, 1911
### Plate 78, fig. 4

*Proceedings of the United States National Museum*, **40**:395; Pl. 54, fig. 4.

Shell similar to *Bittium attenuatum,* but in every way larger and more robust. Sculpture very much stronger than in the typical form; the tubercles well-developed on the last whorl but much weaker than on the earlier ones. Length, 10.1; diameter, 3.5 mm. (Bartsch.)

TYPE in United States National Museum, No. 211586. Type locality, Barkley Sound, Vancouver Island.

RANGE. Queen Charlotte Islands to Vancouver Island, British Columbia.

### Bittium attenuatum multifilosum Bartsch, 1907
### Plate 78, fig. 3

*Proceedings of the United States National Museum,* **33**:179.

Shell similar to *Bittium attenuatum,* but having seven spiral keels between the sutures on the whorls of the spire instead of four; that is, a strong, intercalated, spiral cord between each of the four primary keels. Length, 9.2; diameter, 3 mm. (Bartsch.)

TYPE in United States National Museum, No. 127051. Type locality, Whites Point, San Pedro, California.

RANGE. Monterey to San Pedro, California.

### Bittium attenuatum latifilosum Bartsch, 1911
### Plate 78, fig. 6

*Proceedings of the United States National Museum,* **40**:395; Pl. 54, fig. 6.

Shell similar to *Bittium attenuatum.* Whorls a little higher between the sutures than in the typical form, with four spiral cords between the sutures, which are much broader than in typical *attenuatum.* The axial ribs are strong on the early whorls, but scarcely indicated on the last volutions. Shell usually mottled. (Bartsch.)

TYPE in United States National Museum, No. 109509. Type locality, Terminal Island, San Pedro.

RANGE. Known only from type locality.

## Bittium subplanatum Bartsch, 1911
### Plate 77, fig. 5

*Proceedings of the United States National Museum,* **40**:395; Pl. 57, fig. 5.

Shell broadly elongate-conic, milk-white. Nuclear whorls a little more than one, well-rounded, smooth. The first of the post-nuclear whorls well-rounded, marked by three spiral cords, one of which is at the summit, another on the middle of the whorl, while a third is a little above the suture. The succeeding turns show four spiral cords, of which the one at summit is a little less strong than the rest; the remaining three divide the space between the sutures into four equal parts. Beginning with the fourth whorl, intercalated cords make their appearance between the primary ones, so that on the last whorl we have an intercalated cord and sometimes two between all the primary cords; these, however, are never quite as strong as the principal ones. In addition to the spiral cords, the whorls are marked by decidedly curved, slender, well-rounded, almost vertical, axial ribs, which are scarcely indicated on the first turn, while 14 of them occur upon the second and third, 16 upon the fourth, 18 upon the fifth and sixth, 22 upon the seventh, 24 upon the eighth, and 26 upon the penultimate turn. The intersections of the spiral cords and axial ribs form weakly developed, rounded tubercles which are truncated on their posterior margin, while the spaces inclosed between them are very shallow, quadrangular pits. Sutures strongly constricted. Periphery and base of the last whorl well-rounded, marked by slender, spiral cords, of which those immediately below the periphery are the strongest and are truncated on the posterior margin, sloping gently anteriorly. Of these cords, seven occur on the base of the type. Aperture rather large, irregularly oval, channeled anteriorly; posterior angle acute; outer lip thin, rendered sinuous by the external sculpture; columella decidedly oblique, strongly curved and reflected. Length, 10.9; diameter, 3.8 mm. (Bartsch.)

TYPE in United States National Museum, No. 160076. Type locality, Catalina Island, California.

RANGE. Monterey Bay south to Coronado Island, California.

## Bittium rugatum Carpenter, 1866

*Proceedings of the United States National Museum,* **40**:397; Pl. 56, figs. 4, 5.

Shell elongate-conic, white. Nuclear whorls a little more than one, well-rounded, smooth. Post-nuclear whorls appressed at the summit, decidedly overhanging, the early ones marked by two strong spiral cords on the anterior half between the sutures, and a third less strongly developed cord at the summit. On the third whorl a fourth spiral makes its appearance between the one at the summit and its neighbor; this fourth spiral soon increases in size, so that on the middle of the shell all four cords are practically of equal strength and spacing. On the last whorl a slender, intercalated cord appears between the anterior two. In addition to the spiral sculpture, the whorls are marked by strong, well-rounded, axial ribs, which are merely indicated on the first two whorls; on the third to sixth turn there are 14, on the seventh there are 16, on the eighth, 18, and on the penultimate whorl 24. The intersections of the axial ribs and spiral cords form elongated tubercles, which have their long axes parallel with the spiral sculpture. The spaces between the spiral cords and axial ribs are elongated, squarish pits. Sutures strongly constricted. Periphery of the last whorl marked by a channel which bears a slender cord. Base well-rounded, marked by six spiral cords, which grow successively weaker and more closely spaced from the periphery to the umbilical area. Aperture oval, channeled anteriorly; posterior angle acute; outer lip thin, rendered sinuous by the external sculpture; columella short, twisted, and reflected. Length, 10.5; diameter, 3.5 mm. (Bartsch.)

TYPE in United States National Museum, No. 7154. Type locality, Lower Pleistocene, Santa Barbara, California.

RANGE. San Pedro and Catalina Island, California. Fossil, Lower Pleistocene of Santa Barbara, Terminal Island, San Pedro, and San Diego, California.

## Bittium quadrifilatum Carpenter, 1864

### Plate 73, fig. 4; Plate 79, figs. 2, 3

*Journal de Conchyliologie,* **13**:143.

B. t. satis tereti, pallide, cinerea, tenuisculpta; anfr. nucleosis, primo omnino coelato, sinistrali, dein 2 laevibus, rotundatis, apice quasi mamillato; anfr. normalibus, subplanatis; suturis valde impressis, haud sculptis; costulis radiantibus circ. 15–22, angustis subrectis, anfr. ult. crebrioribus, suturam versus evanidis; filis spiralibus semper aequilibus, supra spiram 4 sagustis, expressis, costulas transeuntibus, haud nodulosis; filis duabus alteris, inter quas sutura sita est; basi tenue striata; columella intorta, parum effusa; apertura ovata; libio parvulo labro tenui, parum arcuato. Long., 26; long. spir., 16; lat., 9 poll. (Carpenter.)

Shell rather large, elongate-conic, chestnut brown, dull. Nuclear whorls two, well-rounded, smooth, the first obliquely tilted. Post-nuclear whorls shouldered at the summit, moderately rounded, ornamented with four equal and subequally spaced spiral cords, which divide the space between the sutures into four subequal parts. The first of these cords is at the summit of the whorl. In addition to the spiral cords, the whorls are marked by moderately strong, rounded, vertical axial ribs, of which 12 occur upon the first, 14 upon the second to fifth, 16 upon the sixth, 18 upon the seventh and the penultimate turn. The spiral cords cross the axial ribs as regular bands, their junctions forming elongated rounded tubercles, having their long axes parallel with the spiral sculpture, while the spaces inclosed between them are well-impressed quadrangular pits with their long axes also parallel with the spiral sculpture. On the last whorl very faint, intercalated spiral threads appear between the cords. Sutures channeled, showing a faint trace of the subperipheral cord. Periphery of the last whorl marked by a rather broad channel, which shows several faint spiral lirations. Base moderately long, somewhat concave, ornamented with seven subequal and subequally spaced, spiral threads. Aperture irregularly oval, channeled anteriorly; posterior angle obtuse; outer lip thin, showing the external sculpture within; columella moderately long, somewhat twisted, and reflected. (Bartsch.)

TYPE in United States National Museum, No. 14849. Type locality, San Diego, California.

RANGE. Monterey, California, to San Ignacio, Lower California.

## Bittium quadrifilatum ingens Bartsch, 1907

*Proceedings of the United States National Museum,* **33**:180.

Shell similar to *B. quadrifilatum,* but in every way stronger and larger and of a white color. The spiral bands in *B. quadrifilatum* do not form strong cusps at their intersection with the axial ribs, but simple nodes, while in the present form these intersections are decidedly cupped. Length, 12.2; diameter, 4.5 mm. (Bartsch.)

TYPE in United States National Museum, No. 32213. Type locality, "Albatross" Station 4475, ten miles off Pinos Light, Monterey, California.

RANGE. Known only from type locality.

## Bittium armillatum Carpenter, 1864
### Plate 76, fig. 6

*Supplementary Report,* British Association for the Advancement of Science, 655. *Proceedings of the United States National Museum,* **40**:391; Pl. 52, fig. 6.

B. testa *B. aspero* simili; anfr. nucl. ii. laevibus, tumentibus, vertice declivi, celato; dein anfr. ix. normalibus planatis, suturis impressis; t. adolescente seriebus nodulorum tribus spiralibus extantibus, supra costas instructis; costis radiantibus circ. xiii. fere parallelis, seriebus, a suturis separatis, spiram ascendentibus; t. adulta, costulis spiralibus, interdum iv., intercalantibus; costulis radiantibus creberrimis; costis suturalibus ii. validis, haud nodosis; basi effusa, liris circ. vi. ornata; apertura subquadratata; labro labioque tenuibus; columella vix torsa, effusa, vix emarginata. (Carpenter.)

Shell broadly conic, rust brown. Nuclear whorls at least three, apparently smooth. Post-nuclear whorls shouldered at the summit, marked by three strong, spiral keels which are considerably narrower than the three spaces into which they divide the whorls between the sutures. In addition to the spiral sculpture the whorls are marked by slightly retractive axial ribs, which are about as strong as the spiral keels. Of these 16 occur upon the fourth, 18 upon the fifth, 20 upon the sixth, 24 upon the seventh, and 32 upon the penultimate turn. The junction of the axial ribs and the spiral cords form well-rounded tubercles, while the spaces inclosed between are well-impressed squarish pits. Sutures strongly constricted. Base moderately long, ornamented by six low, well-rounded, subequally spaced, spiral cords. Aperture channeled anteriorly; posterior angle obtuse; outer lip rendered sinuous by the external sculpture; columella moderately long, oblique, curved, and reflected; parietal wall covered with a moderately thick callus. Length, 9.5; diameter, 3.2 mm. (Bartsch.)

TYPE in United States National Museum, No. 15653. Type locality, Lower Pleistocene, Santa Barbara, California.

RANGE. Santa Barbara to San Pedro, California.

### Bittium purpureum Carpenter, 1864

Plate 76, figs. 1, 3

*Supplementary Report,* British Association for the Advancement of Science, 660. *Proceedings of the United States National Museum,* 40:391; Pl. 52, figs. 1, 3.

C. testa compacta, haud gracili, marginibus spirae parum excurvatis; purpurea seu fusco-purpurea, circa peripheriam pallidiore; anfr. nucl. ii., laevibus; norm. vii., planatis, suturis impressis; seriebus iii. nodulorum minorum supra costulas spirales minores, ad intersectiones costularum radiantium circ. xxiii., lines fere rectis, ad suturas interruptis spiram ascendentium sitis; interstitiis impressis, quadratis; costulis suturalibus ii. haud nodulosis; basi rotundata, antice lirulis paucis expressis inter eas et costulas suturales vix sculpta; apertura subquadrata columella torta, emarginata. Long., 29; long. spir., 19; lat., 1 poll. (Carpenter.)

Shell broadly elongate-conic, wax-yellow, variously mottled and banded with chestnut brown. Nuclear whorls a little more than one, apparently smooth. Post-nuclear whorls marked by three strong spiral cords, which divide the space between the suture and the summit into three almost equal areas. The cord at the summit is very slightly below the summit and renders this strongly shouldered. In addition to the spiral sculpture, the whorls are marked by almost vertical axial ribs which nearly equal the spiral cords in strength. Of these, 14 occur upon the first, 16 upon the second, 18 upon the third, 20 upon the fourth, 22 upon the fifth, and 25 upon the penultimate turn. The intersections of the spiral cords and the axial ribs form well-rounded tubercles, while the spaces inclosed between them are well-impressed, squarish pits on all but the last; on this they are oblong, their long axes coinciding with the axial sculpture. Sutures strongly constricted, showing the peripheral cord on the later whorls. Periphery of the last whorl marked by a slender, smooth cord, the space between which and the first supraperipheral cord is about as wide as that which separates the next two cords posteriorly and is crossed by the continuations of the axial ribs which terminate at its posterior border. Base marked by five subequally spaced spiral cords, of which the strongest is immediately below the periphery and is equal to the peripheral cord, while the next two in strength are at the columella; the two intervening are slender threads. Entire surface of spire and base crossed by numerous, rather strong lines of growth. Aperture broadly oval, decidedly channeled anteriorly; posterior angle obtuse; outer lip rendered sinuous by the external sculpture; columella short, somewhat twisted, curved and reflected; parietal wall covered with a thick callus. Length, 7.3; diameter, 2.5 mm. (Bartsch.)

Doctor Carpenter's cotypes in United States National Museum, No. 14823. Type locality, Monterey and Santa Barbara, California.

RANGE. Monterey to San Diego, California.

Described as *Cerithiopsis purpurea*.

## Bittium challisae Bartsch, 1917
### Plate 74, figs. 2, 6

*Proceedings of the United States National Museum,* **52**:673; Pl. 47, figs. 2, 6.

Shell very large, white. Nucleus and early post-nuclear turns decollated, those remaining slightly shouldered at the summit, weakly rounded in the middle and decidedly contracted immediately above the suture, marked by rather strong, low, well-rounded, axial ribs of which 14 occur upon the second and third, 16 upon the fourth to sixth, 18 upon the seventh, and 20 upon the last turn. The spiral sculpture consists of four cords on the early whorls, of which the first, at the summit, is a little weaker than the rest. These primary cords are truncated posteriorly and slope gently anteriorly. Beginning with the fourth whorl an intercalated thread makes its appearance between all the cords and between the summit and the first cord.

Suture strongly constricted. Periphery of the last whorl rendered angulated by a cord. Base short, slightly concave, marked by five low, broad, well-rounded, obsolete cords, which are subequal and subequally spaced. Aperture broadly oval, rather strongly channeled anteriorly; posterior angle obtuse; outer lip thin, rendered sinuous at the edge by the external sculpture, showing the external markings within; inner lip somewhat sinuous, rather stout, reflected over and appressed to the base; parietal wall covered by a moderately thick callus. Length, 13; diameter, 1.2 mm. (Bartsch.)

TYPE in United States National Museum, No. 272376. Type locality, San Juan Island, Washington.

RANGE. Known only from type locality.

## Bittium sanjuanensis Bartsch, 1917
### Plate 74, fig. 4

*Proceedings of the United States National Museum,* **52**:674; Pl. 47, fig. 4.

Shell very large, rather thin, bluish white. Nuclear whorls decollated. Post-nuclear whorls almost appressed at the summit, well-rounded, decidedly contracted immediately posterior to the suture, marked by strong, broad, heavy, slightly protractive, axial ribs which become enfeebled toward the summit and slightly widened there. Of these ribs 12 occur upon the second and third, 14 upon the fourth and fifth, and 16 upon the last whorl. Intercostal spaces not quite as broad as the ribs. The spiral sculpture consists of five cords, of which the first, which is at the summit, is very slender. The two succeeding this are successively a trifle stronger, while the fourth and fifth are very strong, the last being the heaviest of all. The junction of the axial ribs and the spiral cords form well-rounded, elongated nodules, which have their long axis parallel with the spiral sculpture; the spaces inclosed between them are very shallow, rectangular pits. In addition to the above sculpture, the entire surface of the spire and base is marked by many very slender lines of growth and exceedingly fine microscopic spiral striations, the combination of which gives the surface a somewhat cloth-like texture. Suture strongly constricted; it would be channeled were it not for the fact that the peripheral keel makes its appearance above the summit of the whorl, hence removes the strongly channeled element. Periphery of the last whorl marked by strong spiral cord, which is about as far anterior to the fifth cord of the spire as that is separated from the fourth. Base very short, decidedly concave, marked by three slender spiral cords, of which the first is about as far anterior to the periphery as that is distant from the fifth cord on the spire. The other two cords are very slender, the first being at the base of the columella while the next is a little distance posterior to it. Aperture subquadrate; quite strongly channeled anteriorly; the junction of the outer and basal lip forming almost a right angle; posterior angle obtuse; outer lip very thin, showing the external sculpture within and rendered sinuous by the external sculpture at the edge; inner lip decidedly oblique, slightly curved, slender, reflected and appressed to the base; parietal wall covered by a thin callus. Length, 11; diameter, 3.8 mm. (Bartsch.)

TYPE in United States National Museum, No. 168753. Type locality, off San Juan Island.

RANGE. Puget Sound, Washington.

### Bittium serra Bartsch, 1917
### Plate 74, fig. 1

*Proceedings of the United States National Museum,* **52**:675; Pl. 47, fig. 1.

Shell stout, broadly conic, grayish white. Nuclear whorls decollated. Post-nuclear whorls appressed at the summit, strongly rounded, marked by slender, rounded, somewhat retractive axial ribs, of which 18 occur upon the first and second, 20 upon the third, 22 upon the fourth, and 24 upon the remaining turns. In addition to the axial sculpture, the whorls are marked by four spiral cords, of which the first, which is at the summit, is a little weaker than the rest. The junctions of these cords with the axial ribs form prominent, strongly rounded tubercles, while the spaces inclosed between them are rounded pits. Suture moderately constricted but not channeled. Periphery of the last whorl rendered angulated by a spiral cord. Base short, slightly concave, marked by four, very low, broad, almost equal, spiral cords, which are separated by mere impressed lines. Aperture oval, narrowly twistedly channeled anteriorly; posterior angle obtuse; outer lip rendered wavy by the external sculpture; inner lip decidedly curved, somewhat revolute, reflected over and appressed to the base; parietal wall covered with a thin callus. Length, 6.5; diameter, 2.6 mm. (Bartsch.)

TYPE in United States National Museum, No. 271076. Type locality, U.S. Bureau of Fisheries Steamer "Albatross," Station 4310, off Point Loma Light, California, in 71–75 fathoms.

RANGE. Off Point Pinos to San Diego, California, and San Bertolome Bay, Lower California.

### Bittium tumidum Bartsch, 1907

*Proceedings of the United States National Museum,* 33:179.

Shell of medium size, light yellowish-brown, shining. Nuclear whorls decollated. Post-nuclear whorls somewhat inflated, well-rounded, separated by constricted sutures, and ornamented with strong, tuberculate, axial ribs, of which there are 18 upon the second of the remaining whorls and 22 upon the penultimate turn. In addition to the axial ribs there are four, unequally broad, low, spiral ridges between the sutures, which are much wider than the spaces which separate them, the latter appearing as strongly incised lines. The intersections of these ridges and the ribs form the tubercles. The whorls slope gently from the second spiral ridge toward the summit, and the first row of tubercles, which is only feebly developed, is located on the sloping shoulder. The second set of tubercles is rounded, while the third and fourth rows are decidedly elongated. Periphery of the last turn marked by a strong smooth spiral keel, which is separated from the supraperipheral keel by a mere constriction. Base rather short, without keel, marked only by lines of growth. Aperture suboval, decidedly channeled anteriorly, outer lip rendered sinuous by the external sculpture; columella short, very broad, and slightly expanded at the insertion; a little lighter in color than the rest of the shell; provided with a strong callus on its inner edge, which is reflected over the parietal wall. Length, 4.2; diameter, 1.7 mm. (Bartsch.)

TYPE in United States National Museum, No. 74001. Type locality, Monterey, California.

RANGE. Monterey, California, to San Pedro.

Subgenus LIROBITTIUM Bartsch, 1911

**Bittium interfossa** Carpenter, 1864

Plate 75, figs. 2, 6

*Report,* British Association for the Advancement of Science, **656**:1864. *Proceedings of the United States National Museum,* **40**:401; Pl. 51, figs. 2, 6.

With five sharp keels crossing 14 strong ribs. (Carpenter.)

Shell rather large and robust, white, sometimes light brown. Nuclear whorls small, two; the first obliquely tilted, smooth; the second with two strong spiral cords, which divide the space between the sutures into three equal areas. Post-nuclear whorls appressed at the summit, strongly slopingly shouldered, ornamented with two strong, spiral keels, which divide the space between the sutures into three equal areas, and very strong, vertical, axial ribs, 16 of which occur upon each of the turns. The intersections of the axial ribs and the spiral cords form strong elongated tubercles, the long axes of which coincide with the spiral sculpture. These tubercles slope more abruptly posteriorly than anteriorly. The intersections of the axial ribs and the spiral cords inclose shallow squarish pits. Sutures weakly impressed. Periphery of the last whorl marked by a strong keel to which the axial ribs extend feebly. This keel is a little nearer to the first post-nuclear keel than that is to its posterior neighbor. Base moderately long, marked by two very strong keels, which divide the space between the peripheral keel and the tip of the columella into three equal areas, the spaces between the keels being very deep and a little wider than the keels. Entire surface of spire and base, including the ribs and intercostal spaces, crossed by numerous fine, closely spaced, spiral striations. Aperture irregular, channeled anteriorly; posterior angle obtuse; outer lip thick within, thin at edge, rendered sinuous by the external sculpture; columella stout, strongly twisted, and reflected; parietal wall glazed with a thick callus. Length, 8.3; diameter, 3.1 mm. (Bartsch.)

TYPE specimen in United States National Museum, No. 56906. Type locality, Monterey, California.

RANGE. Monterey to San Diego, California.

Described as *Rissoa interfossa*.

## Bittium catalinense Bartsch, 1907

### Plate 75, fig. 1

*Miscellaneous Collections*, Smithsonian Institution, 50:28; Pl. 57, fig. 13.

Shell elongate-conic, milk-white. Nuclear whorls a little more than one, marked by two, strong, spiral cords, which divide the turns into three equal areas. Post-nuclear whorls shouldered at the summit, marked by three nodulose spiral keels; one of these, which is a little below the summit, is less strongly developed than the other two on all but the last turn; on this turn it is practically equal to the others. In addition to the spiral keels, the whorls are marked by rather strong, well-rounded, axial ribs, which are about two-thirds as wide as the spaces which separate them. Of these ribs, 16 occur upon the first to fifth, 18 upon the sixth and seventh, 20 upon the eighth, and 24 upon the penultimate turn. The intersections of the axial ribs and spiral cords form strong cusp-like nodules, which are suddenly truncated posteriorly and slope gently to the succeeding cord anteriorly. The space between the summit and the truncated end of the first row of tubercles forms a strong shoulder. The spaces inclosed between the spiral keels and the axial ribs are moderately impressed, rounded pits. Sutures strongly constricted, showing the greater part of the peripheral cord on all the turns. Periphery and base of the last whorl marked by five spiral cords which grow successively weaker from the periphery to the umbilical area. These cords are truncated on the posterior margin and slope gently anteriorly until they fuse with the general surface of the shell.

Aperture irregular, channeled anteriorly; posterior angle obtuse; outer lip rendered sinuous by the external sculpture; columella oblique, somewhat twisted; parietal wall glazed with a thin callus. Length, 7.5; diameter, 2.8 mm. (Bartsch.)

TYPE in United States National Museum, No. 165232. Pleistocene. Type locality, Pleistocene at Santa Barbara, California.

RANGE. Catalina Island to San Diego; Pleistocene, Santa Barbara, California.

### Bittium catalinense inornatum Bartsch, 1911

### Plate 75, fig. 3

*Proceedings of the United States National Museum,* **40**:403; Pl. 51, fig. 3.

Shell similar to *Bittium catalinense,* but lacking the plain spiral keel in the suture. Length, 7.2; diameter, 2.3 mm. (Bartsch.)

TYPE in United States National Museum, No. 195153. Type locality, Catalina Island, California.

RANGE. Catalina Island, California. Also Pleistocene.

### Bittium ornatissimum Bartsch, 1911

### Plate 76, figs. 4, 5

*Proceedings of the United States National Museum,* **40**:403; Pl. 51, figs. 4, 5.

Shell elongate-conic, creamy white. Nuclear whorls one and one-half, marked by two slender threads which divide the space between the sutures into three equal parts. Post-nuclear whorls strongly shouldered at the summit, marked by three strong spiral keels which divide the space between the sutures into four equal parts. The space between the first of these keels below the summit and the summit is a little narrower than the rest. In addition to these three strong spiral keels, intercalated keels are present, the first of which is at the summit, while another occurs between each of the other keels. On the last whorl these attain a strength almost equal to that of the primary keels. In addition to this spiral sculpture, the whorls are marked by well-developed, slightly retractive, axial ribs, of which 14 occur upon the first and second, 15 upon the third, 16 upon the fourth, 18 upon the fifth and sixth, 20 upon the seventh, 22 upon the eighth, 28 upon the ninth and tenth, and 34 upon the penultimate turn. The intersections of the axial ribs and spiral keels form strong cusps, which are suddenly truncated posteriorly and slope gently anteriorly, the spaces inclosed between them being small, rounded pits. Sutures channeled.

Periphery of the last whorl marked by a slender cord. Base moderately prolonged, ornamented by six spiral cords, of which the two immediately
anterior to the periphery and the two at the base of the columella are
decidedly stronger than the rest. Aperture rather large, channeled anteriorly; posterior angle stout, twisted, somewhat revolute, and reflected;
parietal wall glazed with a moderately thick callus. Length, 12.1; diameter,
4 mm. (Bartsch.)

TYPE in United States National Museum, No. 194413. Type locality,
Deadman Island, San Pedro.

RANGE. San Pedro, California. Pleistocene, Santa Barbara, California.

## Bittium munitum Carpenter, 1864

*Report,* British Association for the Advancement of Science, **628**:1863, **660**:1864.
*Proceedings of the United States National Museum,* **40**; Pl. 53, figs. 1, 2.

C. testa, *C. purpureae* simili, sed angustiore, marginibus spirae fere
rectis; costis spiralibus magis expressis, testa adulta minus nodulosis;
basi aequaliter lirulata. Long., 34; long. spir., 24; lat., 11 poll. (Carpenter.)

Shell elongate-conic, yellowish white. Nuclear whorls a little more than
one (with the sculpture abraded in all our specimens). Post-nuclear
whorls strongly shouldered at the summit, marked by three, strong, equal
and equally spaced, spiral keels, which are a little wider than the spaces
that separate them. In addition to the spiral keels, the whorls are marked
by somewhat retractive axial ribs, of which 18 occur upon the second, 20
upon the third, 24 upon the fourth and fifth, and 28 upon the penultimate.
The intersections of the axial ribs and spiral keels form strong tubercles,
while the spaces between them are well-impressed, rounded pits. Sutures
subchanneled. Periphery of the last whorl marked by a strong channel,
across which the feeble continuations of the axial ribs extend. Base moderately long, marked by six, strong, spiral cords which grow successively
weaker from the periphery to the umbilical area. In addition to the above
sculpture, the entire surface of the spire and base is crossed by numerous
fine lines of growth and exceedingly fine spiral striations. Aperture subquadrate, channeled anteriorly; posterior angle obtuse; outer lip rendered
sinuous by the external sculpture; columella short, stout, somewhat twisted,
and reflected; parietal wall glazed with a thin callus. Length, 7.8; diameter, 3 mm. (Bartsch.)

TYPE in United States National Museum, No. 15501. Type locality,
Neah Bay, Washington.

RANGE. Sitka Sound to Straits of Juan de Fuca.

Described as *Cerithiopsis munita.*

## Bittium munitum munitoide Bartsch, 1911

*Proceedings of the United States National Museum*, **40**:405; Pl. 53, fig. 4.

This is the southern race of *Bittium munitum*. It differs from *B. munitum* proper in being smaller and in having many more ribs, as many as 40 occurring upon the last whorl. Length, 8.2; diameter, 2.8 mm. (Bartsch.)

TYPE in United States National Museum, No. 152164. Type locality, San Pedro, California.

RANGE. Known only from type locality.

## Bittium asperum lomaënse Bartsch, 1911

*Proceedings of the United States National Museum*, **40**:406; Pl. 56, fig. 2.

Shell similar to *B. asperum* Gabb, but differing in being uniformly smaller, more slender, and in having more ribs. A specimen of *B. asperum* with 10 post-nuclear whorls measures: Length, 8.1 mm.; while one of *B. a. lomaense* of the same number of whorls measures 7.1 mm. *B. a. lomaense* is a living representative of *B. asperum*, which is a Post-Pliocene species. Length, 7.1; diameter, 2.3 mm. (Bartsch.)

TYPE in United States National Museum, No. 195130. Type locality, U. S. Bureau of Fisheries Station 4310, in 71–75 fathoms, off Point Loma Light, California.

RANGE. Catalina Island to San Diego, California.

## Bittium larum Bartsch, 1911

### Plate 77, fig. 4

*Proceedings of the United States National Museum*, **40**:407; Pl. 57, fig. 4.

Shell very regularly elongate-conic, light brown. Nuclear whorls at least two, worn. Post-nuclear whorls appressed at the summit, decidedly overhanging. The early post-nuclear whorls are marked by four equal and equally spaced spiral cords, the first of which is at the summit; these cords divide the space between the sutures into four equal parts. On the sixth whorl intercalated spiral cords make their appearance in the middle, between all the primary cords; these attain a little more than half the strength of the primary cords on the last turn. In addition to the spiral sculpture, the whorls are marked by moderately strong, almost vertical, axial ribs, of which 14 occur upon all but the penultimate turn, which has 18. The intersections of the axial ribs and spiral cords form elongate tubercles, which have their long axes parallel with the spiral sculpture. The spaces inclosed between the axial ribs and the spiral cords are rectangular pits on the early whorls, and broad, incised lines on the later ones. Sutures slightly constricted. Periphery of the last whorl angulated, marked by a spiral cord. Base short, slightly concave in the middle, marked by six spiral cords, which grow successively weaker from the periphery to the umbilical region. In addition to the above sculpture, the entire surface of spire and base is marked by fine lines of growth and numerous exceedingly fine, spiral striations. Aperture quadrangular, channeled anteriorly; posterior angle obtuse; outer lip thin, showing the external sculpture within, rendered sinuous at the edge by the external sculpture; columella moderately strong, twisted, and reflected; parietal wall glazed with a thin callus. Length, 10; diameter, 3.3 mm. (Bartsch.)

TYPE in United States National Museum, No. 195156. Type locality, San Pedro, California.

RANGE. San Pedro, California, to San Bartolome Bay, Lower California.

## Bittium oldroydi Bartsch, 1911

### Plate 75, fig. 5

*Proceedings of the United States National Museum, 40:408; Pl. 51, fig. 5.*

Shell very large, chestnut brown. (Nuclear whorls decollated in all our specimens.) Post-nuclear whorls moderately rounded, ornamented with three spiral keels, which are truncated on their posterior margin and slope gently anteriorly until they fuse with the general mass of the shell. These keels divide the space between the sutures into four almost equal parts, the space between the summit and the first keel being a little narrower than the rest. In addition to the spiral keels, the whorls are marked by slightly retractive axial ribs, of which 12 occur upon the second, 14 upon the third and fourth, 16 upon the fifth to seventh, 18 upon the eighth, 20 upon the ninth and tenth, and 22 upon the penultimate turn. These ribs extend from the summit to the suture. Their intersections with the spiral cords form strong, cusped nodules, which slope more abruptly anteriorly than posteriorly. The spaces inclosed between the spiral keels and the axial ribs are considerably wider than the ribs or cords and form squarish pits. Sutures strongly marked, showing a slender, smooth, peripheral cord (to which the axial ribs extend) on almost all the turns. Periphery and base of the last whorl well-rounded, marked by six well-rounded spiral cords, which grow successively weaker, and a little more closely spaced from the periphery to the umbilicus. Entire surface of spire and base crossed by numerous slender axial lines of growth. Aperture moderately large, channeled anteriorly; posterior angle obtuse; outer lip rendered sinuous by external sculpture; columella stout, flexuose, and reflected;

parietal wall covered with a thick callus. Length, 13.3; diameter, 3.8 mm. (Bartsch.)

TYPE in United States National Museum, No. 196209. Type locality, Lower California, in drift.

RANGE. Destruction Island, Washington, to Lower California.

### Bittium fetellum Bartsch, 1911
### Plate 75, fig. 4

*Proceedings of the United States National Museum,* 40:409; Pl. 51, fig. 4.

Shell moderately large, elongate-conic, light yellow. (Nuclear whorls decollated in all our specimens.) Post-nuclear whorls well-rounded, slightly shouldered at the summit, marked by three, slender, spiral keels, which divide the space between the sutures into four equal areas. (In addition to the three spiral keels, there is a tendency in many of the specimens to have feeble, intercalated cords between the stronger ones.) The axial sculpture consists of decidedly curved, slender ribs, of which 20 occur upon the fourth post-nuclear whorl in the type, 22 upon the fifth, and about 36 upon the penultimate turn. The spaces inclosed between the ribs and the spiral cords are large, shallow, squarish pits on all the turns but the last; on this they are much longer than broad, their long axes coinciding with the axial sculpture. The intersections of the ribs and spiral cords form slender, sharp cusps. Sutures strongly constricted. Base of the last whorl moderately long, slightly curved, marked by four spiral cords, of which the two middle ones are equal and stronger than the others. Entire surface of spire and base marked by numerous strong lines of growth. Aperture large, channeled anteriorly; posterior angle obtuse; outer lip thin, rendered sinuous by the external sculpture; columella very oblique, curved, and reflected; parietal wall glazed with a thick callus. Length, 9.3; diameter, 3.5 mm. (Bartsch.)

TYPE in United States National Museum, No. 198617. Type locality, off Catalina Island in 6 fathoms.

RANGE. Known only from type locality.

### Genus **POTAMIDES** Brongniart, 1810

Shell turriculated, whorls angulated and coronated; aperture prolonged in front into a nearly straight canal; outer lip thin, sinuous; epidermis thick, olive brown. Operculum many-whorled. (Tryon, *Structural and Systematic Conchology.*)

TYPE. *Potomides lamarckii* Brongniart.

DISTRIBUTION. California, Africa, India. In the mud of the Indus they are mixed with species of *Ampullaria.* Fossil, Eocene.

## Subgenus PIRENELLA Gray, 1847

### Potamides cyclus Dall, 1919

*Proceedings of the United States National Museum,* **56**:346.

Shell small, white mottled with brown, probably darker when fresh, with six flattish whorls (excluding the lost nucleus) separated by a rather obscure suture; spiral sculpture of (on the spire and the upper half of the last whorl, three) rows of prominent hemispherical nodules, with subequal interspaces; on the lower half of the last whorl, four undulated or obscurely nodulose flattish cords with narrower interspaces; besides these there is fine spiral threading over the surface and in the interspaces; the nodules are arranged in vertical lines above one another and give the effect of ribs, but the ribs, if any, are very feeble and hardly perceptible in the interspaces; base rounded, aperture rounded, the outer lip slightly thickened, not internally lirate; body and pillar with a thick coat of enamel, pillar shorter than the aperture, with a sharp siphonal sulcus behind it. Height of shell, 6.5; of last whorl, 3.5; diameter, 3.5 mm. (Dall.)

TYPE in United States National Museum, No. 215572. Type locality, Laguna Beach, California.

RANGE. Long Beach, California, to Gulf of California.

## Subgenus LIOCERITHIUM Tryon, 1887

### Potamides sculptum Sowerby, 1855

*Thesaurus Conchyliorum,* **2**:868, figs. 144, 145.

Cerith. testa parva, solida, vertricosa, fusco-nigrescente variegata, spiraliter costellis planulatis, crenulatis, et interstitiis linearibus cincta; anfractibus rotundatis, ultimo magno; apertura rotundata; canali breve; labio externo crenulato fimbriato. (Sowerby.)

A small, solid, darkly variegated, grooved shell of a rather conical form, with rounded whorls. (Sowerby.)

TYPE in British Museum? Type locality unknown.

RANGE. Santa Barbara Islands, California, to Panama and Galapagos Islands.

Described as *Cerithium*.

## Genus CERITHIDEA Swainson, 1840

Shell turriculated, longitudinally ribbed; whorls numerous; summit of spire more or less decollated, aperture rounded, slightly slit anteriorly, outer lip expanded, thickened, broadly rounded below and usually produced into a beak crossing the sinus to the left. Inhabit salt marshes, man-grove swamps, and the mouth of rivers; they are so commonly out of the water as to have been taken for land-shells. Mr. Adams noticed them in the fresh waters of the interior of Borneo, creeping on pontederia and sedges; they often suspend themselves by glutinous threads. (Tryon, *Structural and Systematic Conchology.*)

TYPE. *Cerithidea decollata* Linnaeus.

DISTRIBUTION. India, Ceylon, Singapore, Borneo, Philippines, Port Essington, Pacific Coast.

## Cerithidea californica Haldeman, 1840

*Monograph of the Freshwater Mollusca of the United States,* cover of No. 1.

Epidermis thin, glaucous, under which the shell is chocolate colored, often with a central narrow white band; longitudinally ribbed, and occasionally thickly varicose, crossed by spiral ridges, the crossings varying from nearly smooth to roughly tuberculate; apex usually entire. Length, 1.25 inch. (Tryon, *Manual of Conchology,* 9: 162; Pl. 33, figs. 65, 66.)

RANGE. Baulinas Bay to San Diego, California.

## Family TRICHOTROPIDAE

## Genus TRICHOTROPIS Sowerby, 1829

Testa univalvis, turbinata, carinata, tenuis, apertura longitudinem spire superante, basi integra, columella ad basim oblique truncata, labio externo tenuissimo acuto. Epidermis cornea, super carinas testæ erinacea. Operculum corneum, lamellis ellipticis confertum, nucleo laterali.

The shell may be characterised as being turbinated, and carinated externally; its aperture is wide, but still longitudinal, and it is rather longer than the spire; its base is entire, without any notch, although immediately below the obliquely truncated base of the columella there is an indistinct canal. The whole shell is thin and delicate, its outer lip particularly so. Epidermis horny, forming numerous sharp-pointed bristle-like processes on the edges of the carinæ outside the shell; very strong and by its contraction in drying frequently breaking the edge of the lip. Operculum horny, much smaller than the aperture, composed of elliptical laminæ, its apex or nucleus lateral. (Sowerby.)

TYPE. *Trichotropis borealis* Gould.

DISTRIBUTION. Circumboreal. Fossil, Cretaceous.

## Trichotropis bicarinata Sowerby, 1825

Tankerville, *Catalogue,* Pl. 9.

T. testa anfractibus, quinque, ultimo ventricoso, carinas duabus validis. Long., 1½₀ ; lat., 1¼₀ poll.

Shell with four or five volutions, the last of which is much larger than the others, and ventricose, smooth on the outside, with two prominent keels, which are ornamented with the numerous, strong, sharp-pointed, bristle-like processes formed by the epidermis. The aperture is large, and rather triangular ; being, however, rounded externally, with two obtuse angles, and pointed at the base. The shell has a narrow, linear umbilicus, which is carinated on the outer edge, and its carina is bristly like those on the back ; its inner edge is formed by the elevation of the edge of the inner lip. The columella is rather flattened. Shell white, translucent, epidermis pale horn color. (Sowerby.)

Near the apex of the shell which was in the Tankerville collection, and which is now in the museum of Dr. Goodall, two *Terebratulae* were attached ; on the body-whorl of one of the specimens brought home by Lieutenant Belcher, some small *Balani* have fixed their abode, and two of them are near the lip. Dr. Goodall's specimen was said to have been brought from Newfoundland ; those from which our description was taken were dredged up in from 10–15 fathoms in the Bay between Icy Cape and Cape Lisbon. (Sowerby.)

TYPE in Tankerville Collection. Type locality, Arctic Ocean.

Described as *Turbo bicarinatus,* Sowerby, in Tankerville *Catalogue,* Appendix, p. 12.


## Trichotropis conica Möller, 1842

### See Part II, Plate 31, fig. 12

*Index Molluscorum Groenlandiae,* **12**:

T. conica nob., testa subconica, subumbilicata, lutea ; anfr. 6 angulatis, optimum teretiusculis, postremum ventricosioribus, lincis elatis longitudinalibus cinctis, quarum duæ præcipue eminent, et harum inferior basim a superiore testæ parte acute separat. L. 6‴Rs. (Möller.)

Shell pyramidal, straw-colored, scarcely umbilicated ; spire conical, elevated ; whorls three-keeled with sloped sides ; columella narrow, tortuous ; aperture rather square, flattened anteriorly. (*Conchologia Iconica.*)

TYPE in British Museum. Type locality, Greenland.

RANGE. Forrester Island, Alaska, also Greenland.

## Trichotropis costellata Couthouy, 1838

*Boston Journal of Natural History,* 2:108; Pl. 3, fig. 2.

T. testa ovata, turrito-acuta, fuscescente, epidermide foliacea luteo-albida testa; spira canaliculata; anfractibus costis 5 rotundatis instructis; apertura ovali, lutescente; labro costis indentato; columella arcuata, albescente. (Couthouy.)

Shell ovate, acutely turreted; spire elongated; whorls six, the lowest ventricose and larger (longer) than all the rest together, having four or five and sometimes even six, revolving prominent, rounded costae, with intervening striae; the largest of these costae borders the superior portion of the whorls, forming distinct angulations; numerous fine longitudinal striae, not interrupted by the ribs, cover the whole shell; color a dull yellowish or brownish white; aperture oblong-oval; narrowed and slightly channeled at the base; lip continuous at its superior extremity, and distinctly indented by the costae; columella whitish, arcuated, with a slight projection near its lower third, and abruptly compressed near its base, which inclines to the right, and forms a sharp angle with the outer lip; umbilicus slight, and bordered externally by a strongly marked imbricated ridge. Interior a light brown. Epidermis a dirty whitish yellow, thick, and foliated or laminated by its successive growth, which traversing the costae, gives the shell quite a clathrate appearance. At each stage of the increment, the epidermis is produced upon the costae into two or three, stiff, corneous filaments or bristles, which when the shell is wet, cause it to appear covered with short hairs; the same setae are apparent though less conspicuous upon the ridge bounding the umbilicus. (Couthouy.)

TYPE in Couthouy Collection. Type locality, deep water Massachusetts Bay.

RANGE. Arctic Seas, Bering Sea, Aleutian Islands, and Queen Charlotte Islands, British Columbia. Circumboreal.

## Trichotropis cancellata Hinds, 1843

*Proceedings,* Zoölogical Society of London, 17. *Zoölogy of the Voyage of H. M. S. Sulphur, Molluscae,* Pl. 11, figs. 11, 12.

Tri. testa pyramidata; spira elevata; anfractibus subrotundis, ultimo ventricoso, costellis seto, crenulatis, ciliatis cincto; umbilico parvo; columella tenui, arcuata; apertura rotunda, antice acuta. (*Conchologia Iconica.*)

Shell pyramidal; spire elevated; whorls rather rounded, last whorl ventricose; encircled with eight crenulated, ciliated ribs; umbilicus small;

columella thin, arched; aperture rounded, sharp in front. (*Conchologia Iconica.*)

TYPE in British Museum. Type locality, Sitka, Alaska.

RANGE. Southern part of Bering Sea to Oregon.

### Trichotropis insignis Middendorff, 1849

See Part II, Plate 31, figs. 9, 9a

*Beitrage zu einer Malacologia Rossica,* 107; Pl. 10, figs. 7–9.

Testa ovato-turrita, albescente; anfractibus confertim et grosse carinato-striatis, striis duabus prominentioribus, sub-bicarinatis, angulatis; apertura ampla, ad carinas angulata.

In der Gestalt stimmt diess Art mit der vorhergehenden, in der Skulptur aber nahe mit der folgenden uberuin.

Long., 17; lat., 16; alt. anfr. ult., 13; lat. apertur, 9; colum. ext. long., 8 mm. (Middendorff.)

Shell thick, rather globose, covered with a pale brown epidermis, encircled with undulating crenulated ribs; aperture rather square, with crenulated margin; umbilicus small. (*Conchologia Iconica.*)

TYPE is in Academy, St. Petersburg. Type locality, Bering Strait.

RANGE. Bering Strait, Aleutian Islands, and Cooks Inlet, Alaska. Also Japan.

### Subgenus IPHINOË H. and A. Adams, 1854

### Trichotropis coronata Gould, 1860

See Part II, Plate 31, fig. 7

*Proceedings of the Boston Society of Natural History,* 7:324. Tryon, *Manual of Conchology,* 9: Pl. 7, figs. 53, 54.

T. ovato-rhomboidea, turrita, tenui, cinerea, epidermide fibroso ad carinam in fimbriam cirrosam producto induta; umbilico lato, profundo, acute marginato; anfr. 6 cito crescentibus, postice tabulatis ad angulum carinatis: apertura ovato-triangularis, labro simplici; columella recta, antice vix reflexa, acuta, subcanaliculata; operculum ovatum, corneum, apice terminali. Long., 25; lat., 15 mm. (Gould.)

Umbilicus rather wide and deep, acutely margined; whorls about six, tabulate and carinate at the angle; epidermis ash-colored, fibrously produced at the angle. (Tryon.)

TYPE in United States National Museum. Type locality, Straits of Semiavine, Arctic Ocean, in 20 fathoms.

RANGE. Arctic Ocean and Bering Sea.

## Trichotropis kroyeri Philippi, 1849

### See Part II, Plate 31, fig. 14

*Zeitschrift für Malakozoologie,* 5:175.

Tr. testa oblonga, umbilicata; cingulis obtusis parum elevatis tribus in anfr. supp., sex in ultimo; suturis profundissimis, fere canaliculatis; umbilico infundibuliformi, angulo elevato, obtuso cincto; apertura ovato-oblonga utrinque angulata, angulo inferiore acutiore. Long., 14½; diameter, 10½; alt. apert., 8½; latit, ejus, 5½'''. (Phillippi.)

Shell thick, widely umbilicated; whorls 5, with distinct spiral ribs; umbilicus encircled by a rib; epidermis yellowish brown.

Length, 1.5 inches. (Tryon, *Manual of Conchology.*)

TYPE in ? Type locality, Spitzbergen.

RANGE. Arctic and Bering Sea, eastward to Shumagin Islands.

## Trichotropis kelseyi Dall, 1905

### See Part II, Plate 31, fig. 8

*Proceedings of the United States National Museum,* 34:254.

Shell small, whitish, with a velvety, pale-olive periostracum, and three and a half whorls; spire very short; suture very deep, not channeled, but with the whorl in front of it elevated so as to make a shallow V-shaped trough; nucleus large for the size of the shell, turgid, nor distinctly marked off from the rest of the shell; sculpture of fine, even-rounded closely adjacent, spiral threads, a little more distant on the base, absent from the trough of the suture, with about 22 between the suture and the rim of the umbilical funnel; axial sculpture only of incremental lines; last whorl much the largest, rounded, produced basally, with a deep narrow funicular umbilicus, bounded by a rounded ridge corresponding to a siphonal fasciole; aperture semilunate, rather narrow, produced and almost channeled in front; outer lip thin, arcuate, simple, sharp, not reflected; pillar lip thin, straight, sharp, elevated, connected across the body by a thin layer of callus with the outer lip; pillar absolutely smooth and simple, without any trace of plaits; operculum wanting. Height of shell, 6.2; of last whorl, 5.5; of aperture, 3.5; maximum diameter, 4.0 mm. (Dall.)

TYPE in United States National Museum, No. 110653. Type locality. U.S.S. "Albatross," Station 2936, off San Diego, California, in 359 fathoms.

RANGE. Off San Diego, California, and Point San Quentin, Lower California.

Subgenus PROVANNA Dall, 1918

## Trichotropis lomana Dall, 1918

*Proceedings of the Biological Society of Washington,* 31:7.

Shell thin, white under a dull smooth olive-green periostracum, decollate but indicating more than three whorls; suture deep but not channeled, whorls well-rounded; axial sculpture on the last whorl of about twenty slightly arcuate ribs, with subequal interspaces, ceasing abruptly at the periphery; on the preceding whorls these extend from suture to suture; on the base the axial sculpture is reduced to inconspicuous incremental lines; spiral sculpture, behind the periphery, of obscure close-set threads only visible in the interspaces; in front of the periphery and between it and the canal are four or five strong cords with narrow deep interspaces, slightly crenulating the thin anterior margin of the outer lip; aperture rounded, throat white, showing the impress of the external sculpture; body and inner lip with a thin white layer of callus; pillar arcuate, angulate at the extremity; canal very short and rather wide. Length of (decollate) two and a half whorls, 7.0; of last whorl, 6.0; of aperture, 4.0; diameter, 5.5 mm. (Dall.)

TYPE in United States National Museum, No. 209112. Type locality, U.S. Bureau of Fisheries Station 4354, off Point Loma, San Diego, California, in 650 fathoms.

RANGE. Known only from type locality.

## Family CAECIDAE

## Genus CAECUM Fleming, 1817

*Journal of the Washington Academy of Science,* 10. 1920.

Surface of adult shell marked by numerous axial annulations. Operculum thin, corneous, concave. (Bartsch.)

RANGE. San Diego, California.

Five species are known living on our Pacific shores north of San Diego, California.

TYPE. *Dentalium trachea* Montagu.

## Caecum californicum Dall, 1885

*Proceedings of the United States National Museum,* 8:541. *Supplementary Report,* British Association for the Advancement of Science, 655.

New name for *Caecum cooperi* Carpenter, 1864; not *C. cooperi* Smith, 1862.

Shell small with 30–40 sharp narrow rings. (Carpenter.)

Shell rather narrow, whitish or brownish, with 34–38 close, acute annulations, and subconcave interstices; septum subungulate, apex obtuse, not elevated. Length, .09 inch. (Dall.)

TYPE in United States National Museum, No. 15719. Type locality, San Diego, California.

RANGE. Monterey, California, to Lower California. Fossil: Pleistocene, San Pedro, San Diego, California. Pliocene, San Quentin Bay, Lower California.

This is the most abundant West American *Caecum*. (Dall.)

Paul Bartsch is working on the family *Caecidae* and the following is his classification:

### Caecum dalli Bartsch, 1920

TYPE in United States National Museum, No. 340724. Type locality, San Diego, California.

RANGE. San Diego, California, to Lower California.

### Caecum grippi Bartsch, 1920

Height, 2.5; diameter, 0.7 mm. (Bartsch.)

TYPE in United States National Museum, No. 206961. Type locality, San Diego, California.

RANGE. Known only from type locality.

### Caecum licalum Bartsch, 1920

Height, 2.2; diameter, 0.5 mm. (Bartsch.)

TYPE in United States National Museum, No. 340725. Type locality, San Diego, California.

RANGE. Monterey to San Diego, California.

### Caecum diegense Bartsch, 1920

Height, 2.0; diameter, 0.4 mm. (Bartsch.)

TYPE in United States National Museum, No. 340726. Type locality, San Diego, California.

RANGE. Known only from type locality.

### Genus **MICRANELLUM** Bartsch, 1920

Surface of the shell marked by closely spaced, slender, axial annulations; operculum thin, corneous, concave. (Bartsch.)

Seven species of *Micranellum* are known living in northwestern America. Five of these have the plug at the trunkated apex, forming an attenuated, obliquely placed spur, the base of which is narrower than the diameter of the plug. Of these, *Micranellum pedroënse, catalinense,* and *profundicolum* have the anterior portion of the adult shell bulbously expanded, while in the other two, *M. barkleyense* and *oregonense,* the diameter does not increase at the anterior termination. Of those with the bulbously expanded anterior portion, *M. pedroënse* has very fine, closely spaced annulations, there being about a hundred present in the adult segment of the shell, while in *M. catalinense* and *profundicolum* the annuli are less numerous and more pronounced, there being about seventy-five in the last segment. The shell of *M. catalinense* is shorter and stouter than of *M. profundicolum.* The two species which lack the bulbous anterior expansion, *M. barkleyense* and *M. oregonense,* are distinguished from each other at once by their great difference in size, *M. barkleyense* being both longer and thicker than *Micranellum oregonense.* Two of the seven species, *M. rosanum* and *M. cerbricinctum,* have the spur of the plug expanded basally to cover the entire width of the plug. Of these, *M. rosanum* is easily distinguished from *M. cerbricinctum* by being much longer and having the annuli much more distinct than *M. cerbricinctum.* (Bartsch.)

RANGE. Western North America.

TYPE. *Caecum crebricinctum* Carpenter.

DISTRIBUTION. Alaska to San Diego, California.

## Micranellum crebricinctum Carpenter, 1864

*Proceedings of the California Academy of Sciences,* 3:215. Tryon, *Manual of Conchology,* 8: Pl. 67, figs. 71, 83.

C (Anellum) t., quoad genus, magna tereti, solidiore, rufofusca, interdum radiis intensioribus longitudinalibus ornata; annulis gracillimis, creberrimis, rotundatis, haud elevatis circ. lxxx. cincta; interstitiis nullis; sculptura longitudinali nulla; apertura acuta, vix contracta, vix declivi; septo subangulato, submucronto; margine laterali recto; apice acuto, ad angulum circ. 45° maxime elevato; operculo vix concavo, lira spirali elevata. Long., 0.14; lat., 0.04 poll. (Carpenter.)

This is a shallow water species abundantly distributed from Monterey to Lower California. (Bartsch.) Large, with aspect of *Elephantulum,* but very fine close annular sculpture; plug subungulate.

Described as *Caecum.*

TYPE in State Collection, No. 388. Type locality, San Diego, California.

RANGE. Monterey, California, to Todos Santos Bay, Lower California.

### Micranellum pedroënse Bartsch, 1920

Height, 5.3; diameter, 1.5 mm.
TYPE in United States National Museum, No. 346723. Type locality, San Pedro, California.
RANGE. San Pedro to San Diego, California.

### Micranellum catalinense Bartsch, 1920

Height, 4.5; diameter, 1.3 mm.
TYPE in United States National Museum, No. 211331. Type locality, off Santa Rosa Island, California.
RANGE. Santa Rosa and Catalina Islands, California, in 50 fathoms.

### Micranellum profundicolum Bartsch, 1920

Height, 5.5; diameter, 1.3 mm.
TYPE in United States National Museum, No. 209960. Type locality, off San Diego, California.
RANGE. Off San Diego, in 50 to 199 fathoms.

### Micranellum barkleyense Bartsch, 1920

Height, 6.2; diameter, 1.6 mm.
TYPE in United States National Museum, No. 211589. Type locality, Barkley Sound, Vancouver Island, British Columbia.
RANGE. Known only from type locality.

### Micranellum oregonense Bartsch, 1920

Height, 4.6; diameter, 1.2 mm.
TYPE in United States National Museum, No. 216413. Type locality, Forrester Island, Alaska.
RANGE. Forrester Island, Alaska, to Monterey, California.

### Micranellum rosanum Bartsch, 1920

Height, 6.9; diameter, 1.2 mm.
TYPE in United States National Museum, No. 211859a. Type locality, off Santa Rosa Island, California.
RANGE. Known only from type locality, in 40 fathoms.

### Genus FARTULUM Carpenter, 1856

C. t. laevi, cylindracea, saepius utrinque contracta; apertura saepe declivi. (Carpenter.)

Shell smooth, excepting microscopic incremental lines. (Carpenter.)
TYPE. *Caecum laeve* C. B. Adams.
DISTRIBUTION. Europe, United States, West Indies, Mazatlan, Australia, Japan, Mauritius. Fossil, Eocene, . . . .

## Fartulum orcutti Dall, 1885

*Proceedings of the United States National Museum,* 8:541.
Shell small, smooth, but not polished; light warm brown in color and without sculpture, except very slight lines of growth. Shell slightly curved, the anterior aperture very oblique, about at right angles to the plane of the diameter of the plug, the superior margin being the anterior; plug glandiform, smooth, rounded, without mucro; operculum brown, thin, smooth. Length of shell, 2.0; diameter, 0.75 mm. (Dall.)
TYPE in United States National Museum, No. 20927. Type locality, San Diego, California.
RANGE. San Pedro to Mazatlan, Mexico.

## Fartulum occidentale Bartsch, 1920

TYPE in United States National Museum, No. 152166. Type locality, San Pedro, California.
RANGE. San Pedro to Lower California.

## Fartulum hemphilli Bartsch, 1920

TYPE in United States National Museum, No. 340728. Type locality, San Pedro, California.
RANGE. San Pedro, California, to Lower California.

## Fartulum bakeri Bartsch, 1920

TYPE in United States National Museum, No. 340729. Type locality, San Pedro, California.
RANGE. San Pedro, California, to Lower California.

## Genus ELEPHANTULUM Carpenter, 1858

Surface of shell marked by raised ridges, which coincide with the long axis of the shell; annulations strongly developed; operculum thin, corneous, concave. (Bartsch.)
TYPE. *Caecum hexagonum* Carpenter.
DISTRIBUTION. California and south.

## Elephantulum carpenteri Bartsch, 1920

*Journal of the Washington Academy of Sciences,* **10**:1920.

This is a large thin shell, in which the segments of all stages are marked by very fine spiral sculpture and a little stronger incremental lines, while the last portion of these stages bears well-developed annuli. These increase in number in succeeding stages. Length, 4.8; diameter, 0.9 mm. (Bartsch.)

TYPE in United States National Museum, No. 340726. Type locality, San Diego, California.

RANGE. San Pedro, California, to Point Abreojos, Lower California.


## Family VERMETIDAE

### Genus **BIVONIA** Gray, 1850

Shell affixed, mostly spiral, aperture contracted, circular, with spiral, interruptedly nodulose lirae, and a median elevated line; columella smooth. Operculum small, rudimentary. (Tryon, *Structural and Systematic Conchology.*)

TYPE. *Bivonia triquetra* Bivona.

DISTRIBUTION. Mediterranean, Philippines, west coast, North and South America, and West Africa.


## Bivonia compacta Carpenter, 1864

*Annals and Magazine of Natural History,* ser. 3, **14**:427.

B. testa satis magna, saepe solitaria, purpureo-fusca, spiraliter plerumque satis regulariter contorta, obsoletim cancellata seu sculptura fere evanida; testis tenacissime adhaerente. Long. (plerumque), 7; lat., 3; diameter aperture, 1 poll. (Carpenter.)

Entirely open within; but color and growth alike. (Carpenter.)

TYPE in ? Type locality, Barkley Sound, Vancouver Island, British Columbia.

RANGE. Vancouver Island, British Columbia, to San Pedro, California; Peru?


### Genus **VERMICULARIA** Lamarck, 1799

Shell free, in its early stage regularly coiled like a *Turritella;* subsequently uncoiled, the tube variously twisted, or more or less straight and prolonged. Operculum size of the aperture. (Tryon, *Manual of Conchology.*)

TYPE. ?

DISTRIBUTION. Tropical and subtropical. Fossil, Carboniferous. . . . .

## Vermicularia eburnea Reeve, 1842

*Conchologia Systematica,* 2:46; Pl. 152, fig. 2.

Shell white, loosely whorled, with distant, subobsolete longitudinal sculpture. (Tryon, *Manual of Conchology.*)

TYPE in Cuming Collection. Type locality, South America.

RANGE. West coast, North, Central, and South America; San Pedro, California, to Panama.

## Genus ALETES Carpenter, 1857

Whorls larger than in Section *Macrophragama* Carpenter; columella with a very faint, median, thread-like line; color usually yellowish brown, the early whorls chestnut. Operculum concave externally, with 5 to 6 spiral laminae, the last abrupt; interiorly convex, shining, with irregular spiral lirae; muscular area irregular, opaque. (Tryon, *Manual of Conchology.*)

TYPE. Example: *Vermetus carinatus* Quoy.

DISTRIBUTION. West coast of North America; Peru; Galapagos Islands, Panama.

## Aletes squamigerus Carpenter, 1856

*Proceedings,* Zoölogical Society of London, 226. Tryon, *Manual of Conchology,* 8:181; Pl. 54, figs. 73, 74.

A.t. majore, flavido-albida, solute spirali, plerumque glomerata; superficie costis spiralibus, squamis instructis; costilus pluribus intercalantibus, squamulis minoribus; squamis et squamulis imbricatis, arcuatis; interdum aperturam versus sculptura obsoleta. (Carpenter.)

Yellowish white, usually conglomerated, loosely twisted, becoming erect, loosely longitudinally costate, with intermediate raised lines, scaly decussated, the erect anterior portion marked by rugose incremental striae only. (Tryon, *Manual of Conchology.*)

TYPE in Santa Barbara, California.

RANGE. Monterey, California, to Payta, Peru, and the Galapagos Islands.

## Aletes centiquadrus Valenciennes, 1846

*Voyage de la frégate "Venus,"* Atlas, Pl. 11, fig. 1a. Tryon, *Manual of Conchology,*
    **8**: 174; Pl. 49, fig. 5; Pl. 50, figs. 36–40.

Laterally attached, spirally twisted, earlier whorls rather narrow, rapidly increasing, the last wide, spread out and compressed at the margin;
light yellowish brown, with obscure narrow lines, earlier whorls dark
brown, the interstices of the lirae impressed punctate. (Tryon.)

TYPE in ? Type locality, Gulf of California (?).

RANGE. San Diego, California, to Panama.

Described as *Vermetus.*

## Genus SPIROGLYPHUS Daudin, 1800

Animal excavating a groove on the surface of shells or stones, covering
it over with shelly material, and thus forming a tubular planorbiform case.
When first hatched, the shell is spiral and regular, consisting of one and a
half whorls; it soon attaches itself, the channel it excavates being at first
shallow, afterward deeper; color bright purple to nearly black. The operculum is large, thick, convex exteriorly, with strong concentric laminae,
plane interiorly, concentrically lirate, with central mamilla, and narrowly
elevated margin. (Tryon, *Manual of Conchology.*)

Example, *Vermetus annulatus* Dauden.

DISTRIBUTION. Philippines, Zanzibar, Red Sea, Indian Ocean, California, West Indies on shells, Mediterranean.

## Spiroglyphus lituellus Mörch, 1861

*Proceedings,* Zoölogical Society of London, 154. Tryon, *Manual of Conchology,*
    **8**: Pl. 54, fig. 82.

T. varie torta plerumque lituiformis, profundissime immersa; anfr.
graciles plani, utrinque obtuse angulati, saepe fossula longitudinali mediana, unde carina vel lira obtusa lateris externi; striae et lirae incrementi
arcuatae, approximatae, regulares; apertura circularis interdum obliqua,
nonnunquam soluta; color albescens vel cinereus, interdum fascia obsoleta
fusca, anfr. primi badii. T. infantilis nitida, apice bullata. Diam. ap., 2 m.
(circiter). (Mörch.)

Shell variously twisted, mostly crozier-form, most profoundly im-
mersed; whorls slender and plane, on one side obtusely angulated, often
with a largest median fossula, from which a carina or obtuse riblet comes
on the outer side; striae or lirae of growth arcuate, approximate, regular;
aperture circular, sometimes oblique, occasionally desolved; color whitish
or ashy, sometimes with obsolete tawny dashes, first whorls brownish.

Larval shell polished, with swollen apex. Diameter about 2 m. (Translation.)

Type in Cuming Collection. Type locality, imbedded in the surface of a young Haliotis.

Range. Forrester Island, Alaska, to San Diego, California.

## Genus PETALOCONCHUS Lea, 1843

Animal ignotum. Operculum (speciebus ii.) parvum, corneum, diapanum, tenuissimum; anfractibus paucis, vix apparentibus. Testa extus Bivoniae similis; intus transversim rarissime septata; anfractibus medianis laminis elongatis spiralibus varie dispositis, cameram saepe secantibus; plerumque duabus, plica columellari una.

Shell of corkscrew growth, glomerate, or single, affixed by one side of the whorls; earlier and later whorls open; middle whorls divided by spiral laminae, often of complex structure, which gradually pass away at each end; generally two, nearly meeting, with a third rudimentary, forming a columellar plait. (Lea.)

Example. *Vermetus sculpturatus* Lea.

Distribution. Mare Mediterraneum; Oceanum Atlanticum, Pacificum, Indicum.

## Petaloconchus anellum Mörch, 1861

*Proceedings,* Zoölogical Society of London, 359. Tryon, *Manual of Conchology,* 8: 173; Pl. 49, fig. 34.

T. sinistralis, candida, spira affixa, spirorbiformis, umbilico aperto; anfr. pauci, ultimus solum adspectabilis, rapide crescens, peripheria dilatata, costis transversis acutis valvidis, leviter flexis; apertura soluta, subovalis, latere interno dilatato, recto, superne et inferne angulo recto. Diam. testae fere, 3¼ m.; aperturae, 1 m. (Mörch.)

Shell sinistral, spirorbiform, white, spire affixed, umbilicus open, whorls few, with transverse close ribs. Diameter of shell about 2 to 3.5 mm. (Tryon.)

Type in British Museum (?). Type locality, California on Haliotis.

Range. Monterey, California, to Todos Santos Bay, Lower California. Described as *Vermetis;* in United States National Museum, *Bulletin 112,* as *Vermiculum annellum.*

## Petaloconchus macrophragma Carpenter, 1857

*Mazatlan Catalogue,* 309. Tryon, *Manual of Conchology,* 8: Pl. 48, fig. 16.

P.t. parva, dextrali, dense purpureo-fusca; cylindracea, solute spirali, marginibus spirae saepe subparallelis; plerumque glomerante, interdum solitaria; basi ad conchas, seu altera ad alteram, constricte adhaerente, saepe erodente; superficie rugis irregularibus spiralibus haud extantibus, et rugulis incrementi ornata; intus anfr. promis, et ultimis quoad iv., apertis; medianis laminatis; lamina superiore multo majore, prius conspicua, a columella extante; primum simplici, dein angulo recto reflexa, extus, carinus i–iii. quarum ii. acutissimis; lamina inferiore simplici, a columella extante, altero juxta carinam fere attingente; lamina tertia minima, intercalante, inferiori paene attingente; laminis tenuissimis, albis, diaphanis, lineis incrementi conspicuis; pagina interna maxime nitente, transversim haud septato. (Carpenter.)

The shell is of small diameter; when growing freely taking a tolerably regular spiral, like a *Turritella* squeezed sideways; the whorls enlarging very slowly, and resembling a winding staircase. It is known when fresh by its lustrous purple-brown color and absence of pits on the surface. It occasionally eats into the shell on which it grows, like *Spiroglyphus,* from which it is known by being dextral and cylindrical. Sometimes it clusters in large masses, like *Bivonia glomerata.* At which whorl the internal laminae commence, it is difficult to say; they have been counted running through 10 whorls; and the terminal number of open whorls appears to vary. At each end they commence (the large upper one first) as simple columellar plaits, afterward increasing till they fill the greater part of the cavity and nearly meet. The upper one bends, at right angles, with two sharp keels at the corners, and a third, not always developed, on the upper side. The lower one goes to meet it, forming with the columella a rectangle, only communicating with the remainder by a narrow slit. A small columellar plait supports the inside base of the lower lamina. The delicate texture of these laminae resembles the cup in *Crucibulum,* etc. No transverse partitions have been seen. Long., 65; lat., 13 poll. (Carpenter.)

TYPE in Liverpool and Havre collection. Type locality, Mazatlan.

RANGE. San Diego, California, to Panama.

## Petaloconchus complicatus Dall, 1908

*Bulletin,* Museum of Comparative Zoölogy, **43**:326.

Coil at first closely wound and more or less obliquely bent in conformity with its situs, the first few turns with a subcircular section, very irregularly disposed, those following with a roughly hexagonal section closely coiled around a barely perforate axis, closely coherent to each other and running up to twenty or more in number, after which the tube once more becomes erect, with a circular section, and a slight dextral twist; aperture circular, often with a series of internal concentric lamellae as if the animal had attempted to contract the opening by secreting a succession of smaller tubes within it; sculpture irregular; apart from incremental lines, there is a longitudinal, irregular, but tolerably smooth ridge on the middle of the exposed whorl with strong wrinkles at right angles to it on each side but not crossing it. At resting stages there are sometimes angular projections of the margin of the temporary aperture; the wrinkles are sometimes reticulated by longitudinal subequal threads, which may be entirely wanting on other parts of the same individual; the erect portion is nearly smooth except for lines of growth; the color is a pale, ferruginous brown. Length of the coiled portion may be about 16; of the erect part, 27; the tube with a diameter at the aperture of 2.3 mm. There is one internal basal lamella and a smaller one projecting from the axial wall at about right angles to and a little above the former. (Dall.)

TYPE in United States National Museum, No. 123035. Type locality, U.S.S. "Albatross" Station 3368, near Cocos Island.

RANGE. Puget Sound to Panama.

## Petaloconchus montereyensis Dall, 1919

*Proceedings of the Biological Society of Washington,* **32**:250.

Shell yellowish white, closely irregularly loosely coiled, occurring in masses, the sculpture of crowded concentric wrinkles with, on the earlier part of the shell, an occasional obscure longitudinal ridge; the internal lamellae occur only in the earlier coils, the later portion shows no traces of them. Maximum diameter at the aperture, 2 mm. (Dall.)

TYPE in United States National Museum, No. 334650. Type locality, Monterey, California.

RANGE. Known only from type locality.

## Family TURRITELLIDAE

## Genus **TURRITELLA** Lamarck, 1799

Shell elongated, many-whorled, whorls rounded with revolving striae; aperture rounded. Operculum many-whorled, with a fimbriated margin. The shells are usually brown, with red-brown spots or flames. (Tryon, *Structural and Systematic Conchology.*)

TYPE. *Turritella terebra* Linnaeus.

DISTRIBUTION. World-wide, ranging from the Laminarian Zone to 100 fathoms. West Indies, United States, Britain, Iceland (esp.), Mediterranean, West Africa, China, Australia, West America. Fossil, Triassic; Britain, North and South America, Australia, Java.

### Turritella jewetti Carpenter, 1864

*Supplementary Report,* British Association for the Advancement of Science, 655. *Annals and Magazine of Natural History,* ser. 3, **17**:276.

T. testa satis tereti, haud tenui, cinerea rufo-fusco tincta; anfr. subplanatis, suturis distinctis; lirulis distinctibus (quarum t. jun. duae extantiores) et striolis subobsoletis spiralibus cincta; basi parum angulata; apertura subquadrata; labro tenui, modice sinuato. (Carpenter.)

Shell turreted, with slender, tapering spire; number of whorls variable; whorls flat, with two distinct spiral ridges on anterior portion and three or four less distinct ridges posteriorly; incremental lines distinct and concave anteriorly; suture thread-like and rather indistinct; not usually impressed; aperture angular. (Arnold, *Paleontology of San Pedro, California.*)

TYPE in United States National Museum. Type locality, Santa Barbara, California. Pleistocene.

RANGE. Santa Barbara, California, to Salina Cruz, Mexico. Fossil: Pleistocene—Santa Barbara to San Pedro, San Diego, California; Pliocene, San Diego Well, San Pedro, California.

### Turritella mariana Dall, 1908

*Bulletin,* Museum of Comparative Zoölogy, **43**:327; Pl. 11, fig. 14.

Shell slender, pale pinkish-brown, acute, with about eighteen whorls; suture rather obscure, not appressed; whorls strongly constricted in the middle, a sharp keel on each side of the constriction, which in the last two or three whorls is undulate or obscurely beaded; nucleus lost; on the earlier whorls the keels are entire and the whorl slopes about the same distance from each to the adjacent suture; on the latter whorls there is a single thread behind the posterior keel, and two more are in the trough between the two keels, all somewhat undulate, low, and inconspicuous; base of the whorls with a sharp carina upon which the suture is laid; within the carina deeply concave, with well-marked lines of growth and microscopic spiral striae; aperture rounded-quadrate, the outer lip with the margin at first retractive to a wide sinus nearly coincident with the posterior keel, then prominently protractive to the basal keel, thence in a deeply excavated curve to the pillar, which is thin and arcuate, short, and gyrate, about a narrowly pervious axis. Long. of shell, 25; of last whorl, 6; of aperture, 3; max. diam., 5 mm. (Dall.)

TYPE in United States National Museum, No. 123036. Type locality, U.S.S. "Albatross" Station 3427, near the Tres Marias Islands, in 50 fathoms.

RANGE. Catalina Island, California, to Panama Bay.

### Turritella cooperi Carpenter, 1866

*Proceedings of the California Academy of Sciences*, 3: 216. Tryon, *Manual of Conchology*, 8: 200; Pl. 61, fig. 61.

T. t. valde tereti, tenuiore, cinerea, rufo-fusco tincta; anf. plurimis, angustis, subplanatis, suturis distinctis; liris ii. et striulis crebris spiraliter ornata; basi angulata; apertura subquadrata; labro valde sinuato. Long., 1.80; long. spir., 1.50; lat., 0.45. (Carpenter.)

Whorls about 17, slopingly flattened, excavated at the suture, two-ridged and finely striated spirally, the last whorl doubly ridged below, with a single ridge above; yellowish, longitudinally flamed with chestnut or chocolate. Length, 1.75 in. (Tryon and Pilsbry, *Manual of Conchology*.)

TYPE in State Collection, No. 564. Type locality, San Ped. o, California.

RANGE. Monterey to San Diego, California.

### Turritella nodulosa King, 1831

*Zoölogical Journal*, 5: 347. Tryon, *Manual of Conchology*, 8: 202; Pl. 63, figs. 78, 79.

Testa elongato-turrita; anfractibus striatis; striis duabus maximis sub-nodulosis. (King.)

The two large striae, which are remarkable for the nodules, are not far from the middle of each whorl, and generally are nearer the upper suture: of these the lowest is the larger. (King.)

Whorls 14 or 15, nodulously ridged, a central ridge usually stronger, making an angulation, concave and striate in the middle; light yellowish brown, longitudinally chestnut-flamed between the nodules. Length, 2 inches. (Tryon.)

TYPE in British Museum. Type locality not known to writer.

RANGE. San Diego, California, to Panama.

### Genus TACHYRHYNCHUS Mörch, 1868

Shell like a small *Turritella*, the whorls of the spire grooved across, and rounded; sculptured by spiral and longitudinal lines. Aperture oval, slightly produced in front into a rudimentary channel; lip sharp.

EXAMPLE. *Turritella acicula* Stimpson.

DISTRIBUTION. Arctic to San Pedro, California; also Atlantic.

## Tachyrhynchus reticulatus Mighels, 1842

*Boston Journal of Natural History,* **4**:50; Pl. 4, fig. 19.

T. testa turrito-subulata; anfractibus duodecim, convexis, longitudinaliter plicatis, transversim striatis; sutura valde impressa; apertura suborbiculari.

Shell turreted, very slender, of a dingy white or ash color; whorls eleven to twelve, convex, distinctly, though somewhat irregularly plicate longitudinally, with from three to five delicate, impressed, revolving striae on the five lower whorls; from and above the fifth whorl the transverse striae gradually diminish in number, until they wholly disappear on the upper two or three whorls. The whole surface of the shell has a reticulated appearance. Suture well impressed; aperture suborbicular; labrum thin; operculum horny. Length, 7; breadth, 2 in. (Mighels.)

TYPE in Boston Society of Natural History. Type locality, Bay Chaleur, in the Gulf of St. Lawrence.

RANGE. Arctic Ocean and eastern Bering Sea to the Aleutian Islands, to British Columbia. Also Greenland to Gulf of St. Lawrence.

## Tachyrhynchus erosus Couthouy, 1838

*Boston Journal of Natural History,* **2**:103; Pl. 3, fig. 1.

Shell turreted, with from nine to eleven slightly convex whorls, tapering gradually to a point; sutures deeply impressed; whorls having from three to five transverse sulci, with alternate, slightly elevated, rounded striae or costae, most numerous and strongly marked upon the lower whorl; striae of growth apparent, slightly wrinkling the shell longitudinally; the portion above the last three whorls usually much eroded; aperture rounded, lip thin and impressed by the termination of the costae; columella with a very slight callus, having an angular base; color of shell a reddish horn color. Long., sixteen-twentieths; basal diam., six-twentieths of an inch. (Couthouy.)

TYPE in author's collection. Type locality, Massachusetts Bay, coast of Maine.

RANGE. Arctic Ocean to Kuril Islands, the Aleutian and Kodiak Islands. Also Greenland south to Massachusetts Bay, Maine.

## Tachyrhynchus erosus major Dall, 1919

*Proceedings of the United States National Museum,* **56**:346.

Shell resembling *T. erosus,* but much larger and coarser, of eight or more whorls, the apex as usual much eroded; suture almost appressed, rather obscure; spiral sculpture of (between periphery and suture, five) channeled grooves with low, flattish, wider interspaces; on the flattish base there are four more, closer and diminishing toward the axis; aperture rounded, outer lip thin, deeply, widely, retractively arcuate, inner lip with a glaze of enamel, pillar lip slightly thickened and at the anterior end a little produced; the interior of the aperture is smooth and there is no umbilicus. Height of (decollate) shell, 35; of last whorl, 14; diameter, 13 mm. (Dall.)

TYPE in United States National Museum, No. 224134. Type locality, off the Pribilof Islands, Bering Sea, in 51 fathoms.

RANGE. Mouth of the Colville River, Arctic Alaska, Pribilof and Unimak Islands.

## Tachyrhynchus stearnsii Dall, 1919

*Proceedings of the United States National Museum,* **56**:347.

Shell small, slender, acute, white or pale brownish, with a minute subglobular nucleus and about ten subsequent, well-rounded whorls, regularly increasing in size; suture distinct, not appressed; spiral sculpture of (on the spire, three; on the last whorl, four) low prominent lines about equally spaced above the base, and minute, almost microscopic, close-set spirals in the interspaces and on the convexly rounded base; axial sculpture not perceptible; aperture rounded, simple, the outer lip somewhat arcuate, the body thinly glazed. Height of shell, 13; diameter, 4.5 mm. (Dall.)

TYPE in United States National Museum, No. 74014, in Stearns Collection. Type locality, San Pedro, California.

RANGE. Known only from type locality.

## Tachyrhynchus pratomus Dall, 1919

*Proceedings of the United States National Museum,* **56**:347.

Shell small, yellowish white, acute, with a glassy swollen nucleus of about one whorl and six subsequent whorls; suture distinct, not deep, axis imperforate; early whorls moderately convex, the periphery nearer the succeeding suture, with numerous low, rounded, rather irregular ribs extending from suture to suture, with subequal interspaces; these ribs become gradually obsolete on the later whorls; spiral sculpture of low, irregularly distributed, partially obsolete threads, with minute threadlets between them; this sculpture covers the entire whorl; base convexly regularly rounded with no umbilical depression; aperture ovate, simple, the lips sharp, with no callosities on the body. Height, 10; diameter, 4 mm. (Dall.)

TYPE in United States National Museum, No. 219369. Type locality, Semidi Islands, Alaska, at Station 1152, in 20 fathoms.

RANGE. Semidi Islands, Alaska, to Lower California.

## Tachyrhynchus lacteolus Carpenter, 1865
See Part II, Plate 31, fig. 11

*Proceedings,* Philadelphia Academy of Natural Sciences, 62.

M.t. parva, tereti, tenui, albida; epidermide tenui, flavida, induta; anfr. x. rotundatis, suturis valde impressis, laevibus; costis circ. xii. radiantibus, tumentibus, suturam versus evanidis, interstitiis parvis; costis spiralibus rotundatis, costas radiantes transeuntibus, supram iii. validioribus, aliis interdum intercalantibus; costulis suturalibus parvis, antice ii.; basi rotundata, vix striata; columella recta, paulum effusa; labro tenuissimo, parum arcuato. Long., 33; long. spir., 24; lat., 14. (Carpenter.)

M. with small shell, rounded, slender, whitish; covered with a thin yellowish epidermis; with ten rounded whorls, with suture strongly impressed; with about 12 radiating costae, swelling, and becoming obsolete toward the suture; with small interstices; with rounded spiral costae crowning the radial costae, with the upper three stronger, with others sometimes intercalating; with smallish sutural costae, 2 in front, with rounded base, scarcely striate, with straight columella, spread out gradually; with very thin outer lip, slightly arcuate. (Translation.)

TYPE in ? Type locality, Vancouver district.

RANGE. Chinik Bay, Alaska Peninsula, east and south to Point Abreojos, Lower California.

Described as *Masalia.*

## Tachyrhynchus lacteolus subplanatus Carpenter, 1865

*Proceedings,* Philadelphia Academy of Natural Sciences, 62.

M. t. *M. lacteolae* simili; sed sculptura minus extante anfractibus subplanatis; costis radiantibus pluribus minoribus, costulis spiralibus interdum intercalantibus, aperturam versus saepe obsoletis. (Carpenter.)

M. with shell like *M. lacteolae;* but with sculpture less pronounced on the flattened whorls; with many smaller radiating costae, and sometimes with secondary intercalating costae, often obsolete toward the aperture. (Translation.)

TYPE in ? Type locality, Puget Sound, Washington.

RANGE. Puget Sound to San Diego, California.

Described as *Mesalia lacteola subplanatus.*

## Genus TURRITELLOPSIS G. O. Sars, 1878

Shell, like *Turritella,* but the lingual dentation differs. (Tryon, *Structural and Systematic Conchology.*)

TYPE. *Turritellopsis acicula* Stimpson.

DISTRIBUTION. Boreal.

## Turritellopsis acicula stimpsoni Dall, 1919

*Proceedings of the United States National Museum*, **56**:345. Sars, *Mollusca regionis arcticae Norvegiae*, Pl. 10, fig. 14.

A comparison of the figures of Stimpson and Sars, both elegant and accurate draftsmen, shows at once that there is a wide difference between them. *T. acicula* is more slender, with a looser coil and a much more constricted suture. It has three strong spiral cords, while *T. stimpsoni* has five to seven. The West Coast specimens so far obtained are of the *stimpsoni* type, which is represented in the Jeffreys collection from Lofoten, Spitzbergen, and Maine. The original *acicula* type came from Grand Manan; Portland, Maine; and Newfoundland. On the Pacific side we have the *stimpsoni* from Nunivak Island, Fort Etches; Shoal Bay, British Columbia; and San Diego, California. (Dall.)

TYPE in United States National Museum. Type locality ?

RANGE. Nunivak Island, Alaska, to San Diego, California. Circumboreal.

## Family LITTORINIDAE

### Genus **LITTORINA** Ferussac, 1822

Shell turbinated, thick, pointed, few-whorled; aperture rounded, outer lip acute, columella rather flattened, imperforate. Operculum paucispiral. (Tryon, *Manual of Conchology*.)

TYPE. *Littorina littorea* Linnaeus.

DISTRIBUTION. World-wide. Fossil, Norwich Crag, England, Miocene . . . .

### SECTION ALGARODA Dall, 1918

### Littorina squalida Broderip and Sowerby, 1829

*Zoölogical Journal*, **4**:370. *Zoölogy of Beechey's Voyage*, **139**: Pl. 34, fig. 12.

L. testa obovali, apice acuminato; anfractibus superne depressiusculis; apertura rotunda, labio superne coarctato.

Shell very much like the common Periwinkle, but not so thick and heavy; it is further distinguished by the contraction of the upper part of the lip. Long. 1; lat. 8/10 poll. (Broaderip and Sowerby.)

TYPE in British Museum. Type locality, Oceano Boreali.

RANGE. Cape York, Arctic Ocean, to Pribilof Islands and the Okhotsk Sea.

SECTION LITTORIVAGA Dall, 1918

## Littorina aleutica Dall, 1872

*Proceedings of the California Academy of Sciences,* 4:271; Pl. 1, fig. 3.

Shell depressed, whorls four, the nucleus including one and a half. Last whorl much the largest; spire depressed or nearly flattened. Color variable, from dark brown or purple to waxen white, or banded with white on a darker ground. Nucleus polished, dark brown, translucent. Sculpture consisting of rather coarse lines of growth, and about six or eight nodulous revolving ridges more or less strongly elevated in different specimens, the three middle ones being the most prominent, and faint revolving lines being also traceable occasionally between the ridges. Aperture very oblique, smooth, white or purplish within, outer lip sharp, columella broad, straight, generally with a chink behind it. Anterior margin a little produced. Long., .41; lat., .53 in. of an average specimen. (Dall.)

TYPE in United States National Museum. Type locality, Gull Rocks, Akutan Pass, Aleutian Islands.

RANGE. Pribilof Islands and the Aleutians from Kyska to Chika Islands.

## Littorina sitchana Philippi, 1845

*Proceedings,* Zoölogical Society of London, 140. Tryon, *Manual of Conchology,* 9: 240; Pl. 41, fig. 90.

Subglobose, moderately solid, strongly spirally ridged, usually with subequal intervening furrows; dark chocolate color, often with a broad white peripheral band, interior chocolate. Length, 15–18 mm. (Tryon.)

TYPE in British Museum. Type locality, Vancouver Island, British Columbia?

RANGE. Southern Bering Sea, both coasts and southward; Puget Sound.

## Littorina atkana Dall, 1886

*Proceedings of the United States National Museum,* 9:211. *Ibid.,* 24:551; Pl. 39, fig. 11.

Typical form of shell large, solid, nearly smooth, the whorls flattened next the suture, a few obsolete striations on the base, the general form as figured, the outer lip thin, the pillar broad and white. Alt., 20.0; lat., 17.0 mm. (Dall.)

The most abundant form is of a dark chestnut brown throughout, except on the pillar. The variety figured has white bands at the suture, periphery, and umbilical region. These bands do not vary in position. (Dall.)

TYPE in United States National Museum, No. 108986. Type locality, Bering Island.

RANGE. Kyska Island, Alaska, to Cook's Inlet.

## Littorina rudis Donovan, 1800

*British Shells,* 1; Pl. 33, fig. 3.

Shell somewhat tapering, without umbilicus; volutions of the spire, or turbom swelled; lip thick, and glossy within. (Donovan.)

In the mud appeared a species of *Turbo* which, though very similar to *T. littoreus* (the common Periwinkle), has some characters that seem to authorize its being considered as a different shell. The anfractus are much more swollen, as it were, than in the above species; the spire is more depressed; and, besides, there is no appearance of striae, either transversely or longitudinally. This shell has a sort of distorted or rude contour, that may, perhaps, entitle it to the appellation of *T. rudis.* Its color is greenish.

This shell has been kindly communicated by Dr. Maton, as a new species. It is noticed in the first volume of that gentleman's *Observations on the Western Counties;* but has not, we believe, been hitherto figured. (Donovan.)

TYPE in ? Type locality, Banks of the Tamar in Devonshire.

RANGE. Aleutian Islands to Puget Sound, Washington. Also Atlantic.

## Littorina grönlandica Menke, 1830

*Synopsis Methodica Molluscorum,* **45.**

Litt. testa ovato-ventricosa, tenuiculta, anfractibus convexa, spiraliter obsolete liratis; livido-castanea; columella dilatata.

Shell ovately ventricose, rather thin, whorls convex, spirally obsoletely ridged; livid-chestnut; columella broadly dilated. (*Conchologia Iconica.*)

TYPE in Museum Cuming. Type locality, Greenland.

RANGE. The Okhotsk and Bering seacoasts and eastward to Sitka, Alaska; Puget Sound. Also Greenland.

## Littorina planaxis Philippi, 1847

*Abbildungen und Beschreibungen neue oder wenig gekannte Conchylien,* Littorina, Pl. 4, fig. 16.

L. testa ovata, acuta, solida, laevigata, griseo-fusca, maculis caerulescente-albis variegata; anfractibus parum convexis, ultimo ventricoso, obscure angulato; apertura ovata, fusca, (fascia solita basali alba); columella arcuata, dilatata, depressa. Alt., 8'''; diameter, 6½'''; alt. apert. 5'''. (Philippi.)

Whorls convex, rapidly increasing, smooth or very minutely spirally striate, light chocolate color, shining, under a thin olivaceous epidermis, speckled and spotted irregularly with white; interior chocolate color, with a white band near the base; columella broadly excavated, yellowish brown. (Tryon, *Manual of Conchology.*)

TYPE in British Museum. Type locality, California superior.

RANGE. Puget Sound to Magdalena Bay, Lower California. Guadeloupe and Socorro Islands.

## Subgenus MELARHAPHE (Mühlfeldt) Menke, 1828

### Littorina scutulata Gould, 1849

*Proceedings of the Boston Society of Natural History,* **3**: 83. Tryon, *Manual of Conchology,* **9**: Pl. 45, figs. 98–100-1-3.

Testa parva, ovato-conica, plerumque erosa, castanea vel livida albido inordinatim maculata, striis obsoletis cincta; spira anfr. 5, ultimo ventricoso; apertura late ovata; labro acuto, pallido; columella planata, antrorsum expansa; fauce livido. Long. ⅗; lat. ⅖ poll. (Gould.)

Faintly striate with spiral impressed lines, olivaceous chestnut or chocolate color, including aperture, sometimes not variegated, but usually with longitudinal, zigzag, white markings, sometimes broken up into spots, and frequently with an articulated white and chestnut band on the periphery. Length 9–12 mm. (Tryon.)

TYPE in United States National Museum? Type locality, Puget Sound.

RANGE. Kodiak Island, Alaska, to Turtle Bay, Lower California, and Socorro Island.

## Subgenus ALGAMORDA Dall, 1918

### Littorina newcombiana Hemphill, 1876

*Proceedings of the California Academy of Sciences,* **7**:49. *Bulletin 112,* United States National Museum, Pl. 5, fig. 5.

Shell thin, turbinate, with four or five rounded whorls; apex subacute, last whorl somewhat inflated, subrimate, with or without three or four longitudinal brown bands; aperture ovate, outer lip thin, inner lip appressed to the columella and somewhat thickened; suture deep; epidermis greenish.

Operculum with nucleus sub-central with 2½ whorls. (Hemphill.)

TYPE in Hemphill Collection, Stanford University. Type locality, Humboldt Bay, California.

RANGE. Known only from type locality.

Described as *Paludinella.*

## Littorina subrotundata Carpenter, 1865

*Annals and Magazine of Natural History,* ser. 3, 15:28.

A. testa haud parva, laevi, tenui, fusco-olivacea; anfr. nucl.? . . .
(decollatis); norm. v., rapide augentibus, subrotundatis; marginibus
spirae rectis, suturis valde impressis; basi rotundata, haud umbilicata;
apertura rotundato-ovali, intus fuscescente; peritremate continuo; labro
acuto; labio parum calloso; columella arcuata. Long., 28; long. spir., 13;
lat., 2 poll. (Carpenter.)

TYPE in United States National Museum. Type locality, Neah Bay,
Washington.

RANGE. Known only from type locality.

Described as *Assiminea.*

## Littorina castanea Carpenter, 1865

*Annals and Magazine of Natural History,* 15:28.

P. testa compacta, solidiore, fusco-castanea, marginibus spirae rectiori-
bus; rugulosa, lineis distantibus spiralibus irregulariter insculpta; anfr.
nucleosis? . . . (detritis), vertice late mammillato; norm. iv., rapidius
augentibus, tumidioribus, suturis satis impressis; basi regulariter excurvata, vix rimata; apertura suborbiculari, haud continua; labro acuto; labio
supra parietem obsoleto, supra columellam arcuatam intus calloso; operculo, anfr. iv. haud rapide augentibus. Long., .21, long. spir., .09; lat., .17,
div., 70. (Carpenter.)

Shell compact, rather solid, tawny-chestnut, with rather straight margins of the spire; rugose, sculptured irregularly with distant spiral lines;
nuclear whorl worn, with vertex broadly mamillate; normal whorls 4,
growing rather rapidly, thick, with strongly impressed sutures; basis regularly excurvate, scarcely rimate; aperture suborbicular, not continuous;
with outer lip acute; inner lip obsolete above the wall, within callous above
the arcuate columella; operculum with four whorls, not rapidly increasing.
(Translation.)

TYPE in United States National Museum. Type locality, Neah Bay,
Washington.

RANGE. Known only from the type locality.

Described as *Paludinella castanea.* (Carpenter.)

## Family LACUNIDAE

### Genus **LACUNA** Turton, 1827

Shell turbinated or globular, thin, covered by an epidermis; aperture
semilunar, columella flattened, with a parallel groove behind it ending in

the umbilicus, lip sharp, arcuated. Operculum paucispiral. (Tryon, *Manual of Conchology.*)

TYPE.  *Lacuna puteola* Turton.

DISTRIBUTION.  Japan, Europe, west coast of North America, Mediterranean, North Atlantic. Fossil, Eocene, . . . .

## Lacuna porrecta Carpenter, 1864

See Part II, Plate 32, fig. 9

*Annals and Magazine of Natural History,* series 3, **14**:428.

L. testa *L. puteolo* simili, sed multo majore, spira magis exserta; seu omnino fusca, seu zona pallidiore, seu pallida lineolis fuscescentibus tenuissime spiraliter ornata; epidermide tenuiter striata olivacea seu viridescente induta; tenuiore, spiraliter tenuiter striata; anfr. v., vix planatis, rapide augentibus, suturis impressis, vertice mamillato; apertura tumente; labio tenui, vix parietem attingente, intus subrecto; lacuna maxima, elongata, ad basin arcuata; peripheria expansa. Long., 52; long. spir., 2, lat., 4 poll. (Carpenter.)

Larger than *L. puteola,* with more developed spire, body whorl slopingly expanded; effuse below, thinly spirally striate upon the thin olivaceous yellow or light-brown epidermis, sometimes with a pale band under the obtusely angulated periphery. (Tryon, *Manual of Conchology.*)

TYPE in United States National Museum. Type locality, Vancouver Island.

RANGE. Commander Islands, Bering Sea, to San Diego, California.

## Lacuna porrecta effusa Carpenter, 1864

*Supplementary Report,* British Association for the Advancement of Science, 656. *Annals and Magazine of Natural History,* ser. 3, **14**:428.

L. var. *effusa:* testa *L. porrectae* simili, sed multo majore; spira elevata, satis effusa; anfr. tumidioribus, suturis valde impressis; aperturam versus magis expansa. Long., 65; long. spir., 25; lat., 42 poll. (Carpenter.)

Larger, more effuse, whorls more tumid, with well-impressed suture. (Tryon, *Manual of Conchology.*)

TYPE in United States National Museum. Type locality, Neah Bay, Washington.

RANGE. Straits of Juan de Fuca to San Francisco, California.

## Lacuna porrecta exaequata Carpenter, 1864

*Supplementary Report,* British Association for the Advancement of Science, 656. *Annals and Magazine of Natural History,* ser. 3, **14**:428.

*L.* var. *exaequata:* testa *L. effusae* simili, sed anfr. planatis, suturis parum impressis. Long., 5; long. spir., 2; lat., 42 poll. (Carpenter.)

The form *L. exaequata* is intermediate between the very different *L. porrecta* and *L. effusa.* The lacunae vary so much (*vide* Forbes and Hanley *in loco* that, even with a large multitude of specimens, it is not easy to state what constitutes a species. (Carpenter.)

Whorls flattened, sutures scarcely impressed. Length, 12.5 mm. (Tryon, *Manual of Conchology.*)

TYPE in United States National Museum. Type locality, Neah Bay, Washington.

RANGE. Known only from type locality.

## Lacuna porrecta puteoloides (Carpenter MS) Dall, 1919

*Proceedings of the United States National Museum,* **56**:349.

The only difference between it and the typical *L. porrecta,* to which Doctor Carpenter allied it as a variety, appears to be a considerably narrower pillar and consequently sharper and narrower umbilical groove in the variety. (Dall.)

TYPE in United States National Museum, No. 46984. Type locality, Lobitas, California.

RANGE. Known only from type locality.

## Lacuna glacialis Möller, 1842

*Index Molluscorum Groenlandiae,* 9. Tryon, *Manual of Conchology,* **9**: Pl. 55, fig. 76.

Testa ovato-conica, rufo-fusca; anfr. 5 cylindraceis, plicis membranaceis angustis, cinereis, confertis ornatis. L. 5'''. Rs. (Moller.)

Stout, solid, opaque, epidermis light yellowish; whorls convex, a little tumid or round-shouldered above, epidermis thin, often gathered into obliquely longitudinal folds; under it the surface is minutely spirally striate; pillar lip flattened, expanded; imperforate. Length, 12.5 mm. (Tryon, *Manual of Conchology.*)

The English description is of *Lacuna crassior* Mtg., which is equal to *Lacuna gracialis.*

TYPE locality, Greenland.

RANGE. St. Paul Island, Alaska; also Greenland.

## Lacuna divaricata Fabricius, 1780

*Fauna Grönlandica,* 392. Tryon, *Manual of Conchology,* **9**: Pl. 50, fig. 76.

Trochus testa subumbilicata ousta, infractu onfimo remotiore, umbilico subconsolidato. Long., 5½ ; lat., 3 lin. (Fabricius.)

Testa oblonga, subconica, spira satis obliqua, laeuis, diaphana, glabra non nisi oculo armato subtilissime reticulata. Anfractus 6 convexi, sursum valde decrescentes: insimus enim reliquis simul sumtis dimidio major, ventricosus, et sutura sua versus aperturam remotuor. Apertura sublunaris ampla, labro renui marginato. Columella rimam lanceolatam habet, de plica interiore adiacente fere clausam. Color viridia 3 fasciis ferrugineis in anfractu maiore, quarum duae subtus approximatae, tertia supra remota: hae fasciae in anfractibus reliquis confluunt, ut inde ferruginei fere toti euadant. Intus polita est, concolor, fasciis in apertura transparentibus, columella tamen et suprema pars faucis albae. Operculum natiuum oblongum, planum, saturate viride. Vermem non vidi. (Fabricius.)

Obliquely conical, periphery obtusely angulated, somewhat thin, translucent and glossy, with slight, numerous, spiral striae; white, yellowish, or yellowish brown, often with four brown bands of varying width, sometimes confluent into one band, and so disposed as to exhibit a white band just below the suture. (Tryon.)

TYPE in Museum Copenhagen. Type locality ?

RANGE. Norton Sound, Alaska, and south to Santa Barbara, California; Northern Europe; and Greenland. Fossil, Pliocene.

## Lacuna variegata Carpenter, 1864

*Supplementary Report,* British Association for the Advancement of Science, 656. *Annals and Magazine of Natural History,* ser. 3, **14**:428.

L. testa tenui, plus minusve elevata, soluta, irregulari; adolescente fusco-purpureo; adulta livida, radiatim seu diagonaliter varie irregulariter strigata, strigis fusco-aurantiacis, saepe ziczacformibus; anfr. vi., quorum primi compacti, apice submamillato; dein solutis, postice planatis, antice expansis; basi rotundata seu angulata; apertura subovata; labio postice porrecto; labio saepe partiem vix attingente; columella intus recta, estus valde lacunata. Long., 3; long. spir., 16; lat., 17 poll. (Carpenter.)

Thin, expanded in front, periphery rounded or obtusely angulated, smooth, polished fulvus, irregularly striate with chestnut, with frequently a peripheral band of white spots and sometimes another below the suture. (Tryon, *Manual of Conchology.*)

TYPE in United States National Museum. Type locality, Neah Bay, Washington.

RANGE. Neah Bay, Washington, to Monterey, California.

## Lacuna unifasciata Carpenter, 1856

*Proceedings,* Zoölogical Society of London, 205.

L. t. parva, solida, conica, ad basin angulata; anfr. v. laevibus, parum convexis, sutura distincta; rufo-fusca, linea intensiore ad carinam suturae convenientem, interdum maculis adjacentibus; rima umbilicali a labio subcelata; apertura ovali; apice regulari. Long., 23; long. spir., 11; lat., 15 poll. (Carpenter.)

Small, glossy, generally with a colored keel, sometimes broken into dots. (Carpenter.)

TYPE in Museum Gould. Type locality, Santa Barbara, California.

RANGE. Santa Barbara, California, to Magdalena Bay, Lower California.

## Lacuna unifasciata aurantiaca Carpenter, 1864

*Supplementary Report,* British Association for the Advancement of Science, 656.

Keel obsolete, resembling the chinked *Phasianella.* (Carpenter.)

TYPE in ? Type locality ?

RANGE. Santa Barbara, California, to Point Abreojos, Lower California.

## Lacuna marmorata Dall, 1919

*Proceedings of the United States National Museum,* **56**:348. Tryon, *Manual of Conchology,* **9**: Pl. 50, fig. 58.

Shell small, short, acute, dark brown (fading in the cabinet), of three or four rapidly enlarging whorls; suture distinct; surface, when unworn, with a fine spiral striation which is sometimes feeble; the last whorl has a marked keel at the margin of the base in most specimens, but this region is frequently more or less rounded, and is generally whitish with interrupted brown flecks, which painting occasionally spreads over the upper part of the whorl; a white band in front of the suture is common, and on the base the white may be extended to a brown area bordering the umbilicus. Height, 6; diameter, 5 mm. (Dall.)

TYPE in United States National Museum, No. 47081. Type locality, Monterey, California.

RANGE. Saginaw Bay, Alaska, to San Diego, California.

## Subgenus BOETICA Dall, 1918

## Lacuna vaginata Dall, 1918

*Proceedings of the Biological Society of Washington,* **21**:137. *Proceedings of the United States National Museum,* **56**:349.

Shell small, solid, conical, white, smooth except for faint incremental lines, of about five rapidly enlarging whorls including a minute, subglobular, smooth nucleus; suture distinct, not deep; base rounded, aperture sub-ovate, a distinct sharp groove in the subsutural callus, the outer lip simple, thick; the body with a thick coat of enamel curving into the concavely arcuate pillar lip; umbilicus perforate, the area bounded by a thickened, spirally striated ridge, parallel with the pillar lip, with the area between them excavated; at the anterior end of the lippar is a shallow, narrow sulcus, somewhat as in *Trichotropis*. Height of shell, 4; of last whorl, 3; diameter, 2.5 mm. (Dall.)

TYPE in United States National Museum, No. 209891a. Type locality, U.S. Fish Commission Station 4322, off La Jolla, San Diego County, California.

RANGE. Santa Rosa Island, in 53 fathoms, and off La Jolla, in 199 fathoms.

## Genus **HALOCONCHA** Dall, 1886

New name for subgenus *Lacunella* Deshayes, 1861.

Shell depressed, heliciform, few-whorled, thin, with a strong epidermis; margin of the aperture thin, with a narrow reflexed margin in the adult, continuous with the thin, sharp, unreflected, arcuate columella; umbilicated operculum paucispiral. (Dall.)

EXAMPLE. *Lacunella depressa* Deshayes.

DISTRIBUTION. Alaska and Bering Sea. Fossil: Eocene, Paris basin.

### Haloconcha minor Dall, 1919

*Proceedings of the United States National Museum,* **56**:350.

Shell small, purple-brownish, trochiform with about three and a half rounded, rapidly enlarging whorls, including a minute glassy nucleus; surface smooth except for incremental lines, covered with a glossy, oliva-ceous, translucent periostracum; suture distinct and deep; base rounded with a moderately wide umbilicus; aperture ovate, body with a layer of enamel connecting the two lips. Height, 5.5; longer diameter, 6; shorter diameter, 5 mm. (Dall.)

TYPE in United States National Museum, No. 215073. Type locality, English Bay, St. Paul Island, Pribilof group, Bering Sea.

RANGE. Commander, Pribilof, and Aleutian Islands to Chirkoff Island, Alaska.

### Haloconcha reflexa Dall, 1884

*Proceedings of the United States National Museum,* **7**:344; Pl. 2, figs. 1–3.

Shell thin; light to dark chestnut brown, smooth except for faint lines of growth and wrinkles of the epidermis near the suture or in the umbilicus; whorls two and a half to three, the last very much the largest, inflated; suture distinct, the epidermis sometimes wrinkled close to it; nucleus polished; aperture wide, oblique, rounded, the upper end of the columella and the anterior end of the outer lip approximated, united by a thin glaze of callus; interior polished, brown; base of the aperture rounded, columella arcuate; umbilicus small, narrow, marked by raised wrinkles of epidermis, which sometimes give it a carinated aspect from their abrupt cessation at the umbilical margin; base of the shell smoothly rounded; earlier whorls darker colored than the last. Alt. of shell, 10.0; of aperture, 7.0; maximum lat. of shell, 13.8; of aperture, 7.0 mm. (Dall.)

TYPE in United States National Museum. Type locality, Pribilof and Aleutian Islands.

RANGE. Known only from type locality.

## Genus AQUILONARIA Dall, 1886

Shell lioplaciform, more or less membranous, thin, imperforate, without sculpture, but with a rough, transversely shaggy epidermis. Operculum subspiral, with a raised subspiral rib on the inner side. (Dall.)

TYPE. *Aquilonaria turneri* Dall.

DISTRIBUTION. Arctic Ocean, Labrador.

## Aquilonaria turneri Dall, 1886

*Proceedings of the United States National Museum,* 9:204; Pl. 3, figs. 1, 2, 3.

Shell globose-conic, with five and a half full and rounded whorls regularly increasing. Shell substances white, extremely thin, and wanting near the aperture; covered with a thick, shaggy, more or less hairy, transversely rugose epidermis of a brownish color, of which the outer and anterior margins of the aperture are chiefly formed. This is tough and flexible in life, but dries out of shape when dessicated; sutures with a narrow channel except in the last whorl where the channel gradually becomes obsolete; last whorl forming more than two-thirds of the shell; aperture ovate, margin thin, not reflected; columella smooth, thin, rounding gradually into the anterior margin; inner lip without callus; base rounded, full, without any trace of an umbilicus. Operculum thin, brownish, with about three whorls, slightly transversely undulate and longitudinally finely striate; on the inner side a well-marked raised rib gyrates with the whorls near their inner edge, but does not quite reach the anterior margin of the operculum. Maximum long. of shell, 14.25; of last whorl, 11.00; of aperture, 7.00; of operculum, 7.00; maximum lat. of shell, 10.00; of aperture, 6.00; of operculum, 5.00 mm. (Dall.)

TYPE in United States National Museum, No. 73743. Type locality, Labrador's reef, Ungava Bay.

RANGE. Arctic Ocean near Bering Sea, also Labrador.

## Family FOSSARIDAE

### Genus FOSSARUS Philippi, 1840

Shell perforated, sculptured; inner lip thin; aperture semilunate. Operculum not spiral. (Tryon, *Structural and Systematic Conchology.*)

TYPE. *Fossarus costatus* Brocchi.

DISTRIBUTION. Mediterranean, West America, Polynesia, Japan, Red Sea. Fossil Miocene, Europe.

### Fossarus angiolus Dall, 1919

*Proceedings of the United States National Museum,* **56**:350.

Shell small, yellowish white, with a minute globular nucleus and about four whorls; the suture distinct, not appressed; surface dull; axial sculpture none; spiral sculpture on the upper whorls two, on the last whorl six, strong, elevated cords, with somewhat wider channeled interspaces; umbilicus imperforate, the anterior cord forming its outer boundary; aperture circular, the outer lip thickened but not reflected, the inner lip thin, sharp. Height, 2.25; diameter, 1.75 mm. (Dall.)

TYPE in United States National Museum, No. 271503. Type locality, Todos Santos Bay, Lower California.

RANGE. Known only from type locality.

### Genus ISELICA Dall, 1918

New name for *Isapis* Adams, 1853 (not of Westwood 1851).

Shell umbilicated, spire elevated, cancellated or with revolving ribs, columella with a small median tooth (almost obsolete in *F. anomala*). (Tryon, *Structural and Systematic Conchology.*)

TYPE. *Fossarus anomala* C. B. Adams.

DISTRIBUTION. West Indies, Mazatlan, and west coast North America.

### Iselica fenestrata Carpenter, 1864
### See Part II, Plate 31, fig. 6

*Annals and Magazine of Natural History,* series 3, **14**: 429. Tryon, *Manual of Conchology,* **9**: 272; Pl. 52, fig. 11.

I. testa *I. ovoideae* forma et indole simili; carinis ix. acutis (quarum iv. in spira monstrantur) cincta; interstitiis duplo latioribus, concinne quad-

ratim decussatis, lirulis radiantibus acutissimis; anfr. labro a carinis pec-
tinato; labio parietem parum attingente, medio calloso; umbilico angusto.
Long., 18; long. spir., 13; lat., 19 poll. (Carpenter.)

Umbilicated, spirally lirate, the lirae acute, distant, about 9 on the body
whorl, the wider interspaces with closer longitudinal sculpture; light yellowish brown, epidermis shaggy. Length, 5–6 mm. (Tryon.)

TYPE in United States National Museum. Type locality, Neah Bay,
Washington.

RANGE. Vancouver Island, British Columbia, to Lower California.
Described as *Isapis*.

## Iselica obtusa Carpenter, 1864

*Supplementary Report,* British Association for the Advancement of Science, 656. *Proceedings of the California Academy of Sciences,* 3:217.

I.t. *I. fenestrata* indole simili, sed magis elongata, subnitente, pallide
rosacea; vertice nucleoso decliviter immerso, celato; anfr. norm. postice
subplanatis, suturis obtusis; costis tumentibus rotundatis circ. vii., suturas
et umbilicum minorem versus obsoletis, plus minusve extantibus; interstitiis parvis, irregularibus, haud decussatis; peritremati continuo, labro
tenui, secundum costas variantes undulato; labio medio calloso. Long.,
0.23; long. spir., 0.09; lat., 0.16 poll. (Carpenter.)

Narrower, whorls less convex, less sharply costate than *F. fenestrata;*
whorls flattened behind, ribs swollen, uneven. (Tryon, *Manual of Conchology.*)

TYPE in State Collection, No. 682. Type locality, San Diego, California, in 10 fathoms.

RANGE. Monterey to San Diego, California.

## Iselica obtusa laxa Dall, 1919

*Proceedings of the United States National Museum,* 56:351.

Shell much resembling *obtusa,* but having a longer and more acute
spire, much feebler spiral sculpture, and the whorls more laxly coiled,
giving the suture a well-marked channel in adults. The color is grayish
white, with a yellowish dehiscent periostracum, and the umbilicus is a
narrow chink, partly covered by the reflected inner lip. Height, of shell,
8.5; of last whorl, 6.5; diameter, 5 mm. (Dall.)

TYPE in United States National Museum, No. 219754. Type locality,
Maple Bay, Vancouver Island, British Columbia.

RANGE. Vancouver Island, British Columbia, to Puget Sound, Washington.

## Family LITIOPIDAE

### Genus LITIOPA Rang, 1829

Shell minute, pointed; aperture slightly notched in front; outer lip simple, thin; inner lip reflected; operculum spiral. (Woodward, *Manual of the Mollusca.*)

TYPE. *Litiopa bombyx* Rang.

DISTRIBUTION. Atlantic, Pacific, and Mediterranean, on floating seaweed, to which they adhere by threads.

### Litiopa bombyx Rang, 1829

*Annales des sciences naturelles,* 203. Tryon, *Manual of Conchology,* 9:283; Pl. 53, fig. 74.

Whorls about nine, slightly convex, toward the apex microscopically longitudinally costulate, other whorls very finely spirally striate; light yellowish brown, outer lip often bordered internally with black. Length, 6 mm. (Tryon.)

This is the description of *Litiopa melanostoma* Rang, 1829, which is given as equal to the above.

TYPE in ? Type locality, Pelagic, southern.

RANGE. Southern California to Gulf of California. Pelagic.

### Genus ALABA A. Adams, 1862

Shell smooth, semipellucid, whorls sometimes with a few, irregular varices; outer lip thin, simple; aperture slightly emarginate anteriorly. (A. Adams.)

TYPE. *Alaba trivericosa* C. B. Adams.

DISTRIBUTION. Japan, Australia, Mazatlan, West Indies, California, and Gulf of California.

### Alaba jeanettae Bartsch, 1910

*Proceedings of the United States National Museum,* 39:155; figs. 3, 4.

Shell elongate-conic, semitransparent, with strong varices scattered at irregular intervals. Nuclear whorls four, continuing the general outline of the spire, well-rounded, smooth except for very faint, slender, axial threads, which, in most instances, are only apparent at the summit of the whorls. Post-nuclear whorls well-rounded, appressed at the summit; the early ones smooth, the later ones marked by slender, incised, spiral lines, of which those on the anterior half of the whorls between the sutures and those on the posterior half of the base are usually stronger than the rest. In addition to the spiral sculpture, the whorls are marked at irregular intervals by strong, oblique varices. Suture strongly constricted. Periphery of the last whorl inflated, well-rounded. Base moderately long, well-rounded, and inflated, its posterior half marked like the anterior half between the sutures, bearing the feeble extensions of the varices. Aperture very large, broadly oval; posterior angle obtuse; outer lip thin and transparent; columella very oblique and somewhat curved, slightly reflected over the re-enforcing base; parietal wall glazed with a thin callus. Length, 5; diameter, 2.1 mm. (Bartsch.)

TYPE in United States National Museum, No. 182565. Type locality, Margarita Bay, Lower California.

RANGE. San Pedro, California, to Gulf of California.

## Alaba catalinensis Bartsch, 1920

*Journal of the Washington Academy of Sciences,* **10**:1920.

Shell elongate-conic, milk white, early whorls well-rounded, the succeeding turns a little less so. All whorls polished, appressed at the summit, and marked by fine, retractively slanting lines of growth. Beginning with the second turn, vertical thickenings make their appearance; these are very feeble on the early whorls, but increase steadily in strength until on the last turn they form decidedly raised sinuous ridges. The last whorl, too, shows well-marked malleations. Aperture oval; posterior angle obtuse; inner lip curved and reflected, but not appressed to the base; parietal wall covered by a thick callus. Length, 5.3; diameter, 1.9 mm. (Bartsch.)

TYPE in United States National Museum, No. 213369. Type locality, off Catalina Island, in 40 fathoms.

RANGE. Known only from type locality.

## Family ?

## Genus BARLEEIA Clark, 1855

The color is plain red-brown; smooth or slightly wrinkled; of 4½ to 5½ tumid volutions, which form a rapidly increasing cone. Aperture oval, entire, contracted above, rounded basally; outer margin sharp, without the callous pad of the *Rissoa*. Axis one-tenth; diameter, one-seventeenth of an inch. (Clark.)

TYPE. *Barleeia rubra* Adams.

DISTRIBUTION. West coast of America to Mazatlan.

## Barleeia subtenuis Carpenter, 1864

*Supplementary Report,* British Association for the Advancement of Science, 656. *Proceedings of the California Academy of Sciences,* 3: 218. Tryon, *Manual of Conchology,* 9: Pl. 60, fig. 73.

B.t. parva, tenui, interdum subdiaphana, rufo-cornea, anfr. nucleosis normalibus apice submamillato; normalibus 4, planatis, suturis distinctis; basi rotundata; apertura subovata, peritrenate continuo; labro acuto; labio distincto, lacunem umbilicalem formante, columella subangulata, opercula semilunato, dense rufo-vinoso, subhomogeneo haud spirali, rudi; apophysi praelonga antice columellam versus extante. Long., 11; long. spir., .07; lat., .06 poll. (Carpenter.)

Thin, subpellucid, corneous-chestnut; with four normal whorls, flatly convex, with distinct suture; lip acute. Length, 2.8 mm. (Tryon.)

Type in United States National Museum. Type locality, San Diego, California.

Range. San Pedro, California, to San Ignacio Lagoon, Lower California.

## Barleeia subtenuis rimata Carpenter, 1864

*Journal de Conchyliologie,* 13:145.

B.t. *B. subtenui* simili; sed paulum tumidiore; anfractibus minus planatis; rima umbilicali conspicua. (Carpenter.)

A little more tumid, whorls less flattened, umbilicus slit conspicuous. (Carpenter.)

Type in United States National Museum. Type locality, San Diego, California.

Range. Known only from type locality.

## Barleeia bentleyi Bartsch, 1920

*Proceedings of the United States National Museum,* 58:168; Pl. 13, fig. 2.

Shell small, conic, flesh-colored, excepting the two nuclear turns, which are light brown. Nuclear whorls well-rounded, marked by curved axial rows of closely spaced pits, which are separated by spaces about one and a half times the diameter of the pits. Post-nuclear whorls moderately well-rounded. The appressed summit of these whorls falls slightly below the peripheral keel, and allows this to appear in the suture as a slender thread. The surface of the post-nuclear turns is marked by curved incremental lines and numerous fine, rather closely spaced, spiral striations. A strong cord marks the periphery and renders it decidedly angulated. Base moderately long, well-rounded, marked like the spire. Aperture large, posterior angle obtuse, slightly effuse at the junction of the inner basal lip; outer lip thin; inner lip slender, oblique, and appressed for the greater part to the base, the anterior portion only being free; parietal wall covered by a thick callus, which renders the peritreme complete; operculum typical. Altitude, 2.2; diameter, 1.2 mm. (Bartsch.)

TYPE in United States National Museum, No. 332121. Type locality, Venice, California, on Bryozoa.

RANGE. Known only from type locality.

## Barleeia dalli Bartsch, 1920

*Proceedings of the United States National Museum,* **58**:168; Pl. 13, fig. 10.

Shell rather large, broadly conic, yellowish white. Nuclear whorls two, well-rounded, marked by sinuous axial rows of closely spaced pits which are separated by spaces about four times as wide as the pits. Post-nuclear whorls almost appressed at the summit, moderately rounded, marked by slender, slightly retractively slanting incremental lines and numerous closely spaced spiral striations. Suture but slightly constricted. Periphery of the last whorl marked by a cord, which renders it decidedly angulated. Base moderately long and moderately rounded, marked like the spire. Aperture subcircular, posterior angle obtuse; outer lip thin at the edge; inner lip slender, evenly curved, appressed to the base, except at the extreme anterior portion, where it is free; parietal wall covered by a thick callus, which joins the columella with the outer lip at the posterior angle and renders the peritreme complete. The summit of the last turn bends slightly down below the peripheral cord near the aperture. Altitude, 4.4; diameter, 2.4 mm. (Bartsch.)

TYPE in United States National Museum, No. 209013. Type locality, Bureau of Fisheries Station 4310, in 71–75 fathoms.

RANGE. Known only from type locality.

## Barleeia haliotiphila Carpenter, 1864

*Supplementary Report,* British Association for the Advancement of Science, 656. *Journal de Conchyliologie,* **13**:144. Tryon, *Manual of Conchology,* **9**: Pl. 60, fig. 74.

B.t. parva, turrita, laevi, angusta, tenui, rufo-fusca; marginibus spirae subrectis; anfr. nucleosis normalibus, vertice submamillato; norm. 5 subplanatis, suturis distinctis, basi subplanata, obsolete angulata; apertura ovata, peritremati haud continuo; labro tenui; labro parum calloso; columella vix arcuata; operculo ut in *B. subtenui.* Long., 1; long. spir., .06; lat., .05 poll. (Carpenter.)

Shell elongate-conic, pale chestnut brown. Nuclear whorls almost two, well-rounded, marked by curved, axial rows of pits, which are also arranged in spiral series. Post-nuclear whorls moderately well-rounded, almost flat at the summit, marked by fine incremental lines and exceedingly fine, closely spaced, spiral striations. Suture moderately constricted. Periphery of the last whorl obscurely angulated. Base moderately long and moderately rounded, marked like the spire. Aperture rather small, oval; posterior single, obtuse; outer lip thin; the basal and inner lips meet in a well-rounded curve; inner lip appressed to the base, except at the anterior third, which is free; parietal wall covered by a thick callus, which renders the peritreme complete; operculum typical. Altitude, 2.5; diameter, 1.3 mm. (Bartsch.)

This species closely resembles *Barleeia oldroydi*. It is, however, uniformly smaller, with a decidedly smaller aperture. (Bartsch.)

TYPE in United States National Museum, No. 15558. Type locality, Lower California. Collected off the back of a Haliotis.

RANGE. Mendocino County, California, to Lower California.

## Barleeia sanjuanensis Bartsch, 1920
*Proceedings of the United States National Museum,* 58:170; Pl. 13, fig. 3.

Shell broadly conic, chestnut brown, except the nuclear whorls, which are pale brown. Nuclear turns two, well-rounded, marked by numerous rather strong pits, which are arranged in axial and spiral series. Postnuclear whorls very slightly shouldered at the summit, well-rounded, marked by numerous rather coarse, incised, spiral lines. Periphery obscurely angulated. Suture well-marked. Base well-rounded, marked, like the spire. The summit of the last turn bends decidedly downward behind the aperture. Aperture subcircular; posterior angle decidedly obtuse; outer lip rather thick; inner lip strongly curved, appressed to the base; parietal wall covered by a thick callus, which renders the peritreme complete. Latitude, 2.6; diameter, 1.5 mm. (Bartsch.)

TYPE in United States National Museum, No. 334488. Type locality, Puget Sound.

RANGE. San Juan Islands.

## Barleeia coronadoënsis Bartsch, 1920
*Proceedings of the United States National Museum,* 58:173; Pl. 13, fig. 5.

Shell ovate, white. Nuclear whorls two and a fifth, well-rounded, marked by numerous pits, which are arranged in curved axial lines as well as in spiral series. Post-nuclear whorls, strongly rounded, appressed at the summit, marked by feeble incremental lines and exceedingly fine spiral striations. Suture rather constricted. Periphery of the last whorl strongly rounded. Base moderately long and well-rounded. Aperture large, broadly ovate, rather effuse at the junction of the base and outer lip; posterior angle obtuse; junction of the inner and basal lip well-rounded; outer lip thin; inner lip well-rounded, reflected over and appressed to the base except at the extreme anterior portion where it is free; parietal wall covered by a thick callus, which renders the peritreme complete; operculum typical. Altitude 1.3; diameter, 9 mm. (Bartsch.)

TYPE in United States National Museum, No. 226453. Type locality, off Coronado Islands, northwest coast of Lower California.

RANGE. San Pedro, California, to Point Abreojos, Lower California.

### Barleeia oldroydi Bartsch, 1920

*Proceedings of the United States National Museum,* **58**:171; Pl. 13, fig. 9.

Shell narrowly conic, light chestnut brown. Nuclear whorls one and three-fourths, well-rounded, marked by slightly retractively slanting rows of exceedingly minute pits, which appear to be ranged also in spiral series. Post-nuclear whorls almost flat, appressed at the summit, marked by fine incremental lines and exceedingly fine, closely spaced, spiral striations. Suture only slightly constricted. Periphery of the last whorl obsoletely angulated. Base moderately long, moderately rounded. Aperture moderately large, oval; posterior angle obtuse; outer lip thin; the curved inner lip joins the basal lip in a curve; inner lip appresses to the base for a little more than half its length, the extreme anterior portion only being free; parietal wall covered by a thick callus, which renders the peritreme complete. The summit of the last turn falls slightly below the peripheral angle at the aperture; operculum typical. Altitude, 3.3; diameter, 1.5 mm. (Bartsch.)

TYPE in United States National Museum, No. 32376. Type locality, Monterey Bay, California.

RANGE. Mink Bay, Vancouver Island, to Coronado Islands.

### Genus **DIALA** A. Adams, 1861

Whorls not varicose, sometimes nodulated around the middle; columella straightish, not truncated; labrum not thickened. (Tryon, *Structural and Systematic Conchology.*)

TYPE. *Alaba varia* A. Adams.

DISTRIBUTION. Philippines, Japan, west coast of North America.

### Diala exilis Tryon, 1866

*American Journal of Conchology,* **2**:12; Pl. 2, fig. 18.

Shell subulate, light brown, consisting of six convex volutions, with well-marked suture; aperture oval, proportionately very small; surface covered with slight revolving striae. Length, .135; diameter, .07; length of aperture, .04; breadth, .03 in. (Tryon.)

TYPE in Tryon's Cabinet. Type locality, San Diego, California.

RANGE. San Francisco Bay to San Diego, California.

## Diala acuta Carpenter, 1864

*Supplementary Report,* British Association for the Advancement of Science, 657. *Proceedings of the California Academy of Sciences,* 3:218.

D.t. parva, turrito-conica, cerina, polita, nitente; anfr. nucl. ii. naticoideis, vertice mamillato, apice indistincto; norm. v., omnino planatis, suturis indistinctis; peripheria acute angulata, vix carinata; basi omnino planata; apertura subquadrata; peritremati continuo; labro postice paullum contracto; labio appresso; columella antice angulata, vix sinuata. Long., 0.15; long. spir., 0.09; lat., 0.07. (Carpenter.)

Shell small, turreted-conical, waxy, polished shining; with two naticoid whorls, and mamillate vertex, with indistinct apex; normally five, all flattened, with indistinct sutures; periphery acutely angulate, scarcely carinate; basis altogether flattened; aperture subquadrate; with continuous border, outer lip slightly contracted in the rear; inner lip appressed; columella angulate in front, scarcely sinuate. (Translation.)

TYPE in State Collection, No. 390. Type locality, Catalina Island.

RANGE. Monterey, California, to Coronado Islands.

## Diala marmorea Carpenter, 1864

*Supplementary Report,* British Association for the Advancement of Science, 657. *Proceedings of the California Academy of Sciences,* 3: 218. Tryon, *Manual of Conchology,* 9: Pl. 53, fig. 87.

D.t. solida, exacte conica laevia, nitente; subdiaphane pallida, rufo maculata; anfr. nucl. rotundatis, planorbeis, sub-oblique sitis, apice celato; norm. vi. planatis, suturis indistinctis; basi planata, subangulata; apertura ovali, peritremati continuo, haud varicoso. (Carpenter.)

Solid, glossy, whitish, with faint chestnut longitudinal striations; whorls flattened, periphery faintly angulated. Length, 5 mm. (Tryon.)

TYPE in Rowell Collection, University of California. Type locality, Monterey, California.

RANGE. Queen Charlotte Islands, British Columbia, to San Pedro, California.

## Family RISSOIDAE

## Genus **CINGULA** Fleming, 1828

Shell minute, white or horny; conical, pointed, many-whorled; smooth, or cancellated; aperture rounded; peristome entire, continuous; outer lip slightly expanded and thickened; operculum sub-spiral. (Woodward, *Manual of the Mollusca.*)

TYPE. *Rissoa cingillus* Montfort.

DISTRIBUTION. North America, West Indies, Norway, Britain, Mediterranean.

## Cingula martyni Dall, 1886
### Plate 84, fig. 5

*Proceedings of the United States National Museum,* 9:306; Pl. 3, fig. 9. *Ibid.,* 41:485; Pl. 41, fig. 5.

Shell elongate-ovate, light chestnut brown. Nuclear whorls scarcely differentiated from the succeeding turns, smooth. Post-nuclear whorls well-rounded, smooth, excepting fine incremental lines. Suture well-constricted. Periphery and base of the last whorl inflated, well-rounded, the latter narrowly umbilicated, smooth, excepting fine, incremental lines. Aperture very large, broadly oval, almost circular; outer lip decidedly expanded, thin; columella strongly curved and very slightly reflected; parietal wall covered with a thick callus, which renders the peritreme complete. Length, 5; diameter, 3.2 mm. (Dall.)

TYPE in United States National Museum, No. 213527. Type locality, Kyska Harbor, Aleutians, Alaska.

RANGE. Bering Strait to Aleutians and Chignik Bay, Alaska.

Described as *Cingula robusta martyni.*

## Cingula martyni scipio Dall, 1886
### Plate 84, fig. 6

*Proceedings of the United States National Museum,* 9:306; Pl. 4, fig. 10.

This form is much rarer than the typical, about one per cent of those collected being of this sort, but with a certain number of intermediate grades. There seems to be no other difference than that of form, faint revolving lines being occasionally present in both; both are of the same reddish grape-color with whitish bloom, and whitish border to the aperture. (Dall.)

TYPE in United States National Museum. Type locality, Aleutian Islands.

RANGE. Pribilof Islands to the Aleutians and to Middleton Island, Alaska.

Described as *Cingula robusta scipio.*

## Cingula alaskana Bartsch, 1912
### Plate 84, fig. 4

*Proceedings of the United States National Museum,* 41:486; Pl. 41, fig. 4.

Shell subglobose, light brown. Nuclear turns one and one-quarter, well-rounded, smooth, excepting fine, incremental lines. Post-nuclear whorls inflated, well-rounded, marked by strong, incised, equal and equally spaced spiral grooves, which are about two-thirds as wide as the spaces that separate them; about ten of these grooves occur between the sutures on the second, and thirteen on the penultimate whorl. Suture well-impressed. Periphery of the last whorl strongly inflated. Base rather short, inflated, well-rounded, very narrowly umbilicated, marked like the spire with spiral grooves. In addition to the spiral sculpture, the post-nuclear whorls are marked by rather strong, incremental lines, which extend over the entire surface of the post-nuclear spire. Aperture large, oval, slightly angulated at the posterior angle; outer lip thin, showing the external markings within; columella curved, slightly reflected; parietal wall glazed with a thick callus, which completes the peritreme; operculum horny, paucispiral. Length, 2.6; diameter, 2 mm. (Bartsch.)

TYPE in United States National Museum, No. 160995. Type locality, Amchitka Island, Alaska.

RANGE. Known only from type locality.

### Cingula aleutica Dall, 1886
### Plate 84, fig. 2

*Proceedings of the United States National Museum,* 9:307; Pl. 3, fig. 11. *Ibid.,* 41:487; Pl. 41, fig. 2.

Shell elongate-ovate, light yellow. Nuclear whorls scarcely differentiated from the succeeding turns. Post-nuclear whorls strongly rounded, appressed at the summit, marked by very fine, incremental lines only. Suture strongly constricted. Periphery of the last whorl and the moderately long base well-rounded, the latter narrowly umbilicated, marked like the spire. Aperture large, oval, slightly expanded at the edge; posterior angle decidedly obtuse; peritreme complete, dark brown at the edge; outer lip thin; columella oblique, strongly curved; parietal wall covered with a thick callus, which is appressed to the succeeding whorl, completing the peritreme. Length, 3.2; diameter, 1.7 mm. (Bartsch.)

TYPE in United States National Museum, No. 213525. Type locality, Unalaska, Alaska.

RANGE. Pribilof Islands, to the Aleutians, to Windfall Harbor, Admiralty Island, Alaska.

### Cingula katherinae Bartsch, 1912
### Plate 84, fig. 3

*Proceedings of the United States National Museum,* 41:488; Pl. 41, fig. 3.

Shell subglobose, dark brown. Nuclear whorls about one and one-quarter, well-rounded, smooth. Post-nuclear whorls strongly, roundedly shouldered at the summit, well-rounded on the sides, marked by fine, incremental lines only. Suture strongly constricted. Periphery of the last whorl inflated. Base moderately long, well-rounded, narrowly umbilicated, smooth, excepting incremental lines. Aperture large, oblique, broadly oval; outer lip thin; columella curved and slightly reflected; parietal wall covered with a thick callus which renders the peritreme complete. Length, 2.7; diameter, 2 mm. (Bartsch.)

TYPE in United States National Museum, No. 206103. Type locality, Windfall Harbor, Admiralty Island, Alaska.

### Cingula montereyensis Bartsch, 1912
### Plate 84, fig. 1

*Proceedings of the United States National Museum,* **41**:488; Pl. 41, fig. 1.

Shell elongate-conic, light brown. Nuclear whorls one and one-half, well-rounded, smooth. Post-nuclear whorls appressed at the summit, moderately rounded, with a very narrow umbilical chink, smooth, excepting fine, incremental lines. Aperture broadly oval, somewhat effuse anteriorly; posterior angle obtuse; outer lip thin; columella curved; parietal wall glazed with a moderately strong callus, which joins the posterior angle of the aperture with the insertion of the columella. Length, 4; diameter, 2.1 mm. (Bartsch.)

TYPE in United States National Museum, No. 127547. Type locality, Monterey, California.

RANGE. Known only from type locality.

### Cingula californica Tryon, 1865

*American Journal of Conchology,* **1**:221; Pl. 22, fig. 11.

Shell turbinated, consisting of six rather convex whorls; spire elevated, apex acute, suture well impressed; aperture moderate, ovate; umbilical region not perforate; epidermis minutely striated, polished, dark horn color; operculum thin, shining, dark brown. Length, .13; diameter, .09; length of aperture, .06; breadth, .04 in. (Tryon.)

TYPE in Tryon Cabinet. Type locality, Oakland, California.

RANGE. San Francisco Bay to San Pedro, California.

### Cingula orvieta Dall, 1919

*Proceedings of the United States National Museum,* **56**:351.

Shell minute, olive brown, thin, turbinate, of about three rounded whorls, the surface smooth, not polished, the suture distinct; the base convex, imperforate; the aperture subcircular, simple, a chink behind the inner lip, the peristome continuous. Height, 0.75; diameter, 0.60 mm. (Dall.)

TYPE in United States National Museum, No. 105466. Type locality, San Diego, California.

RANGE. Known only from type locality.

## Subgenus NODULUS Monterosato, 1878

### Cingula cerinella Dall, 1886

*Proceedings of the United States National Museum,* 9:307; Pl. 4, fig. 12. *Ibid.,* 41:289; fig. 1.

Shell small, thin, light yellow. The nucleus consists of a single turn, which is smooth and scarcely differentiated from the succeeding whorls. Post-nuclear whorls high between the sutures, moderately rounded, appressed at the summit, marked by fine incremental lines only. Suture well constricted. Periphery of the last whorl and the rather long base gently rounded. Aperture very oblique, large, broadly oval; posterior angle obtuse; outer lip thin; columella very oblique, slightly curved, inner lip partly reflected over the base; parietal wall covered with a thick callus, which is free at the edge and renders the peritreme complete. Length, 2.9; diameter, 1.3 mm. (Dall.)

TYPE in United States National Museum, No. 213453. Type locality, Atka Island, Alaska.

RANGE. Kyska and Amchitka and Atka Islands, Alaska.

Described as *Onoba cerinella* in 1886.

### Cingula asser Bartsch, 1919

*Nautilus,* 23:138; Pl. 11, fig. 9.

Shell elongate-conic, bluish white, semitranslucent. Nuclear whorls one and one-tenth, smooth, a little less elevated than the succeeding turns. Post-nuclear whorls very high between the sutures, overhanging, moderately rounded, appressed at the summit. The preceding whorl shines through the summit of the succeeding turn and gives this the appearance of having a double suture. Sutures well-impressed. Periphery of the last whorl well-rounded. Base moderately prolonged, well-rounded. Entire surface of spire and base marked by closely placed, exceedingly fine, microscopic, spiral striations. Aperture very broadly ovate; posterior angle obtuse; outer lip thin; peritreme complete. Length, 2.2; diameter, 0.9 mm. (Bartsch.)

TYPE in United States National Museum, No. 208434. Type locality, Port Graham, Alaska.

RANGE. Atka Island, Aleutians, to Cook's Inlet, Alaska.

## Cingula kelseyi Bartsch, 1919

*Proceedings of the United States National Museum,* 41:290; fig. 3.

Shell small, very slender, cylindro-conic, translucent, white. Nuclear whorls two, strongly rounded, smooth. Post-nuclear whorls rather high between the sutures, well-rounded, very narrowly shouldered at the summit and lightly constricted a little anterior to the summit, marked by fine lines of growth and exceedingly fine, spiral striations. Suture moderately constricted. Periphery of the last whorl and the long base gently rounded, marked like the spire. Aperture very oblique, subcircular; posterior angle obtuse; outer lip reflected at the edge; inner lip moderately long, reflected; parietal wall covered with a thick callus, which is also reflected, but free at the edge, thus giving the entire aperture a dished appearance. Length, 1.9; diameter, 0.6 mm. (Bartsch.)

Type in United States National Museum, No. 111369. Type locality, Coronado Island, San Diego, California.

Range. San Diego, California, and to the South Coronado Island.

Described as *Nodulus.*

## Cingula kyskensis Bartsch, 1911

*Proceedings of the United States National Museum,* 41:291; fig. 4.

Shell elongate-oval, thin, translucent, yellow. Nuclear whorls one and one-quarter, smooth, well-rounded, scarcely differentiated from the succeeding turns. Post-nuclear whorls appressed at the summit, with a somewhat concavely sloping shoulder which extends over the posterior third between the sutures. Entire surface marked by numerous fine lines of growth only, surface apparently covered with a thin epidermis which has the finely cracked appearance frequently seen on a surface covered with varnish. Suture strongly constricted. Periphery and the rather short base of the last whorl well-rounded. Aperture very large, very broadly oval, decidedly oblique; posterior angle obtuse; outer lip slightly expanded; inner lip decidedly curved and reflected; parietal wall covered with a thick callus which is disjunct at the edge and renders the peritreme complete. Length, 2.3; diameter, 1.1 mm. (Bartsch.)

Type in United States National Museum, No. 161105. Type locality, Kyska Harbor, Aleutian Islands, Alaska.

Range. Kyska and Atka Islands, Aleutians.

## Cingula palmeri Dall, 1919

*Proceedings of the Biological Society of Washington,* 32:251.

Shell minute, translucent brownish with a blunt apex and four well-rounded whorls; the suture rather deep, the surface smooth except for faint incremental lines; base imperforate, aperture lunate, simple, the margin entire and continuous. Height, 2; diameter, 0.75 mm. (Dall.)

TYPE in United States National Museum, No. 212731. Type locality, St. Paul Island, Bering Sea.

RANGE. Pribilof Islands, Bering Sea.

## Genus SKENEOPSIS

Description of genus not found.

### Skeneopsis alaskana Dall, 1919

*Proceedings of the Biological Society of Washington, 32:251.*

Shell minute, resembling *S. planorbis Fabricius,* but smaller, more elevated, and with a much smaller umbilicus; whorls two and a half, pale greenish white, rounded; suture deep, not appressed; sculpture only of more or less evident concentric fine wrinkles, more conspicuous on the top of the whorl; aperture circular, simple, umbilicus deep, with about one-fifth of the diameter of the last whorl; operculum horny, multispiral. (Dall.)

TYPE in United States National Museum, No. 271717. Type locality, St. Paul Island, Bering Sea.

RANGE. Pribilof and Unalaska Islands, Alaska.

## Genus AMPHITHALAMUS Carpenter, 1865

Testa Rissoidea, nucleo magno; apertura labio producto, labro subpostice juncto, subito in adulta contracto. (Carpenter.)

TYPE. *Amphithalamus inclusus* Carpenter.

DISTRIBUTION. California and Japan.

### Amphithalamus inclusus Carpenter, 1865

*Annals and Magazine of Natural History, ser. 3, 15:181. Proceedings of the United States National Museum, 41:264; 2.*

A. testa minuta, lata, solidiore, pallide rufo-fusca; vertice mamillato; anfr. nucl. uno et dimidio, quoad magnitudinem permagnis, minutissime et confertissime spiraliter et radiatim striolatis; anfr. norm. iii., laevibus, subplanatis, suturis impressis; basi subangulata; costa peripherica rotundata, haud extante, interdum in spira se monstrante; costa altera circa regionem pseudo-umbilicarem; labro acuto, haud contracto; labio testa adolescente normali, dein a pariete separata, sinum posticum suturam versus formante, t. adulta valde separata, regionem quasi umbilicarum magnam formante; ad labrum subito fere perpendiculariter, subpostice juncto; operculo tenuissimo. Long., .04; long. spir., .02; lat., .03 poll. (Carpenter.)

Shell elongate-ovate, purplish brown with the nuclear whorls yellowish white. Nuclear whorls about one and one-half, well-rounded, marked by about 15 slender, equal and equally spaced, spiral threads and numerous very fine, axial threads, lending the surface a very minutely pitted appearance. Post-nuclear whorls subtabulately shouldered at the summit, well-rounded on the middle, marked by fine lines of growth only. Suture well-impressed. Periphery of the last whorl marked by a rather broad, low, rounded, spiral thread, which is separated from the space posterior to it by a narrow, shallow, spiral sulcus. Anteriorly, this cord passes, without demarcation, into the general surface of the moderately long and gently-rounded base. Base broadly umbilicate, having a rather broad, low, spiral cord a little posterior to the insertion of the columella. Aperture oval, very oblique, re-enforced within by an inner peristome, which, in the anterior portion of the outer lip, fuses with it, while the columella portion and the parietal portion stand at a considerable distance from the columella and the parietal wall, being connected with these by a concave bridge of shelly matter; columella moderately long and stout. Length, 1.4; diameter, 0.9 mm. (Bartsch.)

## Amphithalamus lacunatus Carpenter, 1866

*Proceedings of the California Academy of Sciences,* 3:218. *Proceedings of the United States National Museum,* **41**: 263; fig. 1.

A. t. adolescente *A. incluso* simili; nucleo similiter minutissime et confertissime spiraliter et radiatim striulato; sed majore, latiore, anfr. subplanatis; basi late lacunata, haud carinata; adulta?—Long., 0.06; long. spir., 0.03; lat., 0.03 poll. (Carpenter.)

Shell minute, broadly conic, milk-white. Nuclear whorls one and one-half, well-rounded, marked by about 15 slender, equal and equally spaced, spiral threads and numerous, very fine, axial threads, lending the surface a very minutely pitted appearance. Post-nuclear whorls feebly shouldered at the summit, slightly rounded, marked by incremental lines only. Suture well-impressed. Periphery of the last whorl inflated, rounded. Base moderately long, slightly rounded, with a narrow umbilical chink. Aperture incomplete (outer lip fractured); inner lip curved, appressed; parietal wall covered with a thick callus. Length, 1.5; diameter, 1.1 mm. (Bartsch.)

TYPE in United States National Museum, No. 15564. Type locality, San Pedro, California, in the shell washings.

RANGE. Known only from type locality.

## Amphithalamus tenuis Bartsch, 1911

*Proceedings of the United States National Museum,* 41:264; fig. 3.

Shell elongate-ovate, dark greenish horn-color, with the columella and the aperture yellowish-white except the dark edge of the peristome.

Nuclear whorls one and one-half, well-rounded, marked by about 15 slender, equal and equally spaced, spiral threads and numerous, very fine, axial threads, lending the surface a very minutely pitted appearance. Post-nuclear whorls very narrowly subtabulately shouldered at the summit, well-rounded, smooth excepting fine incremental lines. Suture moderately impressed. Periphery of the last whorl well-rounded. Base long, well-rounded, bearing a low, broad, spiral cord on its middle, which bounds the posterior termination of the white area. Aperture oval, very oblique; peristome double, the inner fusing with the anterior portion of the outer lip, while posteriorly it is distinct and at a considerable distance from the columella and the parietal wall; the space between the columella parietal wall and the inner peristome is bridged over by a concave band of shelly matter. Length, 1.1; diameter, 0.7 mm. (Bartsch.)

Type in United States National Museum, No. 213541. Type locality, La Jolla, California.

Range. Monterey to San Diego, California.

## Amphithalamus stephensae Bartsch, 1927

*Proceedings of the United States National Museum, 70: 28; Pl. 4, fig. 5.*

Shell minute, pale brown with an ashy tinge except the columellar region which is flesh colored; the early whorls when they contain the animal are much more dusky. Nuclear whorls 1.5; the first half smooth, the rest marked by rather distantly spaced, poorly developed, rather broad spiral lirations of which nine are present between the summit and the periphery. In addition to this, there are inconspicuous lines of growth. Postnuclear whorls strongly rounded, narrowly shouldered at the summit, the portion appressed to the preceding turn appearing through the substance of the shell as a band. Periphery with a weak keel which is truncated rather abruptly posteriorly but grades gently into the substance of the shell toward the base. Suture well-marked. Base short, inflated, strongly rounded, marked by lines of growth only. There is a heavy callus at the insertion of the columella, which at its posterior termination almost forms a cord. The columella itself is very heavy and oblique. The conformation of the aperture is characteristically Amphithalmid, that is, the aperture which is oval is much contracted by having a shelf extending out from the columellar and the parietal wall toward the outer lip, contracting the aperture. This shelf forms a decided pit behind its edge. The inner and parietal lip of the aperture, therefore, are not in contact with the columellar or the parietal wall, but at some distance from it; the posterior portion of the outer lip, however, extends upward to the preceding turn, which it joins immediately below the peripheral keel as in mollusks with a normal aperture. Operculum thin, paucispiral.

TYPE in United States National Museum, No. 348531. Length 1.1 mm.; greater diameter, 0.9 mm.

This species differs from the two previously known forms, *Amphithalamus inclusus* Carpenter and *Amphithalamus tenuis* Bartsch in being much stouter than *A. tenuis* Bartsch, and in having coarser spiral threads on the early whorls, a less strong peripheral keel, and more rounded whorls than *A. inclusus* Carpenter.

## Genus ALVANIA (Leach) Risso, 1820

Shell oval, turbiniform; spire short, apex sharp; whorls rounded, usually cancellated; aperture subcircular, crenulated within; outer lip with a marginal exterior varix. (Tryon, *Structural and Systematic Conchology.*)

TYPE. *Rissoa abyssicola* Forbes.

DISTRIBUTION. Universally distributed; most abundant in North Temperate zone.

## Alvania castanella Dall, 1886
### Plate 80, fig. 1

*Proceedings of the United States National Museum, 9:307; Pl. 3, fig. 5.*

Shell elongate-ovate, thin, yellowish-white. Nuclear whorls about one and three-quarters, well-rounded, smooth. Post-nuclear whorls strongly, slopingly shouldered at the summit, well-rounded, marked between the sutures by six spiral keels, the first of which adjoins the strongly channeled suture, while the second marks the angle of the shoulder; the remaining four growing gradually and successively weaker and a little more closely spaced. In addition to the spiral sculpture, the whorls are marked by numerous very fine, incremental lines. Suture quite strongly constricted. Periphery of the last whorl well-rounded. Base moderately long, well-rounded, very narrowly umbilicated, marked by six equal and equally spaced, spiral keels. Aperture semicircular; posterior angle obtuse; outer lip thin, showing the external sculpture within; inner lip thin, strongly curved and slightly revolute; parietal wall covered with a moderately thick callus, which renders the peritreme complete; operculum thin, horny. Length, 2.4; diameter, 1.3 mm. (Dall.)

TYPE in United States National Museum, No. 213677. Type locality, Atka Island, Alaska.

RANGE. Kyska to Atka Islands.

## Alvania aurivillii Dall, 1886

### Plate 80, fig. 5

*Proceedings of the United States National Museum,* 9:308; Pl. 4, fig. 8.

Shell elongate-conic, thin, light yellow. Nuclear whorls one and one-half, strongly rounded, smooth. Post-nuclear whorls rounded, shouldered at the summit, marked by two, strong, spiral keels, between the sutures, on the first two turns; the first keel being on the middle and the second a little posterior to the suture; the space between the summit and the first keel slopes regularly. Beginning with the third whorl a third keel appears, and the space between the sutures is here divided into four parts, of which the one at the summit is a little broader than the rest. On the last turn the peripheral keel is completely shown in the suture, and a slender, spiral thread appears immediately below the summit. The spaces between the keels are about four times as wide as the keels. Suture quite strongly constricted. Periphery of the last whorl marked by a strong, spiral keel which is a trifle nearer to the first keel posterior to it than that is to its neighbor. Base moderately long, well-rounded, narrowly umbilicated, marked by three equal and almost equally spaced spiral keels. In addition to the spiral sculpture, the whorls are marked by numerous, very slender, raised, axial threads, which are best developed in the broad spaces between the spiral keels. Aperture subcircular, slightly expanded anteriorly; posterior angle obtuse; outer lip thin, showing the external sculpture within; inner lip strongly curved, somewhat reflected; parietal wall covered with a thick callus, which joins the posterior angle to the columella and renders the peritreme complete; operculum thin, corneous. Length, 4.2; diameter, 2.0 mm. (Dall.)

TYPE in United States National Museum, No. 213680. Type locality, Adakh Island, Alaska.

RANGE. Kyska to Adakh Island, Aleutians.

## Alvania bakeri Bartsch, 1910

### Plate 80, fig. 2

*Nautilus,* 23:137; Pl. 11, fig. 8.

Shell minute, bluish-white. Nuclear whorls one and one-third, forming a moderately elevated spire. Upper half of the well-rounded nuclear whorls marked by about ten, very fine, closely spaced, spiral striations; lower half marked by numerous, closely spaced, depressed, elongate granules, each of which has the long axis decidedly protractively slanted, which lends to this part of the turn a finely blistered appearance. Post-nuclear whorls a little more than two, well-rounded, separated by a strongly constricted suture, marked by three, strong, rounded, spiral keels, of which the strongest is a little anterior to the posterior third between the sutures, while the anterior of the other two, which are of equal strength, is at the periphery and the third halfway between them. In addition to these keels, the whorls are marked by fine, incised, spiral lines between the keels, which are best developed on the well-rounded shoulder, between the summit of the whorls and the strong keel below it. Base of the last whorl slightly protracted, well-rounded, marked by two broad, depressed, spiral cords, of which the basal one, which is a little anterior to the middle, is the broader. In addition to these there are numerous, exceedingly fine, spiral striations. The axial sculpture of the entire spire and base consists of very fine, incremental lines only. Aperture very broadly ovate, almost circular; peritreme continuous; outer lip thick within, beveled at the margin to form a sharp edge, which is rendered slightly sinuous by the external spiral sculpture; inner lip strong and strongly curved; parietal wall of the aperture appressed to the preceding whorl. Length, 1.4; diameter, 1.2 mm. (Bartsch.)

TYPE in United States National Museum, No. 208445. Type locality, Port Graham, Alaska.

RANGE. Port Graham, Cook's Inlet, Alaska.

### Alvania trachisma Bartsch, 1911

### Plate 80, fig. 7

*Proceedings of the United States National Museum,* 41:339; Pl. 29, fig. 7.

Shell elongate-ovate, yellowish white. (Nuclear whorls decollated.) Post-nuclear whorls slightly rounded, marked by slender, well-rounded, almost vertical, axial ribs, of which 26 occur upon the first and second and 22 upon the third and the penultimate turn. These axial ribs are about one-third as wide as the spaces that separate them and extend prominently from the summit of the whorls, where they terminate in rounded cusps, to the umbilical area. In addition to the axial sculpture, the whorls are marked by spiral cords, of which 3 occur between the sutures on the first and second whorls, 5 on the third, and 6 on the penultimate turn. The spiral cords pass over the axial ribs as cords. The spaces inclosed between the axial ribs and the spiral cords are elongate oval pits, having their long axes parallel with the spiral sculpture. Suture broad and very deeply channeled. Periphery of the last whorl marked by a sulcus a little wider than those occurring on the spire, crossed by the continuations of the axial ribs. Base moderately produced, slightly concave in the middle, marked by six, equal and almost equally spaced, spiral cords and the continuations of the axial ribs. Aperture oblique, moderately large, oval; posterior angle acute; outer lip very thick, re-enforced immediately behind the edge by a strong varix, showing about 10 internal lirations within the aperture; inner lip very stout, curved, and reflected over and appressed to base; parietal wall covered with a thick callus which completes the peristome. Length, 3.3; diameter, 1.9 mm. (Bartsch.)

TYPE in United States National Museum, No. 213684. Type locality, Monterey, California.

RANGE. Known only from type locality.

## Alvania californica Bartsch, 1911

### Plate 80, fig. 9

*Proceedings of the United States National Museum,* **41**:340; Pl. 29, fig. 9.

Shell elongate-ovate, the spire posterior to the periphery of the last whorl forming a perfect cone, yellowish-white. Nuclear whorls two and one-half, well-rounded, smooth, separated by a moderately constricted suture. Post-nuclear whorls moderately well-rounded, marked by slender, rounded, slightly protractive, axial ribs, of which 26 occur upon the first and 34 upon the second and the penultimate turn. These axial ribs extend quite prominently from the summit of the whorls, where they form slender cusps, to the periphery of the whorls, becoming evanescent immediately anterior to the periphery. In addition to the axial ribs, the whorls are marked by spiral cords about as strong as the axial ribs. Of these two occur upon the first volution, where they divide the space between the sutures into three equal parts; on the second there are four, while upon the penultimate turn there are six. Suture strongly channeled. Periphery of the last whorl marked by a sulcus as wide as the one which separates the two cords posterior to it. Base moderately long, well-rounded, marked by six almost equally spaced, spiral cords, which grow successively stronger from the periphery to the umbilical area. In addition to these spiral cords, there are feeble extensions of the axial ribs occurring in the space between the first and second sulcus anterior to the periphery. Aperture very broadly oval; posterior angle obtuse; outer lip very thick, re-enforced immediately behind the edge by a strong varix, showing about 10 lirations within; inner lip stout, curved, strongly reflected over and appressed to the base; parietal wall covered with a thick callus, which completes the peristome. Length, 2.5; diameter, 1.4 mm. (Bartsch.)

TYPE in United States National Museum, No. 56347. Type locality, Monterey, California.

RANGE. Known only from type locality.

## Alvania carpenteri Weinkauff, 1885

### Plate 80, fig. 8

*Conchylien Cabinet,* ed. 2, 192. *Proceedings of the United States National Museum,* 41:341; Pl. 29, fig. 8.

A. testa parva, subturrita, rufo-fusca, marginibus spirae rectis; anfr. nucleosis 2 et dimidio, naticoideis, laevibus, tumentibus, apice mamillato; norm. 3 tumidis, suturis impressis; liris angustis distantibus, spiralibus circ. 12 (quarum 4–6 in spira monstrantus) et lirulis radiantibus, supra transeuntibus, haud nodulosis, secundum interstitia incurvatis, eleganter exculpta; interstitiis altis quadratis, peritremate continuo, subrotundato, acutiore. Long., 0.85; diam., 0.04; long. spirae, 0.05." (Carpenter.)

Shell small, elongate ovate, yellowish-white. Nuclear whorls one and one-half, smooth, well-rounded. Post-nuclear whorls well-rounded; the first two slightly, slopingly shouldered at the summit, marked by slender, axial ribs, of which 24 occur upon the first, 32 upon the second, and 34 upon the penultimate turn. In addition to the axial ribs, the whorls are marked by spiral cords which equal the ribs in strength; of these cords, 4 occur upon the first and second turns, the space between the summit and the first cord is considerably wider than that between any two of the cords and forms a sloping shoulder. On the penultimate whorl, this space is marked by two additional cords, one of which is at the summit and the other is halfway between this and the next cord. The spaces inclosed between the axial ribs and the spiral cords are elongated pits, which have their long axes parallel to the spiral sculpture in all cases except the median, where the pits are squarish. Suture strongly channeled. Periphery of the last whorl marked by a spiral sulcus equal to the one separating the first and second supra-peripheral cords. Base moderately long, well-rounded, not attentuated anteriorly, marked by six, equal and equally spaced, spiral cords and very feeble continuations of the axial ribs, which here appear as very slender threads. Aperture broadly oval; posterior angle obtuse; outer lip thin, showing the external sculpture within; inner lip very strongly curved and slightly reflected; parietal wall covered with a moderately strong callus, which renders the peritreme complete. Length, 2; diameter, 1.1 mm. (Bartsch.)

TYPE in United States National Museum, No. 17728. Type locality, Neah Bay, Washington.

RANGE. Forrester Island, Alaska, to Neah Bay, Washington.

### Alvania filosa Carpenter, 1865

*Annals and Magazine of Natural History,* series 3, **14**: 7. *Proceedings of the United States National Museum,* **41**:342; Pl. 30, fig. 7.

A. testa *A. reticulatae* indole et colore, haud sculptura, simili; multo majore, elongata; anfr. nucl.? . . . (detritis), norm. iv.; striis parum separatis circ. xviii. (quorum circ. xii. in spira monstrantur) cincta; rugalis radiantibus posticis creberrimis, haud expressis, circa peripheriam evanidis; peritremate continuo; columella rufo-purpureo tincta. Long., .13; long. spir., .09; lat., .06 poll. (Carpenter.)

Shell elongate-conic, thin, translucent, yellowish-white. Nuclear whorls one and one-half, well-rounded, smooth. Post-nuclear whorls well-rounded, roundly shouldered at the summit, marked by slender, feebly developed, sinuous, axial riblets. In addition to these riblets, the whorls are marked by equal and equally spaced, spiral cords, which are equal to the riblets in strength. Of these cords 10 occur between the sutures on the antepenultimate, and 12 on the penultimate turn. The spaces inclosed between the spiral cords and the axial ribs are shallow, impressed, squarish pits. Suture strongly constricted. Periphery and the rather short base of the last whorl well-rounded; the latter very narrowly umbilicated, marked by eight equal and equally spaced spiral cords, which are a little wider than the spaces that separate them. The sulci between the spiral cords on the base are crossed by the feeble extensions of the axial riblets. Aperture broadly oval; posterior angle obtuse; outer lip thin, showing the external sculpture within; inner lip moderately stout, curved and reflected, the posterior half appressed to the base; parietal wall covered with a thick callus, rendering the peritreme complete. Length, 3.5; diameter, 1.7 mm. (Bartsch.)

TYPE in United States National Museum, No. 36632. Type locality, Neah Bay, Washington.

RANGE. Known only from type locality.

### Alvania alaskana Dall, 1886

*Proceedings of the United States National Museum,* **9**:307; Pl. 4, fig. 9.

Shell very elongate-ovate, white. Nuclear whorls one and one-half, strongly rounded, very finely papillose. Post-nuclear whorls strongly rounded, appressed at the summit, marked by fairly strong, spiral cords, of which three occur upon the first whorl, so arranged that the first below the summit marks the anterior termination of the posterior third between the sutures; the other two divide the space anterior to this into equal parts. The spaces separating these cords are a little narrower than the cords. On the second whorl an additional cord makes its appearance halfway between the summit and the first spiral cord on the preceding whorl, thus dividing the space between the sutures into five equal portions. The penultimate whorl has the same spiral sculpture as its predecessor, but in addition this whorl shows the peripheral spiral cord a little posterior to the suture. In addition to these spiral cords, the whorls are marked by numerous very fine, spiral striations, which are apparent on the cords and in the grooves that separate them. The axial sculpture is reduced to very feeble riblets, which are closely spaced and rather irregularly distributed. Suture very strongly constricted. Periphery of the last whorl well-rounded. Base rather short, umbilicated, marked by six low, rounded, spiral cords, which are situated on the posterior two-thirds of the base, and numerous fine, closely spaced, spiral striations. Aperture subcircular; posterior angle obtuse; outer lip thin, showing the external sculpture within; inner lip slender, curved, reflected, but free from the base; parietal wall covered with a thick callus which renders the peritreme complete. Length, 2.8; diameter, 1.5 mm. (Dall.)

TYPE in United States National Museum, No. 213686. Type locality, Nunivak Island, Alaska.

RANGE. Known only from type locality.

## Alvania montereyensis Bartsch, 1911

*Proceedings of the United States National Museum,* 41:343; Pl. 30, fig. 2.

Shell elongate-conic, yellowish-white. Nuclear whorls one and one-third, well-rounded, marked by about 8 very slender, spiral lirations, of which the three near the summit are weaker than the rest. Post-nuclear whorls well-rounded, marked between the sutures by four strong, narrow spiral cords, which divide the space between the sutures into five almost equal parts, that between the summit and the first cord being a little wider than the rest. In addition to the spiral cords, the whorls are marked by slender axial riblets, which are about as strong as the spiral cords. Of these riblets, about 24 occur upon the first and second, and about 26 upon the penultimate turn. The spaces inclosed between the spiral cords and the axial riblets are elongated pits, having their long axes parallel with the spiral sculpture. In addition to the above sculpture, the entire surface of the shell is marked by numerous very fine, closely spaced striations. Suture strongly constricted. Periphery of the last whorl marked by a sulcus as wide as that which separates the first supraperipheral cord from its posterior neighbor. Base moderately long, scarcely produced anteriorly, well-rounded, marked by seven spiral cords, which grow successively weaker and closer spaced from the periphery to the umbilical region. Aperture moderately large, broadly oval; posterior angle obtuse; outer lip thin, showing the external sculpture within; inner lip slender, curved, and re-flected, the posterior edge touching the body whorl; parietal wall covered with a moderately thick callus, which renders the peritreme complete. Length, 2.3; diameter, 1.1 mm. (Bartsch.)

TYPE in United States National Museum, No. 160114. Type locality, Monterey, California.

RANGE. Sitka, Alaska, to Monterey, California.

### Alvania rosana Bartsch, 1911
### Plate 82, fig. 6

*Proceedings of the United States National Museum,* 41:349; Pl. 31, fig. 6.

Shell broadly ovate, yellowish white. Nuclear whorls two and one-half, smooth, well-rounded. Post-nuclear whorls well-rounded, marked by nar-row, well-rounded, somewhat sinuous, almost vertical, axial ribs, which are about one-third as wide as the spaces that separate them. Of these ribs 24 occur upon the second and the penultimate turn. In addition to the axial sculpture the whorls are marked by low, rounded, equal and equally spaced spiral cords, which are a little weaker than the axial ribs. Of these cords 6 occur between the sutures on the second turn and 7 on the penultimate turn, the first being at the summit. These spiral cords are a little narrower than the spaces that separate them. Suture moderately constricted. Periph-ery of the last whorl inflated, marked by a sulcus which is as wide as the spaces that separate the cords on the spire and, like them, crossed by the continuations of the axial ribs, which terminate at the posterior border of the first basal keel. Base strongly rounded, narrowly umbilicated, very slightly attenuated anteriorly, marked by eight equal and equally spaced spiral cords, which are about as wide as the spaces that separate them. Aperture broadly oval; posterior angle obtuse; outer lip thick; re-enforced immediately behind the edge by a moderately thick callus; inner lip strongly curved and somewhat reflected over and partly appressed to the base; parietal wall covered with a moderately thick callus. Length, 2.6; diameter, 1.5 mm. (Bartsch.)

TYPE in United States National Museum, No. 213688. Type locality, U.S. Bureau of Fisheries, Station 2901, off Santa Rosa Island, California.

RANGE. Santa Rosa and Santa Catalina Islands, California.

### Alvania iliuliukensis Bartsch, 1911
### Plate 82, fig. 2

*Proceedings of the United States National Museum,* 41:350; Pl. 31, fig. 2.

Shell ovate, light purplish brown. Nuclear whorls two (surface eroded). Post-nuclear whorls well-rounded, appressed at the summit, marked by slender, protractively curved axial riblets, of which 24 occur upon the first, 36 upon the second, and about 45 upon the last turn. These riblets are almost as wide as the spaces that separate them. In addition to the axial ribs the whorls are marked by low spiral cords which equal the axial ribs in strength. Of these cords 6 occur upon the first and 7 upon the second and the penultimate turn, between the sutures. The spaces inclosed between the axial ribs and spiral cords are small, squarish pits. Suture moderately constricted. Periphery of the last whorl marked by a sulcus equal to those occurring between the spiral cords on the spire. Base strongly rounded, somewhat inflated, marked by eight equal and equally spaced, somewhat flattened, low spiral cords, which equal the sulci between them in width. In addition, the base is marked by very fine lines of growth, which extend over the cords and interspaces. Aperture broadly ovate; posterior angle obtuse; outer lip thin, with a very faint varix immediately behind its edge; inner lip moderately strong, strongly curved and reflected over and appressed to the base; parietal wall covered with a thin callus. Length, 3; diameter, 1.7 mm. (Bartsch.)

Type in United States National Museum, No. 213690. Type locality, U.S. Bureau of Fisheries, Station 3333, off Iliuliuk Harbor, Alaska.

Range. Iliuliuk Harbor, Unalaska, to Belkoffski, Alaska.

### Alvania compacta Carpenter, 1865
### Plate 82, fig. 7

*Proceedings*, Philadelphia Academy of Natural Sciences, 62. *Proceedings of the United States National Museum*, 41:351; Pl. 31, fig. 7.

R. t. parva rufo-fusca, haud turrita, compacta, marginibus spirae excurvatis; anfr. nucleosis iii. globosis, laevibus, apice mamillato; normalibus iii. subplanatis latis; lirulis spiralibus obtusis circiter xv., quarum circ. vi. in spira monstrantur, interstitiis vix aequantibus; lirulis radiantibus circ. xxx., peripheriam tenus evanidis, anfractibus primis superantibus, anfractu ultimo saepe obsoletis; basi rotundata, haud (nisi testa juniore) umbilicata; apertura suborbiculari, peritremati continuo; operculo tenui, paucispirali, rapidissime augente. Long., .06; long. spir., .04; lat., .045 poll. (Carpenter.)

Shell ovate, light chestnut brown. Nuclear whorls two and one-quarter, well-rounded, smooth. Post-nuclear whorls well-rounded, slightly shouldered at the summit, marked by poorly developed, slightly protractive, axial ribs, of which 24 occur upon the first and second and 30 upon the penultimate whorl. In addition to these axial ribs the whorls are marked by equal and equally spaced, spiral cords, which are almost equal to the axial ribs in strength and of which 6 occur upon the first and 7 upon the second and the penultimate turn between the sutures. The first of these spiral cords is at the summit, which it renders feebly crenulate. The spaces inclosed between the axial ribs and spiral cords are rectangular pits, just a trifle longer than broad, their long axes corresponding to the spiral sculpture. In addition to the above sculpture the entire surface of the spire is marked by exceedingly fine, closely spaced, spiral striations and axial lines of growth. Suture moderately impressed. Periphery of the last whorl marked by a sulcus as wide as those which separate the spiral cords on the spire. Base strongly rounded, feebly produced anteriorly, marked by nine, equal and equally spaced, somewhat flattened spiral cords, which are about as wide as the spaces that separate them, and the fine sculpture noted for the spire. Aperture broadly ovate; posterior angle obtuse; outer lip thin, showing the external sculpture within; inner lip strongly curved, slightly reflected over and appressed to the base; parietal wall covered with a thick callus. Length, 3; diameter, 1.7 mm. (Bartsch.)

TYPE. Cotypes United States National Museum, No. 4338. Type locality, Puget Sound, Washington.

RANGE. Port Etches, Alaska, to Trinidad, California.

## Alvania cosmia Bartsch, 1911
### Plate 82, fig. 4

*Proceedings of the United States National Museum,* 41:352; Pl. 31, fig. 4.

Shell small, elongate-ovate, white, semitranslucent. Nuclear whorls one and one-half, marked by four, moderately strong, spiral lirations, which are separated by strongly impressed lines. Post-nuclear whorls appressed at the summit, strongly, slopingly shouldered, the shoulder extending from the summit to the anterior termination of the posterior third between the sutures, marked by two, strong, nodulose, spiral keels, of which the first is situated on the angle of the shoulder, while the second is about as far posterior to the suture as the first is anterior to the summit. In addition to these spiral keels, the whorls are marked by moderately strong, axial ribs which become enfeebled on the shoulder and anterior to the second keel. Of these ribs, 16 occur upon the first and second, and 18 upon the penultimate turn. The spaces inclosed between the spiral keels and axial ribs are large, deeply impressed, squarish pits, while their junctions form cusplike tubercles. Suture strongly channeled. Periphery of the last whorl marked by a strong, sublamellar, spiral keel. Base moderately long, marked by three spiral keels, which grow successively weaker and closer spaced from the periphery to the umbilical area. The broad spaces between the spiral keels are marked by slender, axial lines of growth. Aperture broadly ovate; posterior angle obtuse; outer lip re-enforced immediately behind the edge by a strong varix, transparent, showing the external sculpture within; inner lip rather stout, decidedly curved, and strongly reflected over and appressed to the base; parietal wall covered with a thick callus which renders the peritreme complete. Length, 2.2; diameter, 1.2 mm. (Bartsch.)

TYPE in United States National Museum, No. 213698. Type locality, San Pedro, California.

RANGE. San Pedro, California, to San Martin Island, Lower California.

## Alvania purpurea Dall, 1871
### Plate 82, fig. 1

*American Journal of Conchology, 7. Proceedings of the United States National Museum,* **41**: Pl. 31, fig. 1.

Shell small, of four rounded whorls rather pointed. Sculpture of six or seven revolving ribs, on the last whorl, only two appearing on the second whorl; apical whorl smooth. These ribs are crossed by about twenty longitudinal riblets, which do not pass the second revolving rib on the last whorl. On the antepenultimate whorl they reach from suture to suture, and are conspicuously angulated at their intersection with the two revolving ribs. Suture deep. Color whitish, the revolving ribs rounded, interrupted by the body whorl; peristome thickened, with a groove behind the columella. Altitude, .07; lat., .04 in. (Dall.)

TYPE in United States National Museum. Type locality, Monterey, California.

RANGE. Monterey, California, to San Martin Island, Lower California.

## Alvania aequisculpta Keep, 1887
### Plate 81, fig. 7

*West Coast Shells,* **65**: 1887.

Shell very elongate-conic, light yellow. Nuclear whorls two, moderately well-rounded, marked by six spiral threads, which are about as wide as the spaces that separate them, and numerous slender, closely spaced, axial threads, which are about one-fourth as strong as the spiral threads between which they occur, giving the entire surface a finely reticulated appearance. Post-nuclear whorls appressed at the summit, with a sloping shoulder which extends over the posterior fourth between the sutures, marked by strong, slightly retractive, axial ribs which are about one-fourth as wide as the spaces that separate them. Of these ribs, 14 occur upon the first, 16 upon the second, and 18 upon the penultimate turn. In addition to the axial ribs, the whorls are marked between the sutures by three strong, spiral cords which are almost as strong as the ribs and divide the spaces between the sutures into four almost equal portions. The intersection of the spiral cords and the axial ribs form strong tubercles, while the spaces inclosed between them are well impressed, rectangular pits, having their long axes parallel with the spiral sculpture. Suture strongly constricted. Periphery of the last whorl marked by a spiral sulcus equal to that which separates the supraperipheral spiral cord from its posterior neighbor and, like it, is crossed by the continuations of the axial ribs, which extend over the first two basal spiral cords and render them tuberculate. Base well-rounded, rather short, produced anteriorly, marked by three strong sublamellar, spiral cords which are about one-third as wide as the spaces that separate them. Aperture very oblique, twisted, ovate; posterior angle obtuse; outer lip thickened at the edge within the lip, re-enforced behind the edge by a strong varix, inner lip very stout, strongly curved, and appressed to the base; parietal wall covered with a very thick callus, which renders the peritreme complete.

TYPE in (not known to writer). Cotypes, United States National Museum, No. 219564. Type locality, San Diego, California.

RANGE. Catalina Island, California, to Todos Santos Bay, Lower California.

### Alvania oldroydae Bartsch, 1911

### Plate 81, fig. 3

*Proceedings of the United States National Museum,* **41**:360; Pl. 32, fig. 3.

Shell minute, broadly ovate, yellowish white. Nuclear whorls one and one-half, well-rounded, smooth. Post-nuclear whorls inflated, weakly shouldered on the posterior fourth between the sutures, marked by numerous slender, rather closely spaced, well-rounded, slightly protractive axial ribs, of which 20 occur upon the first, 24 upon the second, and 28 upon the penultimate whorl. In addition to the axial sculpture, the whorls are marked by slender spiral threads, which are almost equal to the axial ribs. Of these threads, three occur upon the first and second whorl, dividing the space between the sutures into four almost equal portions, the space at the summit being a little wider than the rest; the first spiral thread marks the termination of the sloping shoulder. On the penultimate whorl an additional spiral cord makes its appearance in the space immediately below the summit, a little nearer to the summit than the first spiral cord on the previous whorl. The intersections of the axial ribs and the spiral cords form slender rounded tubercles. The spaces inclosed between the three cords on the early whorls and the same on the last turn are squarish pits, while the spaces between the summit and the first spiral cord and the axial ribs on the first two turns are rectangular pits, having their long axes parallel with the axial sculpture. On the last whorl an additional spiral cord renders the pits between this cord and the next spiral cord and the axial ribs also squarish. Suture strongly constricted. Periphery of the last whorl marked by a sulcus as wide as that separating the suprasutural cord from the one adjacent to it anteriorly. Base well-rounded, strongly umbilicated, marked by four equal and almost equally spaced spiral cords, which are as strong as those occurring between the sutures and the feeble continuations of the axial ribs. Aperture subcircular; outer lip thickened all around by a very thick varix; inner lip stout, decidedly curved, somewhat reflected over and appressed to the base; parietal wall covered with a very thick callus, which renders the peritreme complete. Length, 1.6; diameter, 1.05 mm. (Bartsch.)

TYPE in United States National Museum, No. 152193a. Type locality, San Pedro, California.

RANGE. San Pedro, California, to South Coronado Island.

## Alvania kyskaënsis Bartsch, 1917

### See Part II, Plate 67, fig. 6

*Proceedings of the United States National Museum,* **52**:677; Pl. 46, fig. 6.

Shell elongate-ovate, pale brownish-yellow. Nuclear whorls one and one-half, well-rounded, marked by very slender spiral striations and exceedingly fine incremental lines. Post-nuclear whorls strongly rounded, weakly shouldered at the summit, marked by slender axial riblets which are somewhat sinuous and slightly retractively slanting. Of these ribs 24 occur upon the first, 26 upon the second, and 40 upon the penultimate turn; these are about one-third as wide as the spaces that separate them. In addition to the axial sculpture, the whorls are marked by four spiral cords, of which the first, which is about as far from the summit as it is distant from the second, is very weak on the first two whorls, but on the last assumes almost the strength of the three cords. The other three cords are of equal strength and spacing. The junction of the axial ribs and the spiral cords, which are a little stronger than the ribs, form weak nodules, while the spaces inclosed between them appear as well-impressed, squarish pits. In addition to this sculpture, the entire surface is marked by very fine incremental lines and numerous microscopic spiral striations, the two lending it a clothlike texture. Suture strongly constricted. On the last whorl the first basal keel makes its appearance above the summit of the succeeding turn. Periphery of the last whorl marked by a spiral sulcus, about as wide as those separating the cords on the spire. Base well-rounded, marked by seven equal and equally spaced, low, well-rounded, spiral cords, which are a little wider than the spaces that separate them. Aperture subcircular; posterior angle obtuse; outer lip strongly curved, rendered somewhat sinuous by the external sculpture; inner lip strongly curved; parietal wall covered by a thick callus which renders the peritreme complete. Length, 2.5; diameter, 1.2 mm. (Bartsch.)

TYPE in United States National Museum, No. 271407. Type locality, Kyska Harbor, Aleutian Islands, Alaska.

RANGE. Known only from type locality.

## Alvania almo Bartsch, 1911

### Plate 81, fig. 1

*Proceedings of the United States National Museum*, **41**:359; Pl. 32, fig. 1.

Shell minute, broadly ovate, yellowish-white. Nuclear whorls one and one-half, well-rounded, smooth. Post-nuclear whorls somewhat inflated, slopingly shouldered at the summit, marked by very slender, almost vertical, axial ribs, of which 18 occur upon the first, and 20 upon the penultimate turn. In addition to the axial sculpture, the whorls are marked between the sutures by two spiral cords equaling the ribs in strength. Of these, the first is a little posterior to the middle of the whorls, bounding the sloping shoulder, while the second is halfway between it and the suture. The intersections of the ribs and spiral cords form slender tubercles, while the spaces inclosed between them are well-impressed, squarish pits. The spaces inclosed between the first spiral cord and the summit and the axial ribs are rhomboidal areas, having their long axes parallel to the axial sculpture, while the spaces inclosed between the second spiral cord and the axial ribs and the suture are squarish pits. Suture moderately constricted. Periphery of the last whorl marked by a spiral sulcus, which is crossed by the continuations of the axial ribs. Base narrowly umbilicated, the umbilical chink being bounded by a tumid area, moderately long, well-rounded, slightly produced anteriorly, marked on the posterior half by two spiral cords equaling those between the sutures. Aperture almost circular; outer lip very thick all around, re-enforced by a strong varix; inner lip very stout, partly reflected over, and appressed to the base; parietal wall covered with a very thick callus which renders the peritreme complete. Length, 1.5; diameter, 1 mm. (Bartsch.)

TYPE in United States National Museum, No. 23749a. Type locality, Santa Barbara Island, California.

RANGE. Known only from type locality.

## Alvania acutilirata Carpenter, 1866
### Plate 82, fig. 3

*Proceedings of the California Academy of Sciences,* 3:217. *Proceedings of the United States National Museum,* 41:352; Pl. 31, fig. 3.

R.t. tenui, satis turrita, rufo-cinerea, marginibus spirae parum excurvatis; anfr. nucl. iii. normalibus laevibus, vertice parum mamillato; norm. iii. subrotundatis, suturis valde impressis; liris radiantibus circ. xviii. acutis distantibus, ad peripheriam evanidis; lirulis acutis spiralibus distantibus circ., xv., quarum vi. in spira monstrantur, liris radiantibus et interstitiis latis, undatis, eleganter superantibus, haud nodulosis; basi rotundata, haud umbilicata; apertura ovata, peritremati continuo. (Carpenter.)

Shell small, ovate, yellowish-white, with the base of the columella and the posterior angle of the aperture purplish chestnut brown. Nuclear whorls two and one-half, small, strongly rounded, smooth. Post-nuclear whorls very feebly shouldered at the summit, marked by regular, slender, slightly protractive, axial ribs, which are about half as wide as the spaces that separate them. Of these ribs, 24 occur upon the first, 22 upon the second, and 20 upon the penultimate turn. In addition to the axial ribs, the whorls are marked by slender spiral cords, which are a little more than half as strong as the axial ribs, and about half as wide as the spaces that separate them. Of these cords, 6 occur between the sutures on the first, and 7 on the second and the penultimate turn. The spaces inclosed between the axial ribs and the spiral cords are elongated pits, having their long axes parallel to the spiral sculpture. Suture well-impressed. Periphery of the last whorl marked by a well-rounded, spiral cord, at the posterior edge of which the axial ribs terminate. Base well-rounded, feebly produced anteriorly, marked by 8 spiral cords, which grow successively weaker and more closely spaced from the periphery to the umbilical region. Aperture broadly oval; posterior angle acute; outer lip rather thin at the edge, thinner within, showing the external sculpture within; inner lip moderately stout, strongly curved, and reflected over and appressed to the base; parietal wall covered with a thick callus, rendering the peritreme complete. Length, 2.3; diameter, 1.2 mm. (Bartsch.)

Type. Figured specimen in United States National Museum, No. 153072. Type locality, San Diego, California.

Range. Known only from type locality.

## Alvania dinora Bartsch, 1917
### See Part II, Plate 67, fig. 5

*Proceedings of the United States National Museum,* 52:678; Pl. 46, fig. 5.

Shell small, elongate-ovate, yellowish-white. Nuclear whorls two, strongly rounded, smooth. Post-nuclear whorls strongly rounded, almost appressed at the summit and moderately constricted at the suture, marked by seven very low, flattened, spiral cords between the sutures, which are separated by a shallow impressed line. The axial sculpture consists of numerous very slender threads which are almost vertical. Suture strongly constricted. Periphery of the last whorl somewhat inflated, well-rounded, base moderately long, slightly attenuated anteriorly, marked by six low, ill-defined, rounded spiral cords, which, like those on the spire, are separated by mere impressed lines. The axial ribs also continue over the base. Aperture broadly oval, decidedly effuse at the junction of the basal and the outer lip; posterior angle obtuse; outer lip thick within, thin at the edge, evenly curved from the posterior angle to its junction with the inner lip; inner lip decidedly curved, somewhat reflected, and appressed to the base; parietal wall covered by a thick callus, which practically renders the peritreme complete. Length, 2; diameter, 1 mm. (Bartsch.)

TYPE in United States National Museum, No. 268730. Type locality, Forrester Island, Alaska.

RANGE. Known only from type locality.

## Alvania burrardensis Bartsch, 1921

*Proceedings of the Biological Society of Washington,* **34**: 38.

Shell very broadly ovate, pale yellow. Nuclear whorls decollated in all our specimens. Postnuclear whorls strongly inflated, marked by strong, rather distantly spaced, curved, and slightly protractively slanting axial ribs, of which twenty-four occur upon the next to the last and twenty-two upon the last turn. In addition to the axial ribs, the whorls are crossed by six equal, and equally spaced, broad spiral cords which render the axial ribs obscurely nodulose at their junction. The spaces separating the spiral cords are a little less wide than the cords. Periphery of the last whorl marked by a sulcus, which is crossed by the continuation of the axial ribs which extend partly over the base, but evanesce soon after passing the periphery. Base short, strongly rounded, marked by nine equal and equally spaced prominent spiral cords, which are a little wider than the spaces that separate them. Aperture subcircular; posterior angle obtuse; outer lip reinforced by a callus at the edge; inner lip curved and appressed at the base; parietal wall covered by a moderately thick callus. Length, 2.2; diameter, 2 mm. (Bartsch.)

TYPE in United States National Museum, No. 340938. Type locality, Burred Inlet, British Columbia.

RANGE. Known only from type locality.

## Alvania sanjuanensis Bartsch, 1921

*Proceedings of the Biological Society of Washington,* **34**: 37.

Shell moderately large, chestnut brown excepting the tip, which is a little paler, and the extreme base, which is also lighter. Nuclear whorls one and a half, well rounded. (The sculpture of the nuclear whorls eroded in all the shells seen except in a very small fraction of the last turn in the type, which presents a finely, somewhat wavy, spirally lirate surface. I am not quite certain whether axial threads are present or not.) Nuclear whorls strongly shouldered at the summit, strongly rounded, marked on the first turn by three strong spiral cords, which occupy the anterior half of the turn; on the second turn a fourth cord occurs a little anterior to the median line between the summit and the first strong cord, while on the next turn a fifth slender thread makes its appearance between the summit and this cord. This last cord at the summit never attains a strength as great as the third anterior to it, while the second one is fully as strong on the penultimate turn. In addition to these spiral cords, the shell is marked by rather weak axial ribs, of which twenty-four occur upon the second, twenty-six upon the third, and about thirty-two upon the last turn; on this they are decidedly enfeebled. The junction of the axial ribs and spiral cords forms feeble nodules. The entire surface of the spire between ribs and interspaces is crossed by fine spiral and axial threads which lend it a fine clothlike texture. Suture strongly constricted. Periphery of the last whorl well rounded. Base moderately long, well rounded, marked by seven equally spaced spiral threads, of which the seventh immediately behind the inner lip is feeble. The rest are almost as wide as the spaces that separate them. The axial ribs do not extend over the base, but the fine sculpture described for the spire is also present here. Aperture ovate; posterior angle obtuse; outer lip thin at the edge, strongly curved; inner lip strongly curved, reflected and appressed to the base except at the extreme tip; parietal wall covered by a thick callus, which renders the peritreme complete. Altitude, 3; diameter, 1.5 mm. (Bartsch.)

TYPE in United States National Museum, No. 334487. Type locality, San Juan Islands, Puget Sound, Washington.

RANGE. Puget Sound, Washington.

## Alvania dalli Bartsch, 1927

*Proceedings of the United States National Museum,* **70**: 30; Pl. 3, fig. 6.

Shell small, thin, semitranslucent, bluish white. Nuclear whorls one and one-half, strongly rounded, finely granular. Postnuclear whorls rather inflated, strongly rounded, appressed at the summit, marked by five low, rounded, not quite equal and equally spaced spiral cords, which are about as wide as the spaces that separate them, and numerous, fine lines of growth and microscopic, closely spaced, spiral striations. The incremental lines and fine spiral sculpture give to the surface of the shell a fine cloth-like texture. Suture strongly constricted. Periphery of the last whorl inflated, well-rounded. Base short, strongly rounded, narrowly umbilicated, marked by 14 spiral cords which become a little less strong and closer spaced anteriorly. In addition to this the base is marked by the fine sculpture referred to on the spire. Aperture subcircular, posterior angle decidedly obtuse; outer lip thin, strongly curved, showing the external sculpture within; columella slender and slightly reflected; parietal wall covered by a thick callus which renders the peritreme complete.

TYPE in United States National Museum, No. 362154. Type locality, Shuyak Straits, Afognak Island, Alaska. Length, 2.5 mm.; diameter, 1.3 mm.

## Family RISSOINIDAE

## Genus **RISSOINA** Orbigny, 1840

Rissoiform, ribbed or cancellated, whorls numerous, apex mamillated; aperture semilunar, lip thickened within, a little reflected, anteriorly effuse or faintly channeled. Operculum corneous, thick, semilunar, paucispiral, with a claviform process on the internal face. (Tryon, *Manual of Conchology.*)

TYPE. *Rissoina catesbyana* Orbigny.
DISTRIBUTION. World-wide.

### Rissoina kelseyi Dall and Bartsch, 1902

*Nautilus,* **16**:94. *Proceedings of the United States National Museum,* **49**:49; Pl. 30, fig. 4.

Shell of medium size, elongate-conic, white, variously banded, or uniformly chocolate brown. Nuclear whorls mamillate, smooth. Post-nuclear whorls slightly rounded, ornamented axially by a few broad, depressed, almost obsolete ribs, which are best seen near the summit of the whorls, and many irregular, more or less deeply impressed striations, which extended almost undiminished to the umbilical region. The spiral sculpture, however, is more conspicuous than the axial, and consists of deeply impressed lines which are more closely placed and less strongly developed near the summit of the whorls than at the periphery, grading gradually in this respect between these two regions. Sutures simple, well-marked.

Periphery and base of the last whorl well-rounded, the latter ornamented by spiral sculpture similar to that between the sutures, but a little more distantly spaced and more strongly impressed. Eighteen of the spiral lines appear between the sutures upon the penultimate whorl and ten upon the base. Aperture large, oblique, decidedly effuse anteriorly; posterior angle acute, peristome continuous; columella strong, short, somewhat twisted and slightly revolute. Long., 6.3; diameter, 2.5 mm. (Bartsch.)

TYPE in United States National Museum, No. 168605. Type locality, Pacific Beach, California.

RANGE. San Pedro, California, to Coronado Islands.

## Rissoina bakeri Bartsch, 1902

### Plate 83, fig. 4

*Nautilus,* **16**:9. *Proceedings of the United States National Museum,* **49**:56; Pl. 33, fig. 4.

Shell small, sub-diaphanous to milky white. Nuclear whorls two, quite large, with beveled shoulder, smooth. Later whorls well-rounded, somewhat angulated about one-fourth below the summit, ornamented by about twelve to fourteen quite well-developed axial ribs and a series of prominent axial striations, between them in the intercostal spaces, which are about four times as wide as the ribs; both ribs and striations extend from the summit of the whorls to the umbilical region, which is bordered by a basal fasciole. Sutures simple, well-marked. Aperture large, very oblique, sub-oval, slightly notched at the posterior angle. Outer lip varicose. Long., 2.7; diameter, 1.0 mm. (Bartsch.)

TYPE in United States National Museum, No. 130562. Type locality, San Pedro, California.

RANGE. San Pedro to Coronado Islands.

## Rissoina cleo Bartsch, 1915

### Plate 83, fig. 3

*Proceedings of the United States National Museum,* **49**:58; Pl. 33, fig. 3.

Shell small, elongate-conic, milk-white. Nuclear whorls 2, well-rounded, slightly shouldered near the summit. Post-nuclear whorls well-rounded, very feebly shouldered at the summit, marked by slender, very distantly spaced, somewhat sinuous, decidedly protractive, axial ribs, of which 16 occur upon the first, 18 upon the second and third, and 20 upon the penultimate whorl. Intercostal spaces about four times as wide as the ribs, marked by numerous, fine, irregular wavy markings, which give a watered-silk effect. Suture moderately constricted. Periphery well-rounded. Base of last whorl moderately long, concaved in the middle, having a slender fasciole at its anterior termination, marked by the feeble continuations of the axial ribs and the same wavy sculpture observed in the intercostal spaces of the spire. Aperture large, decidedly effuse, feebly channeled posteriorly; outer lip very thick and effuse, reinforced immediately behind the edge by a thick varix; inner lip slender, curved and appressed at the base; parietal wall covered with a thick callus which renders the peritreme complete. Length, 2.8; diameter, 1.2 mm. (Bartsch.)

TYPE in United States National Museum, No. 226456. Type locality, off South Coronado Island.

RANGE. Known only from type locality.

### Rissoina californica Bartsch, 1915
### Plate 83, fig. 1

*Proceedings of the United States National Museum,* 49:55; Pl. 33, fig. 1.

Shell very minute, elongate-conic, semitranslucent, yellowish-white. Nuclear whorls 27, well-rounded, smooth. Post-nuclear turns well-rounded, marked by strong, decidedly protractive, slightly sinuous, axial ribs which are about half as wide as the spaces that separate them. Of these ribs, 14 occur upon the first, and 16 upon the remaining turns. These ribs extend strongly from the summit to the sutures on each turn and are not at all constricted below the summit. The intercostal spaces are deep, well-rounded, and smooth. Suture strongly impressed. Periphery of the last whorl well-rounded, marked by the continuation of the axial ribs which extend over the somewhat prolonged base to the umbilical chink where they become slightly fused on the tumid area surrounding the umbilical region. Aperture oval; slightly channeled at the posterior angle and at the junction of the outer lip and the columella; outer lip reinforced immediately behind the edge by a thick callus; columella strongly reflected over and appressed to the base; parietal wall covered with a thick callus which renders the peritreme complete. Length, 2.8; diameter, 1.2 mm. (Bartsch.)

TYPE in United States National Museum, No. 271644. Type locality, off South Coronado Island.

RANGE. Catalina Island to South Coronado Island.

## Rissoina dalli Bartsch, 1915

### Plate 83, fig. 2

*Proceedings of the United States National Museum,* **49**:59; Pl. 33, fig. 2.

Shell small, elongate-conic, subdiaphanous. Nuclear whorls almost 2, inflated, well-rounded, smooth. Post-nuclear whorls moderately rounded, appressed at the summit, the appressed portion being somewhat excurved, lending the whorls the aspect of having a double suture. Whorls marked by numerous, very fine, closely spaced, almost vertical, axial threads which are about as wide as the spaces that separate them. Suture feebly impressed. Periphery of the last whorl well-rounded. Base moderately long, well-rounded, marked like the spire. Aperture moderately large, ovate; outer lip somewhat effuse, thick at the edge, thin within where the external sculpture is seen through the substance of the shell; inner lip moderately thick, strongly curved, reflected over and adnate to the base. Parietal wall covered with a thick callus, which renders the peritreme complete. Length, 2.8; diameter, 1 mm. (Bartsch.)

TYPE in United States National Museum, No. 107281. Type locality, San Pedro, California.

RANGE. San Pedro, California, to South Coronado Island.

## Rissoina coronadoënsis Bartsch, 1915

### Plate 83, fig. 8

*Proceedings of the United States National Museum,* **49**:60; Pl. 33, fig. 8.

Shell small, elongate-conic, milk-white. Nuclear whorls a little more than two, well-rounded, smooth. Post-nuclear whorls moderately rounded, appressed at the summit, the appressed portion slightly excurved, marked by slender, slightly protractive, somewhat sinuous, axial threads of which 40 occur upon the first, 42 upon the second, 48 upon the third, 56 upon the fourth, and 52 upon the penultimate turn. These threads are separated by intercostal spaces about as wide as the threads. Suture well-impressed. Periphery of the last whorl well-rounded. Base rather short, slightly concaved, marked by the continuations of the axial threads, which extend undiminished to the umbilical area. Aperture oval, posterior angle acute; outer lip thin at the edge, thickened immediately behind the edge by a varix, thin, deep within where the external sculpture shines through the substance of the shell; inner lip short, strongly curved, reflected over and adnate to the base; parietal wall covered with a thick callus, which renders the peritreme complete. Length, 3.5; diameter, 1.3 mm. (Bartsch.)

TYPE in United States National Museum. Type locality, Coronado Islands.

[ 709 ]

RANGE. San Diego, California, to San Martin Island, Lower California.

## Rissoina newcombei Dall, 1897

### Plate 83, fig. 7

*Bulletin of the Natural History Society of British Columbia,* No. 2, 14; Pl. 1, fig. 12.

Shell small, thin, white, with six and a half whorls; nucleus transparent, polished, smooth; subsequent whorls with close-set, low, fine, slightly flexuous, transverse (or axial) riblets; suture distinct, somewhat appressed; aperture large, patulous, with the lips rather thickened. Long. of shell, 3; diameter, 1 mm. (Dall.)

TYPE in ? Type locality, Cumshewa Inlet, British Columbia.

RANGE. Forrester Island, Alaska, to Vancouver Island, British Columbia.

## Family ANAPLOCAMIDAE

## Genus **ANAPLOCAMUS** Dall, 1895

Shell short-spired, with a thick brown periostracum, with a simple, sharp, outer lip, parietal callus, arched pillar, the anterior extreme of the aperture slightly produced and pointed, as in some *Litorinas;* the base imperforate, the aperture destitute of lirae, teeth, or other projections; operculum, relatively to the size of the animal, large; area of attachment, small; form, U-shaped, the apex without any spiral inclination, rather blunt, the increment being applied to the proximal end, and the edges entire. (Dall.)

TYPE. *A. borealis* Dall.

DISTRIBUTION. Northern.

## Anaplocamus borealis Dall, 1895

### See Part II, Plate 31, fig. 15

*Proceedings of the United States National Museum,* **18**:9 and **24**:550; Pl. 38, fig. 4.

Shell short, rude, of about four and a half whorls (the apex in each specimen eroded), smooth, except for lines of growth and darker lines, which might indicate resting stages; whorls somewhat flattened above and near the apex, more or less appressed at the suture; periphery rounded, or, in the younger shells, obscurely angular; base full, smooth, with no indication of an umbilicus or axial depression; aperture subovate, pointed in front or behind; outer lip thin, sharp, simple; pillar rather thick, white, with a smooth, well-marked callus over the body; operculum dark brown, with strong incremental lines. Height of (somewhat eroded) shell, 17; of last whorl, 15; of aperture, 10; major diameter of shell, 13; of aperture, 7 mm. (Dall.)

TYPE in United States National Museum, No. 122592. Type locality, Pacific Ocean, south of Unimak Island.

RANGE. Known only from type locality.

## Family TRUNCATELLIDAE

### Genus **TRUNCATELLA** (Leach, 1818) Risso, 1826

Shell minute, cylindrical, truncated; whorls striated transversely; aperture oval, entire; peristome continuous. Operculum corneous, subspiral. (Tryon, *Structural and Systematic Conchology*.)

TYPE. *T. truncatula* Drap.

DISTRIBUTION. World-wide, mostly tropical.

### **Truncatella stimpsoni** Stearns, 1872

*Proceedings of the California Academy of Sciences*, **4**: 249.

Shell cylindrical, solid, light reddish horn-color, or amber; shining, slightly decreasing in size toward apex; closely and strongly longitudinally ribbed, the ribs even, regular, and interrupted only by the suture; upper whorls wanting, remaining whorls, 4; aperture oval, somewhat oblique, slightly angulated above; peristome continuous, thickened and moderately angulated at its junction with the body whorl. Length of largest specimen, .22; length of aperture, .06 in. (Stearns.)

TYPE in ? Type locality, False Bay, near San Diego, California.

RANGE. Catalina Island, California, to Magdalena Bay, Lower California.

### **Truncatella californica** Pfeiffer, 1857

*Proceedings*, Zoölogical Society of London, **111**. Binney, *Terrestrial Air-breathing Mollusks of the United States*, **4**: 28; Pl. 79, figs. 20, 22.

T. non-rimata, turrito-cylindracea, truncata, tenuiscula, leviter striata, parum nitens, pallide rubello-cornea; spira sursum vix attenuata; sutura simpliciter marginata; anfr. superst. 4 convexi, sensim accrescentes, ultimus basi non compressus; apertura verticalis, ovalis, superne vix angulata; perist. simplex, continuum, margine dextro expanso, superne sub-repando, columellari adnato. (Pf.) (Binney.)

Shell imperforate, cylindrical, truncated at tip, thin and translucent with light striae, shining, amber-colored; spire in perfect state of the shell composed of about 10 whorls, of which 4 only are not deciduous; these are convex, increasing in size rather rapidly; aperture oval, vertical, rounded above; peristome simple and continuous, slightly expanded, its pillar scarcely attached to the shell. Length, 4⅔; diam., 1⅔ mm. (Binney.)

TYPE in British Museum? Type locality, San Diego, California.

RANGE. Santa Barbara to San Diego and San Martin Island, California.

## Family SYNCERATIDAE

### Genus **SYNCERA** Gray, 1821 (Assiminea)

Shell conic, usually strong. Nuclear whorl smooth, the rest of the shell marked by lines of growth and fine spiral striations only. Outer lip simple; inner lip continuing over the base as a thick parietal callus. Operculum subspiral, thin, horny. (Bartsch.)

TYPE. *Syncera hepatica* Gray, equals *Assiminea grayana* Leach. The *synceras* are littoral forms, frequently inhabiting the brackish reaches of our coast.

DISTRIBUTION. West coast of America, Europe.

### Syncera translucens Carpenter, 1864

*Supplementary Report,* British Association for the Advancement of Science, 613. *Proceedings of the California Academy of Sciences,* 3:219. *Proceedings of the United States National Museum,* 58:164; Pl. 12, fig. 7.

J.t. *Barleeiae subtenui* simili sed tenuiore, tumidiore; cornea, pallide fulva, laevi, nitente, satis diaphana; anfr. nucl. normalibus, apice sub-mamillato; norm. iv. subconvexis, suturis distinctis; basi rotundata, haud umbilicata; apertura ovata, peritremati vix continuo; labro acuto; labio appresso, regione umbilicari parum calloso. (Carpenter.)

Shell broadly ovate, light brown. Nuclear whorls not differentiated from the remaining turns, well-rounded. Post-nuclear whorls strongly rounded, very narrowly shouldered at the summit, marked by decidedly retractively curved axial lines of growth, and exceedingly fine microscopic spiral striations. Suture strongly impressed. Periphery of the last whorl well-rounded. Base inflated, well-rounded. Aperture subcircular; posterior angle obtuse; outer lip thin; inner lip very strongly curved, thick, reflected over and appressed to the base; parietal wall covered with a thick callus which fuses with the reflected inner lip and forms a decided callosity over the umbilical region. Operculum typical. Altitude, 3; diameter, 1.9 mm. (Bartsch.)

TYPE. The specimen figured and described is in the United States National Museum, No. 271483. Type locality, San Diego, California. in shell washings.

RANGE. Vancouver Island to Lower California.

## Family CAPULIDAE

### Genus **CAPULUS** Montfort, 1859

Shell conical, apex posterior, spirally recurved; aperture rounded; muscular impression horseshoe-shaped. (Tryon, *Structural and Systematic Conchology*.)

TYPE. *Capulus hungaricus* Linnaeus.

DISTRIBUTION. West Indies, Europe, India, Australia, West America. Fossil, Silurian . . . .

### Capulus californicus Dall, 1900
#### Plate 91, figs. 15, 16

*Nautilus,* **13**:100. *Bulletin 112,* United States National Museum; Pl. 15, fig. 6.

Shell only moderately elevated, oval or more or less conformable with the object upon which it roosts, the apex small, somewhat laterally compressed, incurved almost symmetrically, nearly concealing the smooth, one-whorled nucleus, situated near the posterior margin; surface nearly smooth, somewhat irregular, mesially with small faint radial not very close-set ridges, covered with an imbricated, dense, soft, glistening periostracum which projects beyond the margins; interior polished, white, with faint rosy rays extending from the apex to the anterior margin. Alt., 10; long., before the apex, 30; behind it, 5.5; total basal length, 36.5; average width, 29 mm. (Dall.)

TYPE in United States National Museum. Type locality, on *Pecten diegensis* from off San Pedro, California, in 20–25 fathoms.

RANGE. San Pedro to San Diego, California.

### Genus **PILISCUS** Loven, 1859

Shell thin, patelliform, with thin epidermis; apex not spiral, somewhat inclined to the right and posteriorly. (Tryon and Pilsbry, *Manual of Conchology*.)

*Piliscus* Loven, 1859, and *Pilidium* Middendorff, not Forbes.

DISTRIBUTION. Boreal Seas.

#### Piliscus commodus Middendorff, 1851

*Sibirische Reise,* **2**:214; Pl. 17, figs. 4–11.

Es sind mir im Okhotskischen Meere nur zwei Schalen dieses Weichthieres aufgestossen, deren grossere zum Glücke auch das Thier enthalt.

Er Schale nach gerath man in Zweifel, ob man dieses Thier zu den Fatellen bringen dürfe, oder ob es nich vielmehr ein *Capulus* sei; von letzterem eschlechte entfernt sich ubrigens die Schale noch mehr als von den Patellen durch ihre grosse Dünne, vermoge welcher sie sogar die Velutinen ubertrifft, und durch den unbedeutender und kaum hakig zuruckgebogenen Wirbel. Die Maassverhaltnisse der Schale sind folgende. Long., 22; latit., 19; altit., 10 mm. (Middendorff.)

TYPE in St. Petersburg? Type locality, Okhotsk Sea.

RANGE. Arctic Ocean to Pribilof Islands, Bering Sea, and the Okhotsk Sea.

Thin, transparent, radiately striate or smooth except growth lines, whitish or yellowish, faintly striate with darker color, under a thin yellowish epidermis, base rounded, apex small, inclined. (Tryon and Pilsbry, *Manual of Conchology*.)

This is the description of *C. radiatus* Sars, which is given as the same as *Piliscus commodus*.

## Family HIPPONICIDAE

## Genus **HIPPONIX** Defrance, 1819

Shell thick, obliquely conical, apex posterior; base shelly, with a horseshoe-shaped impression, corresponding to that of the adductor muscle. (Woodward, *Manual of the Mollusca*.)

TYPE. *Hipponix cornucopia* Defrance.

DISTRIBUTION. West Indies, Persian Gulf, Philippines, Australia, Pacific, West America. Fossil, U. chalk . . . . Britain, France, North America.

### Hipponix antiquatus Linnaeus, 1767

*Systema Naturae*, 12th ed., Tryon, *Manual of Conchology*, **8**:134; Pl. 40, figs. 93–99.

Testa integra oblonga imbricata, vertice postico recurvato. Testa opaca, lactea, multum imbricata. Apertura ovata, apex posticus, recurvus. (Linnaeus.)

Shell entire, oblong, imbricate, recurved in the rear. Shell oblique, milky, strongly imbricate. Aperture ovate, apex posterior, recurved. (Translation.)

TYPE in Linnaeus Society, London. Type locality ?

RANGE. Crescent City, California, to Panama, and the Galapagos Islands. Also Atlantic.

## Hipponix antiquatus cranioides Carpenter, 1864

*Annals and Magazine of Natural History,* series 3, **14**:427. Tryon, *Manual of Conchology,* **8**:135; Pl. 40, figs. 6, 7.

H. testa valde planata, majore, albida; vertice nucleoso? . . . . ; testa adulta apice interdum subcentrali, saepius plus minusve postico; laminis incrementi confertis, undique rapide augentibus; striis radiantibus fortioribus, confertissimis, laminarum margines saepe crenulantibus; margine acuto; cicatr. musc. angusta, margini continua, regione capitis minore, saepe dextrorsum torsa; epidermide? . . . . Long., 85; lat., 75; alt., 3 poll. (Carpenter.)

Rounded, convexly planate, the apex subcentral, radiately striate, concentrically laminate. Length, .85 inch. (Tryon, *Manual of Conchology.*)

TYPE in ? Type locality, Neah Bay, Washington.

RANGE. Vancouver Island, British Columbia, to San Pedro, California.

## Hipponix serratus Carpenter, 1857

*Mazatlan Catalogue,* 296. Tryon, *Manual of Conchology,* **8**:134; Pl. 40, fig. 100.

H.t. conica seu depressa, albida; apice decollato, subcentrali seu valde remoto, interdum valde prominente; lamellis crebris, acutis, extantibus, basi parallelibus, concentrice ornata; apicem versus radiatim costata; supra lamellas radiatim tenuissime sulcata; basi latissima, planata, lamellis creberrimis instructa; lamellis profunde serratis, lobis subquadratis; inter lamellas epidermide fusca copiosissime induta, taeniis incisis conferta; cicatrice musculari transversim corrugata. (Carpenter.)

Differs from *H. antiquatus* in the character of the base of the shell (margin of aperture), which is broad and flat, made up of very numerous close-set lamellae, deeply serrated into large, scarcely rounded lobes; the interstices are filled with epidermis, in irregular ribband-like shreds. Length, 1 inch. (Tryon, *Manual of Conchology.*)

TYPE in the Liverpool collection. Type locality, Mazatlan.

RANGE. Monterey, California, to Panama.

## Hipponix tumens Carpenter, 1865

*Annals and Magazine of Natural History,* series 3, **15**:180. Tryon, *Manual of Conchology,* **8**:135; Pl. 40, fig. 7.

H. testa normaliter fornicata, rotundara, albida; epidermide rugulosa, interstitiis pilulosa; vertice nucleoso nautiloideo, laevi, parum tumente, apice celato, interdum persistente; dein rapidissime augente, expansa, undique regulariter arcuata; liris acutis, subelevatis, distantibus, aliis intercalantibus; lineis incremente minoribus decussantibus; margine acuto; apertura plerumque rotundata: cicatrice musculari a margine parum remota, regione capitis valde interrupta. Long., 7; lat., 46; alt., 33 poll. (Carpenter.)

A somewhat higher form, less worn by attrition, but having the essential characters of *H. cranioides*. (Tryon, *Manual of Conchology*.)

TYPE in Mrs. Boyce's Collection, Utica, New York. Type locality, Santa Barbara, California.

RANGE. Crescent City, California, to San Diego.

### Hipponix barbata Sowerby, 1835

*Proceedings,* Zoölogical Society of London, **5**; 1835. Tryon, *Manual of Conchology,* **8**:135; Pl. 40, figs. 2, 3.

Hipp. testa pallide fulva, subelevato-conica, radiatim confertim striata; margine ventrali producto; epidermide piloso-barbata; margine interno crenulato. (Sowerby.)

Shell depressed, apex suberect, subposterior, whitish, concentrically and radiately striated, with a pilose brownish epidermis, aperture margin smooth. Length, .75 inch. (Tryon, *Manual of Conchology*.)

TYPE in Museum Cuming. Type locality, coral reefs around Toobouai, one of the Society Islands.

RANGE. Crescent City, California, to Panama and the Galapagos Islands.

### Family CALYPTRAEIDAE

### Genus CALYPTRAEA Lamarck, 1799

Shell oval, limpet-like; with a posterior, oblique, marginal apex; interior polished, with a shelly partition covering its posterior half. (Woodward, *Manual of the Mollusca*.)

TYPE. *Calyptraea chënensis* Linnaeus.

DISTRIBUTION. West Indies, Honduras, Mediterranean, West Africa, Cape, India, Australia, West America.

### Calyptraea contorta Carpenter, 1865

*Proceedings of the California Academy of Sciences,* **3**:215.

G.t. parva, tenui, albida; vertice nucleoso planato, extante, minimo, anfr. uno et dimidio planorbi-vormibus, apice conspicuo; dein conoidea, elevata, solute spirali, suturis impressis; superficie rudi, laminis incrementi interdum conspicuis; lamina interna. Long., 0.26; lat., 0.24; alt., 0.15; div., 80°. (Carpenter.)

Shell small, thin, whitish, with nuclear vertex planate, standing out, very small; with one whorl, and in the middle flattened-vermiculate, with conspicuous apex; then conoidal, elevated, with corroded spire, with impressed sutures; surface rough, with laminae of growth sometimes conspicuous; with internal lamina. (Translation.)

TYPE in State Collection, No. 369. Type locality, Monterey, California.

RANGE. Monterey, California, to the Gulf of California. Described as *Galerus contortus* Carpenter.

## Calyptraea fastigiata Gould, 1846
### Plate 93, fig. 5

*Proceedings of the Boston Society of Natural History, 2:161. Mollusca and Shells of the United States Exploring Expedition, 379; figs. 484, a, b.*

Testa parva, tenuis, laevis, rotundata, elevato-conica, concentrice et tenuissime striata, epidermide fuscente induta; spira anfractibus tribus planulatis, apice sub-mediano, acuto; lamella interna spiralis, striata; margine soluto, ad centrum duplicato, haud appresso, umbilicum parvum efformante.

Shell small, thin, smooth, rounded at base, elevated-conical; the apex central, acute; the spire composed of three flattened whorls. Surface faintly marked by very delicate lines of growth, and covered by a thin, pale-yellow epidermis. Interior white, with a spiral lamella (septum) commencing at the vertex, and making half a revolution, terminating at the margin; its free edge folded. Diameter half an inch; altitude five-eights of an inch. (Gould.)

TYPE in United States National Museum. Type locality, Puget Sound.

RANGE. Port Etches, Alaska, to Puget Sound.

## Genus CRUCIBULUM Schumacher, 1817

Differs from *Calyptraea* in the internal cup-shaped lamina, which is entire and attached along a line on one side to the inner wall of the shell. (Tryon, *Structural and Systematic Conchology*.)

TYPE. *C. rudis* Brod.

DISTRIBUTION. Temperate and tropical; world-wide.

## Crucibulum spinosum Sowerby, 1824

*Genera of Shells*, "Calyptraea," figs. 4, 7. *Conchologia Iconica*, 11; Pl. 4, fig. 10.

Cruc. testa orbiculari, interdum subquadrato-ovata, nunc solidiuscula, convexa, nunc depressa, tenuicula, radiatim lirata et irregulariter corrugato-striata, tubulo-spinosa, interdum autem raro inermi, vertice subcentrali, oblique contorto; livida aut lutescente-alba, purpureo contorto-radiata, intus intense cinerea vel alba; appendice interna cyathiformi, ampla, lateraliter compressa, crystallino-alba, medio interdum cinerea.

Shell orbicular, sometimes squarely ovate, rather solid and convex, or depressed and rather thin, radiately ridged and irregularly wrinkle-striated; generally tube-spined, sometimes, but rarely, without spines, vertex nearly central, obliquely twisted; livid or yellowish-white, twistedly rayed with purple, interior dark-ash or white; internal appendage cup-shaped, large, laterally compressed, crystalline-white, sometimes ash-stained in the middle. (Reeve.)

TYPE in British Museum. Type locality, not known to writer.

RANGE. Trinidad, California, to Tome, Chile, and the Galapagos Islands.

## Family CREPIDULIDAE

### Genus **CREPIDULA** Lamarck, 1801

Shell oval, limpet-like; with a posterior, oblique, marginal apex; interior polished, with shelly partition covering its posterior half. (Woodward, *Manual of the Mollusca*.)

TYPE. *Crepidula fornicata* Linnaeus.

DISTRIBUTION. West Indies, Honduras, Mediterranean, West Africa, Cape, India, Australia, West America. Fossil, Eocene . . . . France, North America, and Patagonia.

### Subgenus CREPIDULA s.s.

#### Crepidula grandis Middendorff, 1849

*Beitrage zu einer Malacologia Rossica*, 101. Tryon, *Manual of Conchology*, 8:128; Pl. 37, fig. 33.

Testa ovali, magna, crassa; extus; flavascente aut rufescente, epidermide solida decidua fuscescente obtecta; intus lactaea; apice postico, submarginali, soluto, uncinato, dextrorsum recurvo; septo solido, ceasso, ad dextrum latus impresso; impressione musculari magna, maxime conspicua; aperturae margine postico prominente, subfornicate. (Middendorff.)

Shell oval, large, thick, externally yellowish or reddish, covered with a thick deciduous swarthy epidermis; inside milky; with posterior apex, submarginal, corroded, hooked, curved to the right; with solid septum, thick, impressed broadly to the right; with large muscular impressions, very conspicuous, with posterior margin of aperture prominent, subfornicate. Of this species, which is the largest of its race, I possess only four full-grown individuals; but they agree completely among themselves in their form. (Translation.)

Sprächen nicht geographische Grunde dawider, so müsste ich diese Art für identisch mit der *Cal. pallida* Brod. erklären (Translation, Zoöl. Soc. of London, 1835, Vol. I, p. 204, Pl. 29, fig. 3) Broderip's Beschreibung beschränkt sich leider bloss auf Folgenden: Testa sordide alba, ovata, apice prominente: diam., ⅞ poll.; lat., ⅝; alt., ⅜:—hab. ad insulas Falkland dictas.—Found under stones. (Middendorff.)

TYPE in Academy, St. Petersburg. Type locality, St. Paul Island, Bering Sea.

RANGE. Cape Franklin, Arctic Ocean, south and east to Sitka, Alaska; also Kamchatka.

## Crepidula onyx Sowerby, 1825

*Genera of Shells,* No. 5:51. *Conchologia Iconica,* 8; species 9.

Crep. testa oblongo-ovata, crassiuscula, intus extusque fusca, rufolineata, intus saturatiore, livida; appendice subampla, plana, alba, margine medio emarginato.

Shell oblong-ovate, rather thick, brown within and without, rayed with red lines, interiorly darker, livid; appendage rather large, flat, white, edge notched in the middle. (*Conchologia Iconica.*)

TYPE in Museum Cuming. Type locality, Panama.

RANGE. Monterey, California, to Panama.

## Crepidula norrisiarum Williamson, 1905

*Nautilus,* 19:51.

There is another variety of *Crepidula rugosa* Nutt. found on *Norrisia norrisii* Sby. . . . . these shells are of a light magenta-pink in the interior. These slipper shells are usually much flatter than typical *C. rugosa,* and the form of the septum or deck also varies. Besides variation in color and form the *Norrisia* specimens are more porcellanous than *Crepidula rugosa* (but not so much so as the form found on *Lunatia*) and the texture does not run into layers as in the typical *C. rugosa.* Some years ago this form was often distributed by collectors and labeled *Crepidula adunca* Sby. . . . . While some of the forms found upon *Norrisia* have the remote apex of *C. adunca,* I have never seen one with the "short, deeply sunk and slanting deck, and a hole above it passing up the spire," as described by Philip P. Carpenter in his catalogue of Mazatlan Mollusca in his note on the *adunca* form. . . . . For the *Norrisia* form I would suggest *Crepidula rugosa* Nutt., variety *norrisiarum*. (Williamson.)

TYPE in Williamson Collection in Los Angeles County Museum. Type locality, San Pedro, California.

RANGE. San Pedro, on *Norrisia*.

Described as *C. rugosa* var. *norrisiarum* Williamson.

### Crepidula excavata Broderip, 1834

*Proceedings,* Zoölogical Society of London, 40. *Conchologia Iconica,* 11. *Crepidula,* fig. 4.

Cal. test. crassiuscula, subtortuosa, laevi, albida vel subflava fusco punctata et strigata; intus alba vel alba fusco fuscata, limbo interdum fusco cilliato-strigato; long., 1⅞, lat., 1⅛ ; alt., ⅝ poll. (Broderip.)

Shell oblong-ovate, rather thick, somewhat twisted, compressed at the side, obliquely incurved and beaked at the apex, fulvous-white within and without, stained and lineated with purple and reddish-brown; internal appendage deep, flat. (Reeve.)

TYPE in Museum Cuming. Type locality, not known to writer.

RANGE. Monterey, California, to Payta, Peru.

### Crepidula excavata naticarum Williamson, 1905

*Nautilus,* 19:50.

. . . . this white porcellanous shell with its brown spots might be labelled *Crepidula rugosa* Nutt. var. *naticarum*. (Williamson.)

TYPE in Williamson Collection in Los Angeles County Museum. Type locality, San Pedro, on *Natica*.

RANGE. Known only from the type locality.

Described as *C. rugosa* var. *naticarum*.

### Crepidula exuviata Nuttall, 1859
### Plate 85, figs. 7, 8

*Conchologia Iconica,* 11. *Crepidula,* fig. 28.

Crep. testa elongato-producta, angusta, curvata, sordide alba; appendice interna convexa, diaphano-alba.

Shell elongately produced, narrow, curved, dead-white; internal appendage convex, diaphanous-white. (Reeve.)

TYPE in Museum Cuming. Type locality, Monterey, California.

RANGE. California—south.

## Crepidula adunca Sowerby, 1825
### Plate 93, fig. 6

*Tankerville, Catalogue of Shells,* Appendix 7, No. 828. Keep, *West Coast Shells,* 207; fig. 201.

C. testa subovali, vertice adunco, margine undata, labio interno septiformi.

The internal septiform lip divides the cavity nearly in the middle—the upper being the smaller portion; this is very deep. (Sowerby.)

TYPE in British Museum. Type locality, Monterey, California.

RANGE. Vancouver Island, British Columbia, to Cape San Lucas, Lower California.

## Crepidula convexa glauca Say, 1822

*Journal,* Philadelphia Academy of Natural Sciences, 2:226. Tryon, *Manual of Conchology,* 8: Pl. 36, fig. 9.

Shell thin, convex, glaucus, with minute transverse wrinkles; apex conic, acute, not excurved, but declining and distinct from the margin of the aperture; aperture oval-orbicular; within entirely reddish-brown; diaphragm plain or convex, less than half the length of the shell, edge widely contracted in the middle. Length about half an inch. (Say.)

TYPE in Cabinet of the Academy. Type locality, coast of the United States.

RANGE. Alameda, California. Introduced with Atlantic seed oysters. Described as *C. glauca.*

## Crepidula orbiculata Dall, 1919

*Proceedings of the United States National Museum,* 56:251.

Shell dextral, suborbicular, minutely concentrically wrinkled, white, covered with an olivaceous velvety periostracum; whorls about four counting the (lost) nucleus; the apex curved strongly to the right and elevated (in the type-specimen) about 4 millimeters above the basal margin; back evenly convex; interior subtranslucent white, the edge of the deck prominently produced at the left center with a deep sulcus at the left; there is no cavity under the spire between the deck and the margin; height, 26; length, 20; width, 17 mm. (Dall.)

TYPE in United States National Museum, No. 31100. Type locality, Royal Roads, Victoria, Vancouver Island.

RANGE. Bering Sea to San Diego, California.

## Crepidula aculeata Gmelin, 1792

*Systema Naturae*, 7:3693. Tryon, *Manual of Conchology*, 8:129; Pl. 39, figs. 61–65.

Oval moderately convex, apex lateral, spiral, surface covered with radiating prickly or spinose ridges; whitish, yellowish or brownish, often chest-rayed, interior often splotched or rayed with chocolate, septum white. Length 1–1.5 inches. (Tryon, *Manual of Conchology*.)

TYPE in ? Type locality, not known to writer.

RANGE. Santa Barbara, California, to Valparaiso, Chile.

### Subgenus CREPIPATELLA Lesson, 1830

## Crepidula lingulata Gould, 1846
### Plate 93, fig. 8

Testa parva, depressa, obliqua, rotundata, alba, lineis numerosis crebre divaricantibus radiata, epidermide pallido induta; apice acuto, libero, prope marginem sito: intus lutescens; septo linguiformi, excavato, ad medium carina diviso et oblique protruso, ad latus sinistrum profunde sinuato.

Shell small, depressed, thick at the centre, but flattened and thinning towards the margin, rounded at base, the apex free, acute, slightly curving to the right, the surface near the margin furrowed with numerous, delicate, radiating lines, which branch very frequently, and covered with a very delicate, pale, fugacious epidermis. Interior pale brownish-yellow, with a broadly expanded flat margin. The septum is small, white, excavated; divided by a delicate ridge, deeply detached from the left side, and at the middle projecting far and obliquely. Diameter, ⅛ inch; altitude, ⅜ inch. (Gould.)

TYPE in United States National Museum? Type locality, Puget Sound, Washington.

RANGE. Bering Sea to Panama.

### Subgenus IANACUS Morch, 1852

## Crepidula nummaria Gould, 1846
### Plate 91, figs. 14, 14a, 14b

*Proceedings of the Boston Society of Natural History*, 2:160. *Mollusca and Shells of the United States Exploring Expedition*, 377, figs. 480, a, b.

Testa tenui, explanata, circularis, alba, striis incrementalibus laxis insculpta, epidermide flavo-cornea induta; apice vix conspicuo, marginali: septum internum latius quam longum, deorsum fornicatum; margine flexuoso: cavositas fere nulli.

Shell thin, white, circular, and perfectly flat, externally wrinkled by the loose stages of growth, and covered by a thick, yellowish epidermis. The apex is marginal and scarcely visible. The internal lamella is broader than long, arching, so as to enlarge the cavity, which would otherwise be almost nothing. Its edge is flexuous, retreating near the middle. Diameter, ¾ in. (Gould.)

TYPE in United States National Museum. Type locality, Classet, Straits of Juan de Fuca.

RANGE. Plover Bay, Bering Strait, to Mazatlan, Mexico.

### Crepidula fimbriata Reeve, 1859

*Conchologia Iconica,* 11, fig. 11.

Crep. testa oblique ovata, plana, apice lateraliter incurva, concentrice striata, versus marginem conspicue laminato-fimbriata, extus intusque pellucido-alba, ad latus livido-fusco uniradiata; appendice ad latus emarginato. (Reeve.)

Shell obliquely ovate, flat, laterally incurved at the apex, concentrically striated, conspicuously laminarly frilled toward the margin, transparent-white without and within, stained with a livid-brown ray on one side; appendage notched at the side. (Reeve.)

TYPE in British Museum? Type locality, Vancouver Strait.

RANGE. Vancouver Island, British Columbia, to Gulf of California.

### Crepidula nivea C. B. Adams, 1852
### Plate 93, figs. 7, 9

*Catalogue of Shells Collected at Panama,* 234.

Shell ovate-elliptic: rather thick: within snow white: without dingy white, sometimes with a faint tinge of brown: very irregularly concentrically more or less wrinkled, with very distinct striae of growth: apex turned more or less to the right, moderately prominent, marginal: septum longitudinally subangular, with a deep sinus at the left and a shallow one at the right: margin thick, exhibiting striae of growth. It closely resembles *C. unguiformis,* but constantly differs in character and station. (C. B. Adams.)

TYPE in Amherst. Type locality, Panama.

RANGE. Puget Sound to Panama.

### Crepidula nivea glottidiarum Dall, 1905

*Nautilus,* **19**:26.

Shell uniformly straight, convex, smooth, equilateral, white and posteriorly attenuated. Living on *Glottidia albida* Hinds, and covering the shell, and in some cases there is a specimen on both valves of the *Glottidia*. (Dall.)

TYPE in United States National Museum. Type locality, San Pedro, California.

RANGE. Known only from type locality.

## Family NATICIDAE

## Genus **NATICA** Scopoli, 1777

Shell oval globular, porcellanous, solid, generally smooth, covered by a fine epidermis, which is transparent, and generally not very persistent; umbilicated, or umbilicus more or less filled with callus; aperture semilunar, vertical, the outer lip simple. Operculum large, semilunar, paucispiral, corneous or calcareous. (Tryon, *Manual of Conchology.*)

TYPE. *Natica canrena* Linnaeus.

DISTRIBUTION. Warm seas in all quarters of the globe.

### Subgenus CRYPTONATICA Dall, 1892

### **Natica clausa** Broderip and Sowerby, 1829
### Plate 97, fig. 2

*Zoölogical Journal,* 4:372. Tryon, *Manual of Conchology,* 8: Pl. 9, fig. 65.

N. testa subglobosa, anfractibus ventricosis, superne subdepressis, margine elevatiusculo; umbilico clauso. Long., 10/20; lat., 17/20 poll. (Broderip and Sowerby.)

The umbilical callus completely covers the umbilicus. Operculum testaceous, rather thin and somewhat concave. (Broderip and Sowerby.)

TYPE in Sowerby Collection. Type locality, Arctic.

RANGE. Arctic and Bering Seas to San Diego, California; also Japan.

### **Natica affinis** Gmelin, 1792

*Systema Naturae,* 7:3675. Sars, *Mollusca regionis arcticae Norvegiae,* Pl. 21, fig. 14.

N. testa globosa crassiore: spira submucronata; anfractibus tribus. (Gmelin.)

Shell globose, somewhat thick; spire submucronate; three whorls. (Translation.)

TYPE locality, habitat in Oceano septentrionali, an bujus tribus.

RANGE. Arctic Ocean north of Bering Strait. Also Greenland.

## Natica janthostoma Deshayes, 1841
### Plate 97, fig. 5

*Guérin, Magasin de Zoölogie, Mollusca.*

Testa globulosa, rufa, albido-zonata, apice nigrescente, levigata; anfractibus convexis, ultimo umbilico, clauso callo semi-circulari; apertura ovata semi-lunari, intus violacea, ad margines alba, operculo calcareo albo, simplici clausa. (Deshayes.)

Shell globulous, reddish, zoned with white; apex darkened, smoothish; convex whorls, the last one umbilicate, closed with a semicircular callus; aperture ovate, semilunar, within violet, toward the margins white, closed with plain white calcareous operculum. (Translation.)

TYPE in Museum, Paris. Type locality, Kamchatka.

RANGE. Commander Islands, Bering Sea; Kamchatka and south to Japan.

## Natica russa Gould, 1859

*Proceedings of the Boston Society of Natural History, 7:43.*

T. imperforata, ovato-globosa, tenuis, laevis, epidermide tenui cerina induta; anfr. 4 ventricosis postice quadratis. Apertura ovata, subeffusa, labro tenui, umbilico callo compresso albo obstructo. Operculum osseum. Axis, 18; diameter, 16 mm. (Gould.)

Shell imperforate, ovate-globose, thin, smooth, covered with a thin waxy epidermis; four whorls, ventricose and quadrate behind. Aperture ovate, subeffuse with thin outer lip, umbilicus obstructed with compressed white callus. Operculum osseous. (Translation.)

TYPE in ? Type locality, Arctic Ocean.

RANGE. Bering Strait to Forrester Island, Alaska.

## Natica salimba Dall, 1919

*Proceedings of the United States National Museum, 56:351.*

Shell small, smooth, except for faint incremental lines, yellowish-white with a faintly darker band in front of the suture and another between the periphery and the base; with four evenly rounded whorls slightly flattened in front of a somewhat appressed suture; base rounded, the umbilicus closed by a semilunar convex mass of callus not notched above; the aperture lunate, outer lip thin, the body callus, the pillar lip somewhat thickened; the operculum is white, porcellanous, of about two whorls. Height of shell, 14; of aperture, 10; diameter, 13.5 mm. (Dall.)

TYPE in United States National Museum, No. 209295. Type locality,

U.S. Fish Commission Station 4423, between Santa Barbara and San Nicolas Island, California, in 216–339 fathoms.

RANGE. Off Esteros Bay, California, and south to Gulf of California.

## Natica aleutica Dall, 1919

*Proceedings of the United States National Museum,* **56**:352. *Proceedings of the Biological Society of Washington,* **32**:251.

Shell large, round, slightly flattened in front of the suture, pinkish-white, covered with a light brownish, spirally minutely striated periostracum, a white area surrounding the umbilical region; whorls six or more, the nucleus minute, aperture lunate, simple, the body with a coat of enamel, the umbilicus completely, smoothly filled with a semilunate pad of callus; operculum solidly calcareous, smooth, with an obscure swelling centrally. Height, 46; diameter, 40 mm. (Dall.)

TYPE in United States National Museum, No. 217156. Type locality, Unalaska, Aleutian Islands.

RANGE. North end of Nunivak Island, Bering Sea, to Aleutian chain to Puget Sound; Kamchatka to Japan.

## Genus POLINICES Montfort, 1810

Shell oval or suboval, solid, smooth, spire short, sharp; aperture semi-circular, inner lip oblique, callous, the callus extending into the umbilicus. Operculum corneous. Color usually white, sometimes colored but not banded or spotted. (Tryon, *Manual of Conchology.*)

TYPE. *Polinices uber Valenciennes.*

DISTRIBUTION. West coast of North America, Japan, Philippines, Mauritius.

## Subgenus EUSPIRA Agassiz, 1842

## Polinices acosmitus Dall, 1919

*Proceedings of the United States National Museum,* **56**:352.

Shell white with a minutely spirally banded yellow brown periostracum and about four whorls, the nucleus defective; suture distinct, the whorl in front of it a little impressed; axial sculpture of fine silky incremental lines; there is no spiral sculpture; aperture semilunar, oblique, the outer lip thin, the body with a coat of enamel reflected over the umbilical region in a flat pad as in *Cryptonatica,* the pillar lip somewhat thickened. Height of shell, 16; of aperture, 13; diameter, 15 mm. (Dall.)

TYPE in United States National Museum, No. 207218. Type locality, U.S. Fish Commission Station 3128, off Monterey Bay, California.

RANGE. Monterey Bay, California, to Coronado Islands, in deep water.

### Polinices canonicus Dall, 1919

*Proceedings of the United States National Museum,* **56**:353.

Shell small, white under a pale yellowish periostracum, with more than four whorls, the nucleus eroded; suture distinct, narrow, almost channeled; axial sculpture of incremental lines and numerous small irregular wrinkles radiating from the suture and extending on the average about halfway to the periphery; there is no spiral sculpture; base rounded with a small perforate umbilicus partly overshadowed by a narrow reflection of enamel from the pillar lip; aperture semilunate, simple, the body pillar, and posterior commissure with a liberal coating of enamel. Height of shell, 8; of aperture, 5; of diameter, 7 mm. (Dall.)

TYPE in United States National Museum, No. 209411. Type locality, U.S. Fish Commission Station 2923, off San Diego, California, in 822 fathoms.

RANGE. Off Alaska Peninsula in 2,923 fathoms, and off San Diego, California, in 822 fathoms.

### Polinices monteronus Dall, 1919

*Proceedings of the United States National Museum,* **56**:352.

Shell of moderate size, of a dark greenish-olive color with a whitish base; whorls about five, the nucleus decorticated; the surface smooth except for faint incremental lines, the suture distinct, not appressed; base rounded, whitish, the umbilicus narrow, filled below with enamel, with a minute perforation above forming a notch-like gap between the white enamel of the inner lip and that of the umbilicus; outer lip thin, sharp; body with a thick coat of enamel, pillar lip much thickened, white; operculum dark brown, of about two whorls. Height, 19; diameter, 18 mm. (Dall.)

TYPE in United States National Museum, No. 220856. Type locality, Station 1199, Captains Bay, Unalaska, in 75 fathoms.

RANGE. Arctic Ocean, Bering Sea, and the Aleutians.

### Polinices politianus Dall, 1919

*Proceedings of the United States National Museum,* **56**:353.

Shell small, white, covered with a pale brown dehiscent periostracum; whorls four and a half, the nucleus decorticated, the specimens probably not quite adult; axial sculpture of retractively radiating grooves from the edge of a channeled suture nearly to one-fifth of the arch of the whorl, with wider interspaces, and faint incremental lines; there is no spiral sculpture; aperture ovate, outer lip thin, sharp, inner lip with a glaze of enamel, thicker under the suture and forming a convex mass over the wholly obliterated umbilicus; pillar lip slightly reflected, the periostracum on the base paler than elsewhere. Height of shell, 16; of aperture, 13; diameter, 13 mm. (Dall.)

TYPE in United States National Museum, No. 205653. Type locality, United States Fish Commission, Station 4779, off Petrel Bank, Bering Sea, in 600 fathoms.

RANGE. Known only from type locality.

### Polinices pallida Broderip and Sowerby, 1829
### Plate 97, fig. 9

*Zoölogical Journal,* 4:372. Tryon, *Manual of Conchology,* 8:37; Pl. 14, figs. 26–28; Pl. 13, fig. 15; Pl. 9, figs. 76–78.

Testa subglobosa, albida, tenui, apice breviter acuminato, eraso; anfractibus rotundatis, margine elevatiusculo, sutura distincta; umbilico parvo; long., 13/20; lat., 1 poll. (Broderip and Sowerby.)

Ovately globose, narrowly umbilicated, smooth, dirty white under a yellowish-brown thin epidermis. Length, 28 mm. (Tryon, *Manual of Conchology.*)

TYPE in British Museum. Type locality, Arctic.

RANGE. Arctic Ocean; Forrester Island, Alaska; Puget Sound.

### Polinices gronlandicus (Beck) Moller, 1842

*Index Molluscorum Groenlandiae,* 7. Sars, *Mollusca regionis arcticae Norvegiae,* Pl. 21, fig. 15.

Testa globosa, perforata, solidiori, laeviuscula, solide alba; anfr. 4 convexis; spira mediocriter elata obtusa (plerumque corrosa); peristomata subcontinuo; umbilico angusto, profundo; operculo corneo, tenui. L. 9‴. (Moller.)

Shell globose, perforate, rather solid; smooth, uniformly white, with four convex whorls. Spire moderately high obtuse (mostly corroded); peristome subcontinuous; umbilicus narrow, but with thin corneous operculum. (Translation.)

TYPE in Zoölogy Museum, Copenhagen. Type locality ?

RANGE. Arctic Seas southward in gradually deeper water, to Monterey, California; also Greenland to Cape Cod.

## Polinices lewisii Gould, 1847

*Proceedings of the Boston Society of Natural History,* 2:239. *Mollusca and Shells of the United States Exploring Expedition,* fig. 293.

Testa praegrandis, solidiuscula, conico-globosa, albida, epidermide sordide stramineo induta, lineis minimis confertissimis, flexuosis cincta; spira anfractibus sex ventricosis postice valde constrictis; apertura obovata, ampla, callo copioso albo castaneo-tincto umbilicum modicum simplicem profundum postice circumambiente, ad anfractum ultimum munita; fauce incarnate. Operculum corneum. (Gould.)

Shell very large, not very ponderous, globose-conic; the spire rather elevated and acute; color pale fleshy, covered by a thin, ashy epidermis, which is everywhere marked with very delicate, crowded, undulating, revolving lines. Surface somewhat undulated, by the stages of growth. Whorls six, moderately convex, somewhat flattened near the suture, the last whorl having a very remarkable broad, well-marked constriction, at about its posterior third, which is continued also about the middle of the posterior whorls. Aperture obovate, broad, having a sharp lip, until it rises on the left margin, when it widens, and presents a rounded edge, terminating in a copious white callus, which seems to flow down about halfway around the umbilicus without closing it, and having a furrow running obliquely inward, from the upper edge of the umbilicus; its edges are somewhat tinted with chestnut, and a strong band of callus also fills the upper angle of the aperture. Umbilicus moderately long, deep, nearly round, with a slight encroachment down the right wall. Interior of the aperture shaded with pale flesh-color. Operculum thin, horny. Axial diameter, 4½–5; transverse diameter, 4 in. (Gould.)

TYPE in United States National Museum? Type locality, Discovery Harbor, Puget Sound.

RANGE. Nanimo, British Columbia, to San Pedro, California.

## Polinices caurinus Gould, 1847

*Proceedings of the Boston Society of Natural History,* 2:239. *Mollusca and Shells of the United States Exploring Expedition,* fig. 254.

Testa parva, solida, laevigata globoso-ellipsoidea, albida, epidermide tenui stramineo induta, vix perforata: spira obtusa, erosa, anfractibus quatuor cum dimidio, ultimo sub-tabulato, antice subcontracto: apertura parva, semilunaris; columella recta, postice callo copioso albo induta. (Gould.)

Shell small, solid, smooth, of a round-ellipsoid form, of a dead, dirty milk-white color. Whorls four or five, the upper ones forming a depressed spire, with the apex eroded, and the suture linear; the last whorl very large, full posteriorly, and somewhat tabular at the suture; the last whorl is quite as long as broad, and perhaps attenuated at base. The aperture is about two-thirds the length of shell, semilunar; the outer lip sharp, the pillar lip straight, heavily and broadly loaded with callus posteriorly, and regularly narrowing, and yet nearly covering a small umbilical pit at the middle, so as to leave merely a small, linear chink, or none at all. Axial diameter, ½; transverse diameter, ⅖ in. (Gould.)

TYPE in United States National Museum? Type locality, Straits of Juan de Fuca.

RANGE. Norton Sound, Alaska, to Puget Sound, to San Diego, California, in 822 fathoms.

Described as *Natica caurina*.

## Polinices algida Gould, 1848

*Proceedings of the Boston Society of Natural History, 3:73. Mollusca and Shells of the United States Exploring Expedition, fig. 256.*

Testa parva, tenuis, globosa, glabra, albido-livescens: spira anfr. 4 ventricosis juxta suturam linearem tabulatis, ultimo magno ampullaceo: apertura semilunaris; columella arcuata: basis umbilico modico spirali funiculato, ad aditum angulato perforata. Long., 5/8; lat., 3/5 poll. (Gould.)

Shell small, rather thin, globular, smooth, of a pale bluish-white color. Whorls four, forming a rounded, slightly elevated spire, the last one very capacious, with a narrow shoulder near the suture. Aperture semilunar, outer lip thin; inner margin curved, covered posteriorly by callus, which is not very closely appressed. There is a rather large, open, somewhat spiral umbilical opening, externally margined by an angle, and partially filled by a revolving pillar, on which is an expansion of the callus. (Gould.)

TYPE in ? Type locality, Classet, Oregon.

RANGE. Coal Harbor, Shumagin Islands, Alaska, to Puget Sound, Washington.

Described as *Natica algida*.

## Polinices draconis Dall, 1903
### Plate 99, figs. 3, 6

*Proceedings of the Biological Society of Washington, 16:174. Bulletin 112, United States National Museum, Pl. 14, figs. 4, 6.*

Shell depressed, solid, cream color, sometimes with a ferruginous or livid tinge, with six whorls; nuclear whorls very small, smooth; later ones with an obscure, nearly obsolete, spiral sculpture like flattened-out threads, over which run microscopic, close-set, spiral striae; suture with the whorl in front of it feebly channeled and the excavation bounded by an obsolete thread; top of the whorls flattened, part of the base bordering the umbilicus also flattish, the remainder of the whorl rounded, turgid; umbilicus wide and deep, its walls excavated and closely spirally striated aperture oblique, semi-lunate, outer lip thin, base rounded; the angle where the lip meets the body filled with a smooth white callus, the anterior angle of the pillar lip also thickened. Height of shell, 51.0; of last whorl, 49.0; of aperture, 44.0; maximum width of shell, 50.0 mm. (Dall.)

TYPE in United States National Museum, No. 172859. Type locality, Drake's Bay, California.

RANGE. Port Althorp, Alaska, to Catalina Island, California.

## Subgenus NEVERITA Risso, 1826

### Polinices recluzianus Deshayes, 1841

*Guérin, Magasin de Zoölogie, Mollusca, Pl. 37.*

Testa ovata-conoidea, turgida, levigata, substriatave, griseo-plumbea, basi albescente, ad suturam zona fusca circumdata, umbilico magno, callo maximo semiclauso, callo sulco inaequaliter bipartito, columella superne callosissima, superne, alba inferne macula fusca nautata, apertura occato-semilunari, superne canaculata intus albo-fuscescente. (Deshayes.)

Shell solid, conically globose, columella strongly callous from the superior function of the outer lip, nearly filling the umbilicus; fawn-color, or yellowish-brown, lighter or whitish below, interior chocolate, callus usually white. Length, 3 in. (Tryon, *Manual of Conchology.*)

TYPE in ? Type locality, Mers de Californie.

RANGE. Crescent City, California, to Tres Marias Islands, to Chile.

### Polinices recluzianus imperforata Stearns, 1909

U. S. Geological Survey, *Professional Paper,* **59**:88.

In the variety *imperforata* found living in the vicinity of San Diego, and in the Pleistocene of the upper San Pedro at Deadmans Island, there is an additional thin deposit of callus filling the open part of the umbilicus and overflowing on the adjacent portion of the base. (Dall.)

TYPE in United States National Museum. Type locality, San Diego, California.

RANGE. Crescent City to San Diego, California.

## Polinices recluzianus alta Dall, 1909

U. S. Geological Survey, *Professional Paper,* **59**:88.

There is a variety *alta* Dall, with small narrow shell and exceptionally elevated. (Dall.)

TYPE in United States National Museum. Type locality, California.

RANGE. Monterey, California, to Catalina Island.

### Subgenus POLINICES s.s.

## Polinices nanus Moller, 1842

*Index Molluscorum Groenlandiae,* p. 7. Sars, *Mollusca regionis arctica Norvegiae,* Pl. 21, fig. 16.

Testa globosa, imperforata, alba, anfr. 3½ ; spira parum elata; sutura lineari; peristomate disjuncto. Operculo corneo, tenui. L. 3''' .R. (Moller.)

Shell globose, imperforate, white, with three and one-half whorls, spire little elevated; suture linear, peristome disjointed, operculum corneous, thin. (Translation.)

TYPE in Zoölogy Museum, Copenhagen? Type locality, Greenland.

RANGE. Arctic Ocean to Aleutian Islands, and in deeper water near San Diego, California; also Greenland.

## Genus SINUM Bolten, 1798

Shell ear-shaped, with minute spire and very large aperture, externally with revolving striae, color usually white, with sometimes a thin corneous epidermis. Operculum minute, horny, subspiral.

TYPE. *Sigaretus neritoides* Linnaeus.

DISTRIBUTION. United States, West Indies, China, Peru, Japan. Fossil, Eocene. . . . .

This is the description of genus *Sigaretus,* which is equal to *Sinum*

## Sinum californicum I. S. Oldroyd, 1917
### Plate 92, figs. 13, 14

*Nautilus,* **31**:13.

Shell white, convex, spirally striate above, with epidermis of a rusty yellow; a thin columellar callus reflected nearly over the umbilicus showing only a faint trace of umbilicus; interior snow-white. This has been called *Sigaretus debilis* Gould, but it is not like the specimens from Lower California. It differs from *S. concavum* in not being as convex, and the interior being white, and the early whorls are much smaller; and from

*S. debile* in being convex and larger. *S. debilis* is very flat, the early whorls are smaller and fewer. Length of shell, 38; breadth, 18; height, 18 mm. (I. S. Oldroyd.)

TYPE in Stanford University, Oldroyd Collection. Type locality, San Pedro, California.

RANGE. Monterey, California, to Todos Santos Bay, Lower California.

## Sinum debile Gould, 1852
### Plate 92, figs. 3, 7

*Boston Journal of Natural History,* **6**:379. Tryon, *Manual of Conchology,* **8**: Pl. 24, fig. 65.

T. parva, tenuis, lucida, depressa, mellea, striis numerosissimis obtusis volventibus, supernis majoribus, insculpta; spira superficiei generali congruens; anfr. duobos; apice ad quadrantem diametris sito: angulo ad peripheriam obtuso, versus aperturam sensim evanido; labio antice dilatato. (Gould.)

Shell small, much depressed, thin, almost pellucid, pale honey-yellow color; whorls two, spire almost coincident with the general surface, apex at one-fourth the diameter of the shell; periphery obtuse-angular, becoming more so as it approaches the aperture; ventral surface excavated at the umbilical region, with a slight unappressed lamina at that point; margin of the aperture having a very slight advance in the outline, as it approaches the peripheral angle; surface with very numerous and very delicate, obtusely excavated, revolving striae, much finer on the ventral than on the dorsal face. Long diameter, ⅞; short diameter, not quite ½; height, ⅕ in. (Gould.)

TYPE in Albany State Museum, Albany. Type locality, La Paz, Lower California.

RANGE. Catalina Island, California, to Gulf of California.

## Sinum keratium Dall, 1919

*Proceedings of the United States National Museum,* **56**:354.

Shell small, white, convex, suborbicular, with a minute subglobular nucleus and three subsequent whorls; suture distinct and deep; spiral sculpture of sharp, rather regular grooves with wider interspaces crossed by fine incremental lines, which in places give a punctate effect to the grooves under the lens; the grooves on the base closer and finer; pillar lip slightly reflected over an umbilical chink, body with a thin glaze. Long diameter, 6.5; shorter diameter, 5; height, 3.5 mm. (Dall.)

Type in United States National Museum, No. 206152. Type locality, Catalina Island, California.

RANGE. Known only from type locality.

## Sinum pazianum Dall, 1919
### Plate 92, fig. 10

*Proceedings of the United States National Museum,* 56:354.

Shell small, grayish-white, depressed, with a translucent smooth nucleus of a whorl and a half and two subsequent whorls; suture distinct, outline oval, axial sculpture of faint incremental lines; spiral sculpture of fine incised lines, sometimes close, sometimes with wide interspaces, but barely perceptible except with a lens; last whorl produced, body with a slight glaze of enamel; pillar lip reflected with an umbilical chink behind the reflection, base slightly flattened. Longer diameter, 9; shorter diameter, 7; height, 2.5 mm. (Dall.)

Type in United States National Museum, No. 211406. Type locality, U.S. Fish Commission Station 2823, off La Paz, Lower California, in 26½ fathoms.

RANGE. Catalina Island, California, to Panama.

## Genus **EUNATICINA** Fischer 1855

Shell umbilicated, oval oblong, thin ventricose; spire sharp; inner lip straight, thin anteriorly, with a median callus. (Tryon, *Manual of Conchology.*)

TYPE. *Sigaretus papilla* Gmelin.

DISTRIBUTION. Moluccas, Japan, Philippines, Madagascar, Australia, California.

## Eunaticina oldroydii Dall, 1897
### Plate 92, figs. 11, 11a

*Nautilus,* 11: No. 8, 85. *Bulletin 112,* United States National Museum, Pl. 14, figs. 1, 2.

Shell large, thin, naticoid, with a short spire and 3–4 inflated whorls; color pale brown, livid on the spire, fading to waxen on the base; surface sculptured with extremely fine wavy spiral striae; aperture ample, oblique, the outer lip thin, a little patulous, the body covered with a thin callus, the pillar lip obliquely cut away, wide near the junction with the body, the basal part of the margin receding; umbilicus large, pervious, its walls covered with a thin, silky, brown wrinkled epidermis. Alt., 35; diameter, 37 mm. (Dall.)

Type in Oldroyd Collection, Stanford University. Type locality, off Catalina Island, California.

Range. Oregon to San Diego, California.

## Genus ELACHISINA Dall, 1918

Shell minute, elevated, with naticoid spiral sculpture, umbilicate. (Dall.)

Type. *Elachisina grippi* Dall.

Distribution. California, deep water.

### Elachisina grippi Dall, 1918

*Proceedings of the Biological Society of Washington,* 31:137. *Proceedings of the United States National Museum,* 56:354; 1919.

Shell minute, having somewhat the aspect of a short *Cingula,* white, with a very thin periostracum, and four well-rounded whorls, exclusive of a small glassy subglobular slightly depressed nucleus; the suture constricted but hardly appressed; spiral sculpture of fine uniform striae with slightly wider interspaces, covering the whole shell; aperture subovate, the outer lip thin, the body with a continuous coat of enamel, the pillar arcuate; the umbilicus naticoid, narrow twisted, with an internal spiral ridge. Height of shell, 3; of last whorl, 2; diameter, 2 mm. (Dall.)

Type in United States National Museum, No. 250230. Type locality, off San Diego, California.

Range. Known only from type locality.

## Genus BULBUS Brown, 1839

Shell globular, spire very short; columellar margin incurved, columella twisted; lip fragile. (Tryon, *Manual of Conchology.*) This is the description of *Acrybia* H. and A. Adams, 1853.

Type. *Natica flava* Gould.

Distribution. Maine, Newfoundland, Finmark, Nova Zembla, Arctic Ocean to Aleutian Islands, and Shumagin Islands.

### Bulbus fragilis (Leach, 1819) apertus (Lovèn) Middendorff, 1851

*Journal de physique,* 88:464. Middendorff, *Sibirische Reise,* Pl. 11, figs. 1–3.

Testa rimata, valde tenui, pellucida, ovato-globosa; anfr. 4, ultimo, inflato, suturae appresso; spira brevi obtusa; apertura ampla, longitudinali ovata, postice emarginata anfractu penultimo; lamina columellari tenuissima supra umbilicum omnino tectum incrassata, effusa, revoluta deinde sensim angustata, producta, et in labrum continuata, acutum, tenuissimum, arcuatum. Long., 35; lat., 34; altit. anfr. ult., 31; platit. apert., 20; col. ext. long., 17 mm. (Middendorff.)

Shell fissured, quite thin, pellucid, ovate-globose; with four whorls, the last inflated, appressed at the suture; with short obtuse spire; with aperture broad, longitudinally ovate, emarginate behind on the next to the last whorl; with very thin columella lamina, covering the entire umbilicus, effuse, revolute, then perceptibly narrowed, produced, and extending to the acute, very thin, arcuate outer lip. (Translation.)

TYPE in Academy, St. Petersburg. Type locality, Sudküste des Ochotskischen Meeres.

RANGE. Icy Cape, Arctic Ocean, south to the Aleutian Islands, and eastward to the Shumagin Islands.

## Genus AMAUROPSIS Morch, 1857

Shell with canaliculated sutures. Scarcely distinct from *Amaura*.

TYPE. *Natica canaliculata* Gould.

DISTRIBUTION. Arctic, Point Barrow, Norway, Halifax, to Massachusetts Bay. Fossil at Montreal.

### Amauropsis purpurea Dall, 1871

*American Journal of Conchology*, 7:124; Pl. 15, fig. 16.

Shell elongate-globose, covered with a yellowish epidermis; color purple brown, when weathered, purplish white. Whorls four, inflated globose. Suture deeply canaliculated. Aperture more than half as long as the shell; outer lip thin; columella white, thickened, rounded; a thin callus on the inner lip. Umbilicus closed, or a mere chink. Spire very short, bluntly rounded. Sculpture of numerous close, very fine revolving grooves, and a few inconspicuous ridges. Long., 1.0; lat., .08 in. (Dall.)

TYPE in United States National Museum. Type locality, St. Michael's, Norton Sound, Alaska.

RANGE. Arctic coast westward from Bernard Harbor, to Point Barrow and south to Plover Bay and Nunivak Island, Bering Sea.

## Family VANICOROIDAE

### Genus MEGALOMPHALUS Brusina, 1871

Coquille mince, naticiforme ou auriforme; spire courte; tours arrondis; sommet régulier; ombilic très grand, plissé; opercule corné, spiral. (Fischer.)

TYPE. *Megalomphalus azonus* Brusina.

DISTRIBUTION. Mediterranean, west coast of North America.

## Megalomphalus californicus Dall, 1903

### Plate 97, fig. 7

*Proceedings of the Biological Society of Washington,* **16**:175. *Bulletin 112,* United States National Museum, Pl. 14, fig. 7.

Shell small, elevated, with a wide umbilicus and whitish color; whorls two and a half, the last much the largest, rounded above with a prominent suture, below with a wide funicular umbilicus bordered externally by an obtuse carina; surface sculptured axially with numerous coarse oblique threads separated by narrower interspaces and crossed by fine partially obsolete spiral striation; aperture semilunate, entire, very oblique; the pillar lip straight, but the whole peristome simple and thin. Height, 5.5; of aperture, 3.5; max. diameter, 5.5 mm. (Dall.)

TYPE in United States National Museum, No. 109307. Type locality, Santa Barbara Islands, California.

RANGE. Known only from the type locality.

Described as *Macromphalina californica* Dall.

## Family LAMELLARIIDAE

### Genus **LAMELLARIA** Montagu, 1815

Shell internal, ear-shaped, thin pellucid; spire lateral, very small; aperture large, patulous, both lips regularly arcuated; axis imperforate. No operculum. (Tryon, *Manual of Conchology.*)

TYPE. *Lamellaria perspicua* Montagu.

DISTRIBUTION. Norway, Britain, Mediterranean, New Zealand, Philippines, west coast of North America. Fossil, Pliocene. . . . .

## Lamellaria stearnsii Dall, 1871

### Plate 92, fig. 6

*American Journal of Conchology,* **7**:122; Pl. 15, fig. 6.

Shell pure white, suborbicular, depressed, of three whorls. Columella sharp, thin, widely arcuated; loosely twisted so that the apex is discernible from below. Spire hardly elevated above the last whorl; suture distinct, sharply defined; aperture very effuse, rounded. Exterior marked by lines of growth, crossed by microscopic fine revolving striulae. Interior polished. Lat., 26; long., 2; alt., 12 in. (Dall.)

TYPE in United States National Museum. Type locality, Monterey, California.

RANGE. Puget Sound to San Diego, California.

## Lamellaria stearnsii orbiculata Dall, 1871

### Plate 92, figs. 4, 5

*American Journal of Conchology,* 7:122; Pl. 15, figs. 2, 3.

Shell resembling the last (*L. stearnsii*) ; of two whorls and a half; but larger than *L. stearnsii;* whorls more inflated, aperture very oblique, suture deeper, spire more elevated and proportionately larger, columella rather thicker and more drawn out. Lat., 34; long., .32; alt., .24 in. (Dall.)

TYPE in United States National Museum. Type locality, Monterey Bay, California.

RANGE. Sannakh Islands, Alaska, to Gulf of California.

## Lamellaria diegoënsis Dall, 1885

*Proceedings of the United States National Museum,* 8:538; Pl. 24, figs. 1, 2, 3.

Shell calcareous except the immediate margins of the aperture, grayish waxen white, slightly iridescent with a mucilaginous polish like dry glue inside and out, somewhat malleated, with indications of the lines of growth by obscure slightly elevated transverse waves and faint irregular spiral tracings; three-whorled, very much inflated, though the form varies slightly between individuals and probably between the sexes; nucleus small, smooth; suture deep, not channeled; spire pervious from below; columella less calcareous than the shell in general, without callus or any thickening, twisted into an open spiral, continuous with the outer lip in front only; aperture very oblique, subquadrate. Alt., 15; max. lat., 17; long. of aperture, 12; lat. of aperture, 10 mm. (Dall.)

TYPE in United States National Museum. Type locality, False Bay, near San Diego, California.

RANGE. San Pedro and the islands to San Diego, California.

## Lamellaria digueti Rochebrune, 1895

P.—Corpus rotundato ovatum, turgidum; pallio ovato, antice subtruncato, marginibus undatis; latis; pede subangusto, circulariter crenulate; regione buccali proboscidea; tentaculis duobus, rotundatis canaliculatis; branchiis subabsconditis; superne miniaceum; inferne albo luteum. Long., 0.022; lat., 0.016; Crass., 0.012 mm. (Rochebrune.)

TYPE in Museum, Paris. Type locality, Mogote, Baie de la Paz.

Described as *Pleurobranchus digueti.*

## SECTION MARSENINA Gray, 1850

## Lamellaria rhombica Dall, 1871

### Plate 92, figs. 1, 2

*American Journal of Conchology,* 7:122; Pl. 15, figs. 4, 5.

Shell pure white, subrhombical in shape, moderately elevated, of three whorls. Columella thickened, stout, reflected, narrow, with a groove behind the callus. Whorl appressed against, and slightly flattened below, the suture; spire very small, not elevated; apex not above the level of the last whorl. Aperture subquadrate, outer lip very much produced, slightly angulated above and below. Suture deep appressed. Nucleus smaller than in the last; surface of the whorls smooth, without striulae. Lat., .46; long., .32; alt., .2 in. (Dall.)

TYPE in United States National Museum. Type locality, Monterey Bay, California.

RANGE. Neah Bay, Washington, to Magdalena Bay, Lower California.

## Genus ONCHIDIOPSIS Berg, 1853

Shell entirely inclosed by the animal, thin, slipper-like, without spire, margin entire. Animal verrucose, with a lanceolate foot. (Tryon, *Structural and Systematic Conchology.*)

TYPE. *O. glacialis* M. Sars.

DISTRIBUTION. Norway, Great Britain, New Zealand, Mediterranean, Philippines, and North Pacific Ocean.

## Onchidiopsis hannai Dall, 1916

*Proceedings,* Philadelphia Academy of Natural Sciences, 376.

The base of the arc of the shell measures 40 mm. long by 32 mm. wide. It has much the shape of the bowl of a deep oval ladle and its depth is about 15 mm. when in normal position. Its structure is concentric, not in circles but in a rounded-quadrate fashion. On the edge of the left side behind is a knot-like nucleus. On the inner surface near this nucleus and extending for a length of about 12 mm. away from it are two elevated straight ridges, which at half their length from the nucleus join to form a single stronger ridge, which gradually diminishes and becomes obsolete on the inner surface of the disk. The appearance of these ridges suggests that if the shell was spirally coiled they would form a columella. The outer surface of the shell, to which the mantle adheres tenaciously, is smooth, but undulated by more or less irregularly disposed concentric wrinkles. It is attached to the body only by a small area at the edge near the nucleus. There is no indication of a periostracum, and the cartilaginous shell is nearly transparent. (Dall.)

TYPE in United States National Museum, No. 215162. Type locality, St. Paul Island, Bering Sea.

RANGE. Known only from type locality.

## Family VELUTINIDAE

## Genus **VELUTINA** Fleming, 1822

Shell thin, with a velvety epidermis; spire small; suture deep; aperture very large, rounded; peristome continuous, thin. No operculum. (Woodward, *Manual of the Mollusca.*

TYPE. *Velutina laevigata* Linnaeus.

DISTRIBUTION. Britain, Norway, North America, Icy Sea to Kamchatka. Fossil, Triassic, Cretaceous, Pliocene.

## Velutina cryptospira Middendorff, 1849

*Beitrage zu einer Malacologia Rossica,* pt. 2, 106. *Sibirische Reise,* Pl. 25, figs. 8, 9, 10. Tryon, *Manual of Conchology,* 8:67; Pl. 27, figs. 45–47.

Testa transversim ovata, subauriculata, cartilagineo-coriacea, tenui (exsixxata, membranacea et tenuissima fit) fusca; spira laterali, subemarginali immersa et plane inconspicuus; anfractu ultimo maximo; apertura ampla, orbiculari-ovata; columella angusta, subacuta, interdum canaliculo obsoletissimo submarginata. (Middendorff.)

Shell narrower, more elongated than *V. laevigata,* the columellar lip forming an angle with the outer lip below; coriaceous, or with a very slight calcareous deposit. Length, 10 mm. (Tryon, *Manual of Conchology.*)

TYPE in Academy, St. Petersburg. Type locality, Gulf of Alaska.

RANGE. Known only from type locality.

## Velutina granulata Dall, 1919

*Proceedings of the United States National Museum,* 56:355.

Shell small, white, naticoid, covered with a thin yellowish dehiscent periostracum, with a minute smooth nuclear and three subsequent rounded rapidly expanding whorls; suture distinct, not appressed; sculpture of quite evident but not elevated incremental lines and very minute granulation over the whole surface; aperture ovate, produced, simple; axis imperforate, twisted, and the pillar lip hardly thickened, slightly reflected. Height, 5; diameter, 4.5 mm. (Dall.)

TYPE in United States National Museum, No. 214455. Type locality, U.S. Fish Commission, Station 4441, off Point Pinos, Monterey Bay, California, in 35 fathoms.

RANGE. Known only from type locality.

## Velutina conica Dall, 1886

*Proceedings of the United States National Museum,* 9:305; Pl. 3, fig. 10.

Shell solid, strong, white, with an imperceptible or extremely thin epidermis; finely striate in each direction, four-whorled, the last much the largest; suture narrow, but channeled; aperture oblique; axis within the columella not pervious to the eye; columella narrow, strong, with a light wash of callus, and no umbilical chink. Alt. of shell, 10.0; of aperture (parallel to the axis), 7.5; maximum lat. of shell, 7.6 mm. (Dall.)

TYPE in United States National Museum. Type locality, Unalaska.

RANGE. Aleutian Islands to Kodiak and Forrester Island, Alaska.

## Velutina zonata Gould, 1841

*Report on the Invertebrata of Massachusetts,* 242, fig. 160.

Shell oval-orbicular, compressed, pellucid, covered with a striped, calcareous incrustation; inner lip flattened and channeled. (Gould.)

Shell thin, opaque, white, and in some places pellucid, minutely striated both ways; whorls less than three, the first two minute, and not seen when the shell is viewed in front; the last, widening with great rapidity, becomes large, though it is not tumid, but appears rather depressed as it lies upon the aperture; the surface is covered with a chalky incrustation deposited by the animal, apparently instead of an epidermis; it is white or flesh-colored, and generally with numerous zones of brown, of various widths; when this is removed, the shell is left pellucid; aperture ovate, ample, nearly the whole length of the shell, more than double the size of the body of the shell; outer lip sharp and spreading; inner lip sharp-edged, but margined by a flattened, crescent-shaped, white, channeled space; the sharp edge is lost as it revolves within the shell, and a thin plate of enamel covers the space between it and the junction of the outer lip. Length, $\frac{9}{20}$; breadth, $\frac{13}{40}$ in. (Gould.)

TYPE in State Collection, No. 126. Type locality, Chelsea Beach.

RANGE. Icy Cape, Arctic Ocean, to Monterey, California. Also Atlantic.

## Velutina prolongata Carpenter, 1865

*Annals and Magazine of Natural History,* series 3, **15**:32. Tryon, *Manual of Conchology,* **8**:66.

V. testa majore, subplanata, tenuiore, carnea, spira minima; anfr. iii. et imidio, rapidissime augentibus; vertice vix conspicuo; anfr. ult. antice valde porrecto; regione columellari incurvata; labio valido; axi haud, rimata; eperdimide tenui, rugis incrementi ornata, spiraliter haud striata. Long., 1; long. spir., 15; lat., .95 poll. (Carpenter.)

Pink, under a smooth, thin yellowish epidermis, the growth-lines crossed by very fine slight spiral impressed striae; whorls 3½, the last very large, suture deeply impressed; aperture long oval, junction of columellar and outer lip somewhat angulated. Length, 10 mm. (Tryon, *Manual of Conchology.*)

TYPE in United States National Museum. Type locality, Neah Bay.
RANGE. Bering Strait to Monterey, California.

## Velutina laevigata (Linnaeus) Müller, 1776
### Plate 92, fig. 8

*Systema Naturae,* 12th ed., 1250. *Prodromus Zoölogiæ Danicæ,* Pl. 101, figs. 1–4. Tryon, *Manual of Conchology,* **8**:65; Pl. 27, figs. 41–44, 48, 59.

H. testa imperforata obovata obtussima pellucida laevissima. Testa magnitudine Pisi, pellucida, laevissima, nitida, obovata, obtussima. Umbilicus vix ullus. Apertura magna, sublunata, postice elongata. Color cornu striis obsoletis, longitudinalibus, obscurioribus; labio interiore albo. (Linnaeus, Müller.)

Shell thin, translucent, whitish, or usually light pink, with numerous fine revolving striae crossing the minute growth-lines; epidermis thin, transparent horn-color, somewhat tufted on the revolving striae, whorls three and a half, suture deeply impressed. (Tryon, *Manual of Conchology.*)

TYPE in Zoölogy Museum, Copenhagen. Type locality, England.
RANGE. Icy Cape, Arctic Ocean, to Monterey, California; also Atlantic.

## Velutina coriacea Pallas, 1788

*Nova Acta Acad. Petropolitana,* **3**:237. Tryon, *Manual of Conchology,* **8**:67; Pl. 27, figs. 49, 50.

Testa transversim ovata, subauriculata, dum humet cartilagineo-cornea (sicca membranaceo-cornea), lutescente, subpellucente, incrementi, striis, imo (in adultis) rugis, subimbricata; labri margine hirsutie quadam exasperato; spira laterali, submarginali simplici, et interius, praesertim in junioribus, crusta calcarea tenui obducta; anfractibus circit 2½, ultimo; apertura ampla suborbiculari; columella angusta, arcuata. (Pallas.)

Membranaceous, corneous, subpellucid, rugosely striate; whorls two and one-half, the last very large, expanded. (Tryon, *Manual of Conchology.*)

TYPE in Academy, St. Petersburg. Type locality, Arctic.

RANGE. Bering Sea and the Okhotsk Sea.

### Velutina sitkensis A. Adams, 1851

*Proceedings,* Zoölogical Society of London, 225.

V. testa nigro-fusca, epidermide liris elevatis transversis confertis obtecta, longitudinaliter valde sulcata, sulcis subdistantibus; apertura ovli, intus sulcata; labro margine reflexo, nigro, incrassato; postice non-producto supra anfractum ultimum. (A. Adams.)

TYPE in Museum Cuming. Type locality, Sitka, Alaska.

RANGE. Bering Sea to Sitka, Alaska.

### Genus TORELLIA Loven, in Jeffreys, 1867

Coquille étroitement perforee, renflee, globuleuse, converte d'un épiderme poilu; spire courte, deprimee; ouverture arrondie; sinus basal a peine perceptible; labre mince, arque; columella concave, subtronquee à la base. (Fischer.)

TYPE. *Torellia vestitis* Jeffreys.

DISTRIBUTION. North America and Europe.

### Torellia vallonia Dall, 1919

*Proceedings of the United States National Museum,* 56:355.

Shell small, the earlier part of it with a calcareous lining under a leathery periostracum, the later part entirely flexible; the general form and aspect that of *Vallonia gracilicosta* on a larger and somewhat less depressed scale; the color is yellowish-brown, with about four whorls, the nucleus missing, the rest of the shell with elevated axial lamellae, regularly disposed with much wider interspaces; the aperture entire, with a simple, not thickened, margin; umbilicus open, moderately wide. Diameter, about 4; height, about 3 mm. (Dall.)

TYPE in United States National Museum, No. 219130. Type locality, Nazan Bay, Atka Island, Aleutian, in 14 fathoms.

RANGE. Known only from type locality.

## Torellia ammonia Dall, 1919

*Proceedings of the United States National Museum,* 56:355.

Shell large, cartilaginous or leathery, depressed, brown, of somewhat more than two whorls; the nucleus, comprising a little more than one whorl, has spiral elevated lamellae with wider interspaces, much as in the larval envelopes of *Lamellaria,* but on a larger scale; the remainder of the shell has close-set axial fringed lamellae; the suture is deep, the aperture entire, wider than high; the umbilicus is pervious, moderately wide, as in *Planorbis trivolvis,* the animal is distinctly *Velutinoid,* and carries no operculum. Height, of dry shell, 12; greatest diameter, 24 mm. (Dall.)

TYPE in United States National Museum, No. 111367. Type locality, southwest of Sannakh Island, Alaska, U.S. Fish Commission Station 3213, in 41 fathoms.

RANGE. Known only from type locality.

## Family LEPETIDAE

## Genus **LEPETA** Gray, 1847

Shell patelliform, the embryonic nucleus spiral, lost in the adult, apex in front of the middle; no internal septum.

TYPE. *Lepeta caeca* Müller.

DISTRIBUTION. Arctic Ocean, North Atlantic, south to Massachusetts Bay, Scotland, Denmark, Point Barrow, Alaska, Sitka, Aleutian Islands, Plover Bay, East Siberia, eastern part of Straits of Magellan, Patagonia.

## Subgenus CRYPTOCTENIDIA Dall, 1918

## **Lepeta concentrica** Middendorff, 1851

*Sibirische Reise,* 183; Pl. 16, fig. 6.

Testa extus incrementi vestigiis lamellulosis, erectiusculis, concentricis et confertis ornata; vertice subinflexo; intus nitidissima, vernicosa. Long., 13; lat., 9; alt., 5 mm. (Middendorff.)

Shell depressed conical, apex directed forward; front slope one-third the length of the shell or a little less; surface faintly radiately striate (more distinctly so in young specimens), not decussated or granulose; light-brownish or greenish tinted. The outline is ovate, a little narrower in front; front slope convex. The fine thread-like radiating striae are larger on the longer slope of the shell; they are not interrupted by concentric growth-lines, the latter being inconspicuous, or sometimes strongly impressed at intervals. Epidermis very thin, yellowish-brown, deciduous.

Inside polished, white, the anterior terminations of the muscle-scar a little behind the apex. Edges of shell level, narrowly bordered with gray, especially in the 'young. (Tryon and Pilsbry, *Manual of Conchology*.)

TYPE in Academy, St. Petersburg. Type locality, Okhotsk Sea.

RANGE. Icy Cape, Arctic Ocean, Okhotsk Sea, Shantur Island, Forrester Island, Alaska, to Puget Sound, Washington.

### Lepeta caecoides Carpenter, 1865

*Proceedings,* Philadelphia Academy of Natural Sciences, **60.**

L. t. *L. caecae* simili; t. alba, ancyloidea, tenui, juniore subdiaphana; apice obtuso, antice verso; parte postica parum excurvata; lateribus haud compressis; margine regulariter ovato; tota superficie sublente minutissime striata, striulis valde distantibus, haud elevatis, haud granulatis, subobsoletis; cicatrice musculari haud impressa. Long. (t. adolesc.), .45; lat., .37; alt., .19 (speciminis multo majoris pars solum superest: long., .94; lat., .73; alt., .55). (Carpenter.)

Shell like *L. caecae;* with white, thin ancyloid shell, in youth subdiaphanous; with obtuse apex, turned toward the front; the hinder part little excavated; with sides not compressed, with margin regularly ovate; the whole surface gently and very minutely striate, with striae strongly distant, not elevated, not granulated, subobsolete; with muscle-scar not impressed; (of much the larger specimen only a part remains).

TYPE in ? Type locality, Puget Sound.

RANGE. Arctic and Bering Seas to Hakodate, Japan, the Aleutian Islands, south to the Farallon Islands, California.

### Lepeta alba Dall, 1869

*American Journal of Conchology,* 5:145; Pl. 15, figs. 3, a–d.

Shell pure white, smooth, or with extremely faint striae; solid; interior pure white, apex anteriorly directed, inconspicuous; shell arcuate before and behind. Length of adult, .96; width, .70; alt., .40 in. (Dall.)

TYPE in Smithsonian Institution. Type locality, Seniavine Strait.

RANGE. Bering Strait and Sea south to Straits of Juan de Fuca.

Described as *Cryptobranchia alba* Dall.

### Lepeta alba instabilis Dall, 1869

*American Journal of Conchology,* 5:145; Pl. 15, fig. 6.

Shell patelliform, depressed, broad, smooth or lightly striate (young). Apex inconspicuous, in the anterior fourth. Sculpture in the adult only of the concentric lines of growth, which are occasionally impressed. Plane of the base of the shell curved upward anteriorly and posteriorly, without compression of the sides. Shell very thick, solid, muscular impression deeply impressed. Long., .56; alt., 20 in. Defl. apex, 120°. (Dall.)

TYPE in United States National Museum? Type locality, Sitka, Alaska.

RANGE. Unalaska to the Shumagins and south to Sitka Sound, Alaska. Described as *Cryptobranchia instabilis* Dall.


## Family ACMAEIDAE

### Genus **ACMAEA** Eschscholtz, 1830

Shell conical, patelliform, apex more or less anterior. The shells may generally be distinguished from *Patella* by the different texture and margin border of the inside. (Tryon and Pilsbry, *Manual of Conchology*.)

TYPE. *Acmaea mitra* Eschscholtz.

DISTRIBUTION. North Atlantic and Arctic, western coast of North America and South America, West Indies, Japan, and Indo-Pacific.

### **Acmaea mitra** Eschscholtz, 1833
### Plate 85, fig. 5

*Zoölogical Atlas,* 18: Pl. 23, fig. 4.

Testa ovato-rotundata elevata, intus alba, extus lutescente calcarea scabriuscula, vertice subcentrali erecto, obtuso, margine integro. (Eschscholtz.)

Die Schale ist ziemlich dick. Der längste Durchmesser ihrer Oeffnung beträgt ungefähr 1 Zoll, und eben so viel beträgt auch ihre Höhe-Ihre aussere Fläche wird von einer mässig dicken Lage einer fast Kreideartigen aber festen Masse gebildet. An kleinen Exemplaren ist die Oeffnung der Schale ovaler, als an den grössern, ja mitunter beinähe elliptisch. (Eschscholtz.)

Shell dull-white, aperture nearly circular, wider behind, in some young examples somewhat elongated, oval; form conical, apex erect, nearly central, blunt, smooth, posterior surface usually straight, but occasionally a little convex; exterior smooth, marked with very faint concentric lines of growth, devoid of epidermis; margin entire, polished, with a narrow semipellucid rim inside. Internally smooth or furnished with grooves radiating from the apex more or less strongly marked. Muscular impressions deep, strong, horseshoe-shaped, with the marks of the anterior ends of the adductors rounded and broader than the rest, connected by a slender impressed line marking the attachment of the mantle. Young shells are often furnishéd with irregular riblets more or less strong, many or few in number, radiating from the apex, but stronger toward the margin. Color dead-white inside and out, often livid or tinged a fine pink or pea-green from *Nullipore,* never wax-yellow or horny-pellucid as in the normal state of *Scurris scurra.* Length, 35; breadth, 31; alt., 23 mm. (Tryon and Pilsbry, *Manual of Conchology.*)

TYPE in St. Petersburg? Type locality, Bering Sea.

RANGE. Pribilof Islands, Bering Sea, to San Diego, California.

## Acmaea mitra funiculata Carpenter, 1864

*Supplementary Report,* British Association for the Advancement of Science, 650.

Shell small, whitish, regularly conical, apex acute, elevated, a little in front of the middle; sculptured with strong rounded riblets, sometimes a little nodulous; sometimes single, sometimes gathered into two's and three's; with wide interspaces in which intercalary riblets appear. Length, 6; breadth, 4½; alt., 3 mm. (Tryon and Pilsbry, *Manual of Conchology.*)

The measurements are from the type.

TYPE in United States National Museum. Type locality, Monterey, California.

RANGE. Shumagin Islands, Alaska, to Magdalena Bay, Lower California.

## SECTION COLLISELLA Dall, 1871

### Acmaea cassis Eschscholtz, 1833

### Plate 94, figs. 5, 6

*Zoölogical Atlas,* 19: Pl. 24, fig. 3.

Testa ovato-oblongata, elevata intus alba, extus rutilo-fusca, costis latis rotundatis fornicatis inaequalibus, vertice excentrico, erecto obtuso, margine repando. (Eschscholtz.)

Die äussere Flache ist, wie auch gewöhnlich bei fast allen folgenden Arten, etwas rauh anzufühlen, und hat das Aussehen, als sei eine Deckfarbe in einer ziemlich dicken Schicht auf die Schale aufgetragen. Die Rippen entspringen ziemlich tief unterhalb des Gipfels, werden gegen den Rand Schale immer breiter, sind an Lange und Dicke unter einander etwas ungleich und erscheinen als Falten der Schale, so dass also der Konvexität, die jede nach aussen zeigt, inwendig eine Konkavität entspricht. Lange, 1

Zoll 9 Linien; grosste Weite, 1 Zoll 3 Linien; Höhe, 1 Linie. (Eschscholtz.)

Bei einer nur 5 Linien langen und hierher gehörigen Schale kommt inwendig unter dem Gipfel ein sehr verwischter bräunlicher Flecken vor, am Rande der Oeffnung befindet sich inwendig ein sehr schmaler bräunlicher Saum, auswendig ist gegen den Rand hin stellenweise etwas Gold unter die rothbraune Grundfarbe untergemischt und die Breite der Schalenoffnung ist im Verhältniss zur Lange derselben etwas geringer. (Eschscholtz.)

Solid, strong, having stout radiating ribs about 25–27 in number, those in front narrower or obsolete. Dark spot of the inside small or obscured; margin with a mere dark line, or a series of scollops between the ends of the ribs. Outside dull gray. (Tryon and Pilsbry, *Manual of Conchology*.)

RANGE. Aleutian Islands to San Francisco, California.

#### Acmaea cassis pelta Eschscholtz, 1833

*Zoölogical Atlas*, 19. Tryon and Pilsbry, *Manual of Conchology*, 13: Pl. 8, figs. 90, 91.

Testa ovato-oblonga depressiuscula, extus fusco-olivacea, intus alba aut caerulescente, macula sub fornice fusca irregulari, striis transversis subtilissimis, costa parum elevatis latis, vertis, excentrico subinflexo. (Eschscholtz.)

Die Schale ist auswendig glatt, sehr dunkel, olivengrün, und in der Nahe des Randes zwischen den Rippen mit einigen gelblichen Langsstreifen versehen. Die vordere Hälfte des breunen Fleckens im Innern der Schale ist bei allen mir vorliegenden Exemplaren verwischt. Am Rande kommt inwendig ein schmaler brauner oder braun und gelb gefleckter Saum vor. Lange I Zoll 1 Linie, grosste Breite 10 Linien, Höhe 5½ Linien.

Hieher glaube ich als blosse Spielart eine andere Schale zahlen zu mussen, die etwas grösser und auswendig nicht glatt, sondern etwas rauh ist, bei der auch die Rippen nicht so deutlich hervortreten, die gelben Langsstreifen fehlen und der Gipfel nicht im mindesten übergebogen ist. Ich habe mehrere Schalen in Handen, die deutlich einen Uebergang in diese Varietät machen. (Eschscholtz.)

Shell oval, conical, apex a little in front of the middle. Surface having rather coarse low ribs. Dark border of the inside very narrow, or reduced to a series of dark scollops. (Tryon and Pilsbry, *Manual of Conchology*.)

TYPE in St. Petersburg. Type locality ?

RANGE. Okhotsk and Southern Bering Sea, Nushsgak, Alaska, Aleutian Islands to Lower California.

## Acmaea cassis olympica Dall, 1914

Plate 94, figs. 12, 13

*Nautilus,* **28**:14. Tryon and Pilsbry, *Manual of Conchology,* **13**: Pl. 8, figs. 92–94.

Shell small, conical, elevated, having much the shape of *A. mitra.* The color outside is gray, pink, or light purple, painted with a few or many black stripes. A dark spot is inside. Ribs obsolete. Length, 25–30 mm. (Tryon and Pilsbry, *Manual of Conchology.*)

TYPE in United States National Museum. Type locality, Olympia, Washington.

RANGE. Alaska Peninsula to San Diego, California.

## Acmaea cassis nacelloides Dall, 1871

Plate 94, figs. 7, 8, 9

*American Journal of Conchology,* **6**:247; Pl. 17, figs. 36, *a–c.*

A very distinct variety of this species (*Collisella pelta* Esch.) has exactly the aspect of *Nacella instabilis* externally. It is of a blackish-brown, with sharp, radiating grooves sometimes obsolete near the apex. Several shells beginning in this way have a margin with the normal characters of *C. pelta.* It is quite distinct from the variety *monticola* Nutt., as described by Dr. Carpenter in the *American Journal of Conchology,* **2**:33, and might readily be taken for a distinct species, as the sculpture differs entirely from that of the normal *pelta,* which is sparsely furnished with prominent bulging ribs. (Dall.)

TYPE in United States National Museum. Type locality, California.

RANGE. Attu Island, Aleutian, to San Diego, California.

Described as *Collisella pelta nacelloides.*

## Acmaea scutum Eschscholtz, 1833

Plate 85, figs. 12, 17

*Zoölogical Atlas,* pt. 5, **19**: Pl. 23, figs. 1–3.

*A. scutum.* Testa ovato-oblonga convexa, transversim striata, extus olivacea, lineis flavescentibus radiantibus inaequalibus picta, intus livida, macula sub fornice fusca spathulaeformi magna, vertice subcentrali obtusissimo, margine crenato.

Der Gipfel ist etwas abgerieben. Um ihn herum befindet sich ein bräunlicher, verwischter Saum, an dem übrigen Theile ist die Grundfarbe ein schmutziges Olivengrün, das gegen den Rand ins Schwarze übergeht. Viele gelblich weisse verschiedentlich länge, im Ganzen aber nur kurze

Streifen verlaufen auf jenem Gründe von oben nach unten und stellen einige concentrische Reihen dar. Der bräune Fleck an der innern Seite der Schale ist ziemlich scharf begrenzt und Ziemlich regelmässig. Am Rande der Schale befindet sich inwendig ein mässig bretter schwarzbräuner Saum, mit vielen kleinen grünlich gelben und in ungleichen Entfernungen von einander abstehenden Querstreifen. Länge, der Oeffnung 1½ Zoll, grösste Weite derselben 1 Zoll 3 Linien, Höhe der Schale 6½ Linien. (Eschscholtz.)

Shell large, oval, or rounded-oval, depressed-conic, the apex rounded and near the middle; slopes slightly convex. Surface obsoletely radiately striated, olive-gray, tessellated, or more rarely striped, with black. Inside white with an irregular brown central area and a rather wide dark or tessellated border. (Tryon and Pilsbry, *Manual of Conchology*.)

TYPE in St. Petersburg. Type locality ?

RANGE. Southern Bering Sea, Pribilof and Aleutian Islands, to Gulf of California.

This is what has been called *A. patina* by west coast collectors.

### Acmaea scutum patina Eschscholtz, 1833
### Plate 85, figs. 4, 10; Plate 94, figs. 3, 4

*A. patina.* Testa ovato-oblonga, depressa, tenui, decussatim subtilissime striata, extus pallide-olivacea, maculis et taeniis flavescentibus decurrentibus inaequalibus picta, intus livida, macula sub fornice fusca elliptica obsoleta, vertice excentrico prominulo, subinflexo, margine acutissimo crenulato.

Shell ovate-oblong, depressed, thin, crosswise very finely striate; outside pale olive, ornamented with unequal yellow spots and bands running downward, within pale, with a tawny obsolete elliptical blotch under the arch, with slightly prominent excentric vertex, somewhat bent inward, margin acutely crenulate. (Translation.)

Die schale hat viele Aehnlichkeit mit dem gleichen Theile der vorigen Art. Doch ist sie kieiner, viei flacher und dünner. Ihr Gipfel dagegen ragt deutlich hervor, und die gelblichen Striche in dem bräunen Saum der innern Fläche sind zahlreicher, als in jener Art. Hierzu kommt, dass bräune Flecken in der innern Fläche der Schale selbst an sehr kleinen Exemplaren mehr oder weniger verwischt erscheint, die bräunliche Grundfarbe ab der äusseren Fläche der Schale aber lichter ist, und einen Uebergang ins Graue zeigt. Länge de Oeffnung 1 Zoll 2 Linien, grösste Weite 1 Zoll, Höhe 4 Linien. Bei kleineren Exemplaren ist die Oeffnung der Schale beinahe elliptisch. (Eschscholtz.)

Length, 53; breadth, 48; alt., 18 mm. (Tryon and Pilsbry, *Manual of Conchology.*)

TYPE in St. Petersburg. Type locality ?

RANGE. Southern Bering Sea to Tugar and the Shentar Islands on the west, Pribilof and Aleutian Islands to Gulf of California.

## Acmaea scutum pintadina Gould, 1846

### Plate 94, fig. 14

*Proceedings of the Boston Society of Natural History,* 2:151. Tryon and Pilsbry, *Manual of Conchology,* 13: Pl. 9, fig. 6.

Testa diversissime figurata et picta, tenuis, ovato-rotundata, radiatim et confertim striata, smaragdina, interdum fuscescens vel cinerascens, maculis crebis albidis tessellata aut radiata; apice plerumque sub-centrali, plus minusve elevata: facies interna coeruleo-albida, nitens; fundo piceo; limbo lato, piceo et albido tessellato. Long., 2; lat., 1⅗; alt., ⅗ poll. (Gould.)

Shell large, flat, open, apex subcentral; tessellated white and dark. (Tryon and Pilsbry, *Manual of Conchology.*)

TYPE in ? Type locality, Straits of Juan de Fuca, Puget Sound, Washington.

RANGE. Aleutian Islands to Monterey, California.

## Acmaea scutum cribraria Carpenter, 1866

*American Journal of Conchology,* 2:335.

This shell when young is dark olive closely dotted all over with white, the eroded apex black; when adult it is usually uniform dull slate-color outside with a ring of light around the black apical spot; inside it has a wide dark border, a large, irregular central dark patch, and generally is suffused with dark brown all over. Sculpture obsolete. (Tryon and Pilsbry, *Manual of Conchology.*)

TYPE in ? Type locality, California.

RANGE. Neah Bay, Washington, to Santa Barbara, California.

## Acmaea scutum ochracea Dall, 1871

*American Journal of Conchology,* 6:249; Pl. 17, fig. 35.

There is another very well marked and pretty variety which I should refer to this species (*patina*), and as it does not appear to have been described, I would propose for it the name of *ochracea;* externally it is of a very light yellowish-brown, without spots or rays; internally white with the characteristic dark brown stain of *patina* in the visceral area. The exterior is covered with fine, regularly radiating, close, equal, thread-like riblets, which pass from apex to margin without bifurcation, imbrication, or asperities of any kind. These riblets will serve to distinguish it from any of the other limpets of the coast; otherwise it approaches very close to some varieties of *scabra,* and can be traced right into varieties of *patina.* The variations of these limpets appear to be absolutely without limits; you may describe seven hundred forms as easily as seven. The only guide to specific identity is a certain habit of growth, easier seen than described, and very easy to overlook. (Dall.)

TYPE in United States National Museum. Type locality, Monterey, California.

RANGE. Known only from type locality.

## Acmaea apicina Dall, 1879

*Proceedings of the United States National Museum,* **1**:341. Tryon and Pilsbry, *Manual of Conchology,* **13**:24; Pl. 7, figs. 66, 67.

Testa parva, conica, tenui, rotundata, plus minusve elevata; albida seu isabellina, apice erecto, luteo; intus luteo, albido, seu fusco, laevi; extus striulis incrementis subobsoletis, munito. Lat., 5; long., 6; alt., 4 mm. (Dall.)

Shell small, conical, thin, rounded, more or less elevated; whitish or isabelline, the apex erect, buff; inside buff, whitish or brown, smooth; provided with subobsolete lines of growth outside. (Dall.)

TYPE in United States National Museum. Type locality, Pribilof Island.

RANGE. Pribilof, Aleutian, and Shumagin Islands, Alaska.

## Acmaea peramabilis Dall, 1872

*Proceedings of the California Academy of Sciences,* **4**:302. Tryon and Pilsbry, *Manual of Conchology,* **13**: Pl. 33, figs. 80–82.

Shell thin, delicate, ovate; externally of a uniform dark rose color, with a few scattered irregular blotches of light or dark brown, nucleus pale. Within polished, bluish white, with a chestnut brown spectrum with sharply defined edges, outside of which for a short distance the white is unsullied, but further toward the margin in adult specimens, radiating brown blotches may be observed forming a more or less interrupted band around the shell, which is wanting in the young. The margin is of the same deep rose as the exterior. Shell moderately elevated, with the apex well marked subacute and situated in the central third. Nucleus smooth, pale, sharply decurved with a chink beneath it, in front. Sculpture of fine elevated threads which extend from the vertex to the margin without bifurcating. These are crossed by very fine sharp lines of growth slightly elevated. Long., 1.03; lat., 0.8; alt., 0.33 in. Posterior slope slightly arched. (Dall.)

TYPE in United States National Museum. Type locality, Shumagin group of islands, Alaska.

RANGE. Known only from type locality.

### Acmaea digitalis Eschscholtz, 1833
### Plate 85, figs. 6, 9; Plate 94, figs. 10, 11

*Zoölogical Atlas,* 5:20; Pl. 23, figs. 7, 8.

Testa ovato-elliptica, convexa, intus albida, macula sub fornice fusca, irregulari obsoleta, costis nonnulla latis tuberculosis, vertice submarginali compresso. (Eschscholtz.)

Der Gipfel befindet sich an dem vordern End, und nicht gar fern von dem Rande der Schale. Seitwärts von ihm fällt die Schale fast ganz gerade und sehr abschussig ab, nach hinten aber steigt die Schale vor ihm erst etwas, obschon an einigen Exemplaren mehr, an andern weniger in die Höhe, und fällt dann allmählig und mit einer starken Wölbung ab. An dem vordern Theile der Schale sind zwischen dem Gipfel und dem Rande Keine Rippen zu bemerken, nach hinten aber werden sie allmählig stärker, und die jenige, welche von Gipfel zu dem hinterne Ende des Schalenrandes verläuft, ist nicht blos die längste, sondern auch die höchste und dickste. Den grossern Rippen entsprechen schwache rinnenformige Vertiefungen an der innern Seite der Schale. Einige Schalen sind in der Gegend des Gipfels stärker zusammengedruckt, als andere, und danach richtet sich denn auch hauptsächlich die Form des nur mässig grossen bräunen Fleckens, der sich an der innern Fläche der Schale befindet. Doch ist in manchen Exemplaren statt dieses Fleckens nur ein leichter bräuner Anflug vorhanden. Am Rande ist inwendig ein schmaler bräunlicher oder auch braun und gelb geflecter Saum. Die Grundfarbe der äussern Fläche ist ein unreines Weiss. An einigen Exemplaren liefen auf diesem einige dunkelbräune Binden von dem einen Seitenrande über die Wölbung der Schale, etwas schief zu dem andern Seitenrande quer herüber. Bei den meisten aber waren die Binden oder auch nur der mittlere Theil derselben abgerieben oder auch gleichaam vervischt. Die Länge der grössten Exemplare betung 11 Linien, ihre grösste Weite 8 Linien und ihre Höhe 6 Linien. (Eschscholtz.)

This is the most usual form found north of San Francisco Bay. It is dull, lusterless, whitish, with stripes and zigzags of blackish-brown. The apex is usually decidedly anterior and elevated; the front ribs are obsolete, the posterior ribs strong, rounded, often uneven. Inside margin conspicuously tessellated; central area generally dark and rather narrow.

This is the *oregona* of Authers, and probably *radiata* of Eschscholtz. It resembles the striped variety of the Chilean *A. ceciliana* so closely that it would be absolutely impossible to separate a mixed lot. (Tryon and Pilsbry, *Manual of Conchology.*)

TYPE in St. Petersburg. Type locality ?

RANGE. Aleutian Islands to Monterey, California.

## Acmaea digitalis umbonata Reeve, 1855

*Conchologia Iconica,* **8:** fig. 107.

Shell globosely convex, apex altogether anterior, sharply hooked, radiately ridged, ridges few, irregular, obtusely crenated; ash-black, sculptured with white impressed punctate, punctures more numerous toward the margin, interior bluish-white, chestnut in the middle. (Reeve.)

TYPE in Museum Cuming. Type locality, Upper California.

RANGE. Saginaw Bay, Alaska, to California.

## Acmaea digitalis textilis Gould, 1846

*Proceedings of the Boston Society of Natural History,* 2:152. *Mollusca and Shells of the United States Exploring Expedition,* Pl. 29, fig. 456.

Testa depressa, oblique conica; apice anteriori, acuto; costis latis, elevatis, subplanulatis, nodosis ad 15 radiantibus; epidermide viridi, punctis albidis quadratis reticulato; basi ovato-rotundato, limbo marginali albido et fusco-viridi fimbriato: facies interior albida; fundo piceo. Long., 1; lat., ¾; alt., ³⁄₁₀ in. (Gould.)

TYPE in State Museum, Albany, No. 6347.

RANGE. Crescent City, California, to Cedros and Socorro Islands.

## Acmaea limatula Carpenter, 1864

*American Journal of Conchology,* 2:340. Tryon and Pilsbry, *Manual of Conchology,* 13:14; Pl. 3, figs. 38–40.

Extus sculptura normali; seu intensiore, lirulis quibusdam majoribus, valde nodosis; intus albida, nitida; limbo lato nigro, seu rarius tessellato; spectro saepius fusco maculato. (Carpenter.)

Outer layer of the shell black, covered with an olive-green, or some-times light bluish, epidermis; inside border black; a deep brown central spot. (Tryon and Pilsbry, *Manual of Conchology*.)

TYPE in ? Type locality, San Pedro, California.

RANGE. Crescent City, California, to Cerros and Socorro Islands.

Described as *Acmaea scabra limatula*.

## Acmaea limatula mörchii Dall, 1878

*Proceedings of the United States National Museum,* 1:47.

Shell conical, much elevated, with a subcentral recurved apex resembling that of *Helcion pectinatus* covered with close-set, rough, imbricated ribs and riblets, the coarse, imbricated, sharp lines of growth forming with the other sculpture a close reticulation in some specimens. Interior with a brown-mottled spectrum and margin, otherwise white; exterior dull grayish or greenish speckled. The imbrications on the principal ribs very strong, in some specimens forming small spines concave beneath. Lat., 16; long., 20; alt., 10 mm. (Dall.)

TYPE in United States National Museum, No. 31268. Type locality, Tomales Bay, California.

RANGE. Known only from type locality.

## Acmaea scabra Gould, 1846

*Proceedings of the Boston Society of Natural History,* 2:152. Tryon and Pilsbry, *Manual of Conchology,* 13: Pl. 3, figs. 41–44.

Testa parva, scabra, per-inequilateralis, oblique arcuato-conica, pallide virens, nigro diversemode virgata, costis radiantibus scabrosis ad 15 ornata; basi ovata: facies interna albida; fundo piceo, limbo marginali viridulo et piceo fimbriato. Long., ⅞; lat., ⅝; alt., ⅜ poll. (Gould.)

Apex rather anterior; slopes rather straight; sculptured with very strong, close, rough ribs, with smaller intervening riblets; center of the inside white, with dark spots and bars. Normally it is solid, rather depressed, with from 20–30 very strong, rounded ribs not evanescent anteriorly, the interstices being occupied by intercalary riblets. The color is white, with fine lines of brown (not striped as in *pelta* and *persona*) between the principal ribs, which delicately dot the otherwise uniform livid in some specimens, especially the young, to a very faint pink. Apex white margin. Sometimes the principal ribs are rather sharp, palmating the margin; occasionally they are small and crowded, becoming faint at the margin, when the shell presents the internal aspect of *A. mitella;* at other times assuming that of *Patella pediculus.* Generally the apex is at the anterior third; rarely at the anterior fourth, with very elongated outline; but sometimes is nearly central with a rounded shell. (Gould.)

TYPE in ? Type locality, San Francisco, California.

RANGE. San Francisco, California, to Socorro Island.

Described as *Patella scabra.*

This has been called *A. spectrum* by collectors.

## Acmaea sybaritica Dall, 1871

*American Journal of Conchology,* 6:257.

Shell depressed, thin; apex subcentral, more anterior in the young. General shape rounded oval, hardly more narrow before than behind. Surface nearly smooth, with rounded concentric lines of growth; in young specimens a few faint, hardly noticeable elevated radiating lines or riblets may be observed near the margin, which is entire. Internally smooth, border polished and also the cavity of the apex above the muscular impressions. Color a clear rose pink, varying from quite deep and a little white, even in very young specimens entirely uneroded, rather blunt and inconspicuous; sides of the shell ornamented with rays of a darker shade of pink, more or less gathered in groups, and more or less evident, according to the shade of the remainder of the shell. Internally, the visceral area is bluish white, usually washed with a faint yellowish-brown, often hardly evident; in which case the area is whitish; the successive layers of brown sometimes appear externally around the apex when eroded. The inner margin, and to some extent the whole interior, exhibits the external markings or rays, through the somewhat pellucid shell. Texture hard and brittle. Epidermis exceedingly thin, usually evanescent; translucent brownish. Long., .6; lat., 1.46 in.

TYPE in United States National Museum. Type locality, Bering Sea.

RANGE. Plover Bay, Eastern Siberia, to Hakodate, Japan, Pribilof and Aleutian Islands to Chirikoff Island.

## Acmaea persona Eschscholtz, 1833
### Plate 85, figs. 13, 14

*Zoölogical Atlas,* 5:20; Pl. 24, figs. 1, 2.

Testa ovato-elliptica, convexa, extus rufescente, intus albo-caerulea, macula sub fornice fusca, ovato-elliptica alata, vertice excentrice inflexo. (Eschscholtz.)

Der Gipfel ist weit aus dem Centrum herausgerückt, so dass der vordere Theil der Schale von ihm sehr abschüssig gegen den Rand hin abfällt. Der bräune Flecken im inner der Schale ist nicht bei allen Exemplaren verhältnissmässig gleich gross. Auch hat er nicht bei allen so deutlich 2 Seitenlappen oder kurze Seitenäste, als es in der gegebenen Abbildung (fig. 2) der Fall ist. Dicht an dem Rande der Schale befindet sich inwendig ein schmaler und gleichmässig dunkelbraun gefärbter Saum. Zwischen ihm und dem bräunene Flecken hat die innere Fläche der Schale nur einen leichten Anflung von bläurlicher Farbe. Länge, 1 zoll; grösste Weite, 10 Linien; Höhe, 5 Linien. (Eschscholtz.)

Shell oval, apex pointing forward; posterior slope long, convex; anterior slope short. Sculptured with strong, rounded ribs, usually nodulous, but sometimes obsolete. Whitish, with strips and zigzags of blackish-brown or olive-green variegated and speckled with white. Margin crenated by ribs. Inside white or stained with yellowish-brown, with a large central deep brown area, rarely absent; border articulated black and gray. (Tryon and Pilsbry, *Manual of Conchology.*)

TYPE in St. Petersburg. Type locality, not known to writer.

RANGE. Shumagin Islands, Alaska, to Socorro Island.

## Acmaea persona strigillata Carpenter, 1866

*Supplementary Report,* British Association for the Advancement of Science, 1863, 618.

A. testa *A. mesoleucae* simili, sed minore, haud viridi; striolis minimis, confertissimis, plerumque erosis tenuissime sculpta; albida, strigis olivaceofuscis, plerumque radiantibus, interdum confluentibus picta; apice saepius nigro; intus albida, margine satis lato, strigis tessellato. Long., 9; lat., 74; alt., 3 poll. (Carpenter.)

The shell is ovate, a little wider behind, elevated; apex at the front fourth of the length. Young with excessively fine close radiating striae crossed by growth-lines, largely worn off in adult specimens. Apex very acute in young, eroded, dark brown and polished in old shells. Coloration: Marked with irregular, forking black stripes on a white ground, interspersed around the apex when not eroded, with dots and small narrow or needle-shaped white streaks. Inside bluish-white, with the central area indistinctly irregularly clouded with brown. Border wide, vividly tessellated with blackish-brown. Length, 19; breadth, 14; height, 7 mm. (Tryon and Pilsbry, *Manual of Conchology.*)

TYPE in United States National Museum. Type locality ?

RANGE. Sitka, Alaska, to Magdalena Bay, Lower California.

This equals *A. strigatella* Carpenter.

## Acmaea rosacea Carpenter, 1866

*American Journal of Conchology*, 2:341. Tryon and Pilsbry, *Manual of Conchology*, 13: Pl. 7, figs. 71–73.

A. t. parva, conica, tenui, laevi; t. jun. pallide rosacea, elegantissime maculis albis et fuscis subradiantim sparsis; t. adulta strigis fusco-rosaceis et albidis picta; apice elevato, parum antico; intus rosacea. Long., .2; lat., .16; alt., .08 poll. (Carpenter.)

Shell small, conical, thin, smooth or with very obsolete ribs. The young are pale roseate, with few white and brown subradiating spots; the adults have rosy brown and whitish streaks or are dotted with pale rose. Apex elevated, a little anterior; inside white or rosy. (Tryon and Pilsbry, *Manual of Conchology*.)

TYPE in ? Type locality, San Diego, California.

RANGE. Neah Bay, Washington, to Acapulco, Mexico.

## Acmaea instabilis Gould, 1846

### Plate 94, figs. 1, 2

*Proceedings of the Boston Society of Natural History*, 2:150. Tryon and Pilsbry, *Manual of Conchology*, 13: Pl. 6, figs. 32, 33.

Testa olivaceo-cervina, elongata, elevata, ad latera compressa, creberrime radiatim striata, apice sub-centrali, obtusa, castanea: apertura oblongo-elliptica; margine integro: facies interna lactea. (Gould.)

Shell narrow and oblong, the basal margin elevated at the ends; texture thin; slopes convex or bulging. Surface finely radiately striated; dark brown or black. Inside white or bluish, with or without a faint brown spot in the cavity. Largest specimen measures 1½ inches in length by ⅞ in breadth. (Tryon and Pilsbry, *Manual of Conchology*.)

TYPE in State Museum, Albany, No. 6356. Type locality, Neah Bay, Washington.

RANGE. Kodiak Island to San Pedro, California.

Described as *Patella instabilis*.

## Acmaea incessa Hinds, 1842

*Annals and Magazine of Natural History*, 10:82; Pl. 6, fig. 3.

Testa conica, ovali, fusca, tenue transversim striata, intus alba; apice maculis albis ornato.

A small, horny, brown shell, remarkable for the white markings on the apex, usually three, but sometimes four in number, the central being rather the larger. It was always found imbedded in the fronds of a *Laminaria*, which it was often necessary to cut with a knife before the shell could be liberated. (Hinds.)

TYPE in ? Type locality, on seaweed, San Diego, California.

RANGE. Trinidad, California, to Magdalena Bay, Lower California.

Described as *Patella incessa*.

## Acmaea depicta Hinds, 1842

*Annals and Magazine of Natural History,* 10:82; Pl. 6, fig. 4.

Testa minima, lineari, diaphana, alba, lineis rufis apice radiantibus; lateribus compressis; longa quadruplo quoad longitudinem.

This is a small, delicate shell, white, with irregular brown rays diverging from the apex, about eight in number on each side, sometimes disposed to fork; clouded with a dark spot anteriorly; and the sides much compressed, so as to make the shell four times longer than broad. Length, $\frac{4}{10}$ of an inch. (Hinds.)

TYPE in British Museum? Type locality, on seaweed, San Diego, California.

RANGE. Santa Barbara, California, to Lower California on *Zostera*.

Described, *Patelloida depicta*.

## Acmaea asmi Middendorff, 1849

### Plate 94, figs. 15, 16

*Beitrage zu einer Malacologia Rossica,* 2:39; Pl. 1, fig. 5.

Testa minuta, crassa, ponderosa, elliptica, elata atque inflata, vertice antico clivo postico multum extus; epidermide fusca vestita, sub microscopico irregulariter rugosa, saepe (margine) erosa, calcarea, albida; intus: albida, margine concolore et centro toto coeruleo. Long., 1; lat., $\frac{3}{4}$; alt., $\frac{5}{8}$; vertice ad, $\frac{5}{8}$; totius longitudinis sito: longitudo adulti, 0.08 Decim. (Middendorff.)

Shell small, thin but strong and solid, elevated, conical, the base short-oval, apex erect, a little in front of the middle; slopes of the cone somewhat convex. Surface lusterless, usually corroded, smooth except for very fine radiating striae visible with the aid of a lens, but obsolete in adult shells. Color rusty black. Inside black, with a brown zone just outside the muscle-scar. Length, 10; breadth, 8½; alt., 7 mm. (Tryon and Pilsbry, *Manual of Conchology.*)

TYPE in Academy, St. Petersburg. Type locality, Sitka, Alaska.

RANGE. Sitka, Alaska, to San Diego, California, and Socorro Island, Lower California.

## Acmaea paleacea Gould, 1851

*Mexican and California Shells,* 3: Pl. 14, fig. 5.

T. parva, tenuis, elongata, lateraliter valde compressa et triangularis, ad dorsum utroque costis obtusis ornata; apice acuto prope trientem anteriorem longitudinis sito; color straminea, vel cinnamomea.

Shell minute, delicate, thin, elongated, laterally compressed, so that the two sides are parallel; dorsal aspect a long, narrow oval; lateral aspect low triangular; apex at about the anterior third, acute, antrorse; surface with delicate lines of growth, and a few obtuse radiating ridges along the dorsal slope at each end; color straw yellow, or cinnamon brown.

Length, $\frac{1}{4}$; height, $\frac{1}{10}$; breadth, $\frac{1}{16}$ in.

On kelp or stems of Zoöphytes, Santa Barbara. (Col. Jewett.)

This curious little shell can be compared with no other species, unless it be *A. alveus,* which is still comparatively much broader. It has much the size and shape of a glume of wheat.

Type in ? Type locality, Santa Barbara, California.

Range. Trinidad, California, to Lower California.

## Acmaea triangularis Carpenter, 1866

*Proceedings of the California Academy of Sciences,* 3:213.

N. t. *N. palleacea* simili, sed multo minus elongata; apice elate marginibus rectangulatim devergentibus; albida, maculis fuscis perpaucis ornata; striulis subobsoletis. Long., 0.28; lat., 0.12. (Carpenter.)

The solitary shell sent by Dr. Cooper is shaped like a right-angled triangle, with five brown spots near the base. (Carpenter.)

Type in State Collection, No. 416c. Type locality, Monterey, California.

Range. Sitka, Alaska, to Gulf of California.

Described as *Nacella.*

## SECTION RHODOPETALA Dall, 1920

### Acmaea rosea Dall, 1872

*Proceedings of the California Academy of Sciences,* 4:270; Pl. 1, fig. 2.

Shell small, egg-ovate, of a deep rose color, externally smooth, except for very faint radiating ridges divaricating from the apex, and for lines of growth. Margin entire; apex minute, produced before the anterior margin. Interior smooth, white, except the margins, which are polished and of the same color as the exterior. Nacre, especially when weathered, silvery. Long., .35; lat., .27; alt., .12 in., of the largest specimen. (Dall.)

TYPE in ? Type locality, Alaska.

RANGE. Kyska Island, Aleutians, to Shumagin Islands, Alaska, to California.

## Genus **LOTTIA** Gray, 1834

Shell patelliform, depressed, the apex anterior. The typical and only species is among the largest and most active of limpets, the shell attaining three inches in length.

TYPE. *Lottia gigantes,* Sowerby.

DISTRIBUTION. West coast of North America.

## Lottia gigantea Gray, 1843

*Journal de Conchyliologie,* **13**:141.

L. t. magna, crassiore, planata, expansa, textura saepius extus spongiosa; nucleo minore, corneo, nigro-fusco, ancyliformi, vertice mamillato, subelevato; dein elongata, postice grisea, undulata; t. adolescente verrucosa, radiis obscuris, antice haud verricosis; t. adulta plus minusve lata, plus minusve radiata seu verricosa; apice plus minusve a margine remoto; parte antica seu haud exstante, seu circiter per quintam totius longitudinis projiciente; parte postica plus minusve elevata, convexa; extus ut in *Acmaea pelta* picta, albido-grisea, fusco-olivaceo copiose irregulariter strigata: intus, plerumque testudinaria, margine lato, nigro; spectro definito, seu rarius albido; cicatrice musculari fortiore, interdum purpurei seu violaceo tincta. Long., 2.6; lat., 2.05; alt., 7 poll. (Gray.) (Carpenter.)

Shell large, solid, oval, depressed, the apex near the front margin; outer surface eroded, of a spongy texture, dull brown, gray toward the summit. Inside having a black rim around the margin, deep chestnut brown outside of the muscle-impression, which is strong, bluish or purplish-white. Central area chestnut brown, more or less mottled with white, rarely entirely white. Length, 75; breadth, 55–60; alt., 17–20 mm. (Tryon and Pilsbry, *Manual of Conchology.*)

TYPE in ? Type locality ?

RANGE. Crescent City, California, to Guadalupe and Cerros Islands.

## Family COCCULINIDAE

## Genus **COCCULINA** Dall, 1881

Shell patelliform, apex posteriorly inclined, with a deciduous spiral nucleus; margin entire.

The shell resembles that of the *Patellidae,* but the animal is more nearly allied to the *Fissurellidae.* (Dall.)

TYPE. *Cocculina rathbuni* Dall.

DISTRIBUTION. In nearly all parts of the deep sea which have been explored.

## Cocculina agassizii Dall, 1908

*Bulletin,* Museum of Comparative Zoölogy, **43**: No. 6, 340.

Shell small, white, covered with a strong light, olive-colored periostracum, beneath which it is chalky, ovate-quadrate, high, with the apex about the posterior third, and the anterior longer slope roundly arcuate; the periostracum is finely, closely, radiately threaded, the threads seem to bear very short projecting hairs, but neither the threads nor the hairs appear to correspond to any sculpture of the shell; on drying, the periostracum immediately detached itself from the upper part of the shell, showing beneath it only very fine, irregularly concentric lines; toward the margin it seemed to be more closely attached to the shell and by its contraction in drying, began at once to split the shell, obliging me to return it at once to the liquid from which it had been taken, or it would have gone to pieces entirely; interior of the shell smooth, white, with a broad, short, horseshoe-shaped muscular impression with a wide anterior hiatus at about the anterior third of the length; nucleus small, bulbous, produced hardly spiral, but decurved; the shell enlarges suddenly on entering the nepionic stage. Alt., 2.0; length, 3.5; width, 2.5 mm. (Dall.)

TYPE in United States National Museum, No. 110660. Type locality, U.S.S. "Albatross" Station 4630, Gulf of Panama, in 556 fathoms.

RANGE. Off Queen Charlotte Islands, British Columbia, in 150 fathoms, to Gulf of Panama.

## Cocculina cazanica Dall, 1919

*Proceedings of the United States National Museum,* **56**:356.

Shell small, brownish or whitish, darker toward the apex, which is at the posterior third of the shell; both slopes slightly convex, the nuclear point always eroded; sculpture near the apex minutely equally reticulate, the sculpture coarser and the radial threads more prominent than the concentric ones toward the margin; interior polished, the muscular impression perceptible but not conspicuous. Length, 5.5; width, 4.5; height, 2.25 mm. (Dall.)

TYPE in United States National Museum, No. 222069. Type locality, U.S. Fish Commission Station 4245, in Kasa-an Bay, Alaska.

RANGE. Kasa-an Bay and Stephens Passage, Alaska, in 95–160 fathoms.

## Family PHASIANELLIDAE

### Genus **PHASIANELLA** Lamarck, 1804

Shell bulimiform or subglobose, polished, without epidermis or nacre, variegated with bright colors; operculum heavy, calcareous, internally paucispiral, with nucleus near the basal margin, externally convex, white.

TYPE. *Phasianella bulimoides* Lamarck.

DISTRIBUTION. Australia, India, Philippines, Mediterranean, Britain, West Indies, west and east coasts of North America. Fossil Devonian . . . . ; Europe and California.

### Subgenus TRICOLIA Risso, 1826

### **Phasianella compta** Gould, 1855

H. R. Doc. 129, Preliminary Report, 25: *Pacific Railroad Surveys, Report,* 5; Pl. 11, figs. 25, 26.

Testa parva, solida, ovato-conica, imperforata, polita, cinerascens, lineis minutis oblique volventibus olivaceis ornata; spira acuta; anfractibus quatuor rotundatis; ultimo ad peripheriam obtuse angulato, et interdum tessellatem fasciato; apertura circularis; labro tenui, alvo; columella alba, compressa; faucibus callo albo incrassatis. (Gould.)

Shell small, solid, ovate, imperforate, smooth and shining, ashy white, minutely and closely lineated in an obliquely spiral manner with olive green. Whorls four, well-rounded, forming an acute spire, the outer one obtusely angular at periphery, where there is sometimes a delicate catenated range of white and olive spots, aperture nearly circular; lip very thin, showing the lineations on the inner margin; throat coated with bluish·white enamel; pillar flattened; white. Operculum patelliform, ivory-like, the outer convex surface marbled black and white; the inner surface black. Length, ¼; diameter, ⅕ in.

*Locality.* Found at San Diego by Mr. Blake and also by Mr. Webb. This pretty little shell is usually more or less coated with cretaceous matters, but when cleaned exhibits a beautifully lineated surface, peculiar on account of the lines running so nearly in the ordinary direction of revolving striae. The coloration, however, sometimes consists of olive and white tessellations and blotches. It would accord pretty well with *P. perforata* Philippi, did it not lack the distinguishing mark of that species, its perforation.

TYPE in United States National Museum? Type locality, San Diego, California.

RANGE. Monterey, California, to Gulf of California.

## Phasianella compta punctulata Carpenter, 1865

*Annals and Magazine of Natural History,* **15**:179.

P. testa *P. comptae* simili, sed elatiore; suturis impressis; anfractibus tumentibus; omnino minutissime fusco punctata; columella lacunata. Long., 24; long. spir., 12; lat., 14 poll. (Carpenter.)

*P.* with shell like *P. comptae,* but more lofty; impressed sutures; with tumid whorls; most minutely punctate all over with tawny spots; columella lacunate. (Translation.)

TYPE in the Boyce Collection, Utica, New York. Type locality, San Diego, California.

RANGE. San Diego, California, to Cape San Lucas, Lower California.

## Phasianella pulloides Carpenter, 1865

*Annals and Magazine of Natural History,* ser. 3, **15**:180.

P. testa *P. pullo* simillima; solida, compacta, spira breviore; suturis distinctis. Long., 2; long. spir., 1; lat., 13 poll. (Carpenter.)

Somewhat similar to *P. pulla;* solid, compact, with shorter spire; suture distinct. (Tryon and Pilsbry, *Manual of Conchology.*)

TYPE in Boyce Collection, Utica, New York. Type locality, San Diego, California.

RANGE. Monterey, California, to Lower California.

## Phasianella pulloides elatior Carpenter, 1865

*Annals and Magazine of Natural History,* ser. 3, **15**:180.

P. testa perparva; spira elongata, ut in *P. pullo* picta; anfractibus subplanatis; suturis haud impressis; columella haud lacunata. Long., 19; long. spir., 12; lat., 11 poll. (Carpenter.)

Very small, spire elongate, painted as in *P. pulla;* whorls subplanulate, suture scarcely impressed, columella lacunate. (Tryon and Pilsbry, *Manual of Conchology.*)

TYPE in the Boyce Collection, Utica, New York. Type locality, Santa Barbara, California.

RANGE. Monterey to San Diego, California.

## Subgenus EULITHIDIUM Pilsbry, 1898

## Phasianella typica Dall, 1908

### Plate 91, fig. 9

*Proceedings of the United States National Museum,* **34**:256. Tryon and Pilsbry, *Manual of Conchology,* **10**:177.

A new name for *Phasianella variegata* Carpenter, 1864, not *Phasianella variegata* Lamarck.

E. testa parva, laevi, turbinoidea, nitente, marginibus spirae valde ex-curvatis; rosaceo et rufo-fusco varie maculata; anfr. nucleosis regularibus, vertice mamillato; normalibus iv., valde tumentibus, rapide augentibus, suturis impressis; anfr. ultimo antice producto; basi rotundata; umbilico carinato; apertura vix a, pariete indentata; peritremate pene continuo, acuto. Long., 1; long. spir., .05; lat., .07 poll. (Carpenter.)

Shell small, smooth, turbinate, bright, outlines of spire convex, vari-ously masculated with rose color and reddish brown; whorls normally four, very convex, rapidly increasing, the last one produced anteriorly, separated by well-impressed sutures; nuclear whorls regular, apex mamil-lated; base rounded; umbilicus carinated; aperture scarcely indented by parietal margin, peristome nearly continuous, acute. Alt., 1; diameter, .07 in. (Tryon and Pilsbry, *Manual of Conchology*.)

TYPE in United States National Museum. Type locality, Cape San Lucas, Lower California.

RANGE. San Luis Obispo, California, to Cape San Lucas, Lower Cali-fornia.

Described as *Eucosmia variegata*.

## Phasianella lurida Dall, 1897
### Plate 91, fig. 10

*Bulletin*, Natural History Society of British Columbia, No. 2, 15; Pl. 1, fig. 11.

Shell small, solid, turbinate, of four whorls, of a lurid purple color, slightly paler on the base and apex; whorls rounded, sculptured only by feeble lines of growth, polished; suture distinct; base rounded with feeble spiral striation, aperture rounded, peritreme sharp-edged, smooth within, the lips united over the body by a wash of callus; umbilical region im-perforate. Height of shell, 3.75; diameter, 3 mm. (Dall.)

TYPE in Natural History Society of British Columbia, Victoria? Type locality, Skidegate Channel, Queen Charlotte Islands, in 20 fathoms.

RANGE. Vancouver Island, British Columbia, to Mendocino County, California.

Described as *Eucosmia lurida*.

## Phasianella substriata Carpenter, 1864

*Annals and Magazine of Natural History*, ser. 3, **13**:475. Tryon and Pilsbry, *Manual of Conchology*, **10**:177.

E. testa E. *variegatae* simillima, sed anfr. circa basin et supra spiram (nisi in anfr. nucl. laevibus), interdum tota superficie tenuiter et crebre striatis; striis anfr. penult. circ. x. (Carpenter.)

Form as in *E. variegata;* but whorls, except the nuclear, very delicately striate, the last with about ten striae. (Tryon and Pilsbry, *Manual of Conchology.*)

TYPE in United States National Museum. Type locality, Cape San Lucas, Lower California.

RANGE. Catalina Island, California, to Panama.

Described as *Eucosmia substriata.*

## Family TURBINIDAE

## Genus **ASTRAEA** Bolten, 1798

### Subgenus POMAULAX Gray, 1850

Shell trochiform, elevated conic, angulated and nodose at the periphery, obliquely ribbed, not umbilicated; inner lip arcuated, with a wide callus, which is channeled, anteriorly truncated. Operculum with three radiating ribs and perforated axis. (Tryon and Pilsbry, *Manual of Conchology.*)

This is the description of the subgenus *Pomaulax.*

TYPE. *Astraea undosa* Wood.

DISTRIBUTION. North Pacific, America and Japan.

### Astraea undosus Wood, 1828
### Plate 102; Plate 103

*Index Testaceologicus,* Supplement 16; Pl. 5, *Trochus,* fig. 1.

Shell large, conic, imperforate, white, covered with a strong obliquely lamellose corneous epidermis; whorls six to eight, planulate, with oblique radiating tuberculate costae above; periphery with an undulating nodose carina; base flattened, with three to five concentric corrugations; aperture subovate, very oblique, angular, pearly within; columella dilated, with a semicircular groove at the position of umbilicus, the umbilicated tract bounded by a white grooved ridge. Alt., 80; diam., 110 mm. (Tryon and Pilsbry, *Manual of Conchology.*)

TYPE in ? Type locality, California.

RANGE. San Pedro, California, to Cerros Island, Lower California.

### Subgenus PACHYPOMA Gray, 1850

### Astraea inaequalis Martyn, 1784
### Plate 108, figs. 1, 2

*Figures of Nondescript Shells, Universal Conchology,* tablet 1, Pl. 31.

Shell conic, imperforate, rather solid, with a chestnut brown cuticle, lighter beneath; whorls six to seven, planulate above, sutures slightly impressed, bordered below by a series of obliquely descending corrugations, which are cut into granules by from one to five spiral furrows; periphery carinate, subspinose on the upper whorls, usually nearly smooth on body-whorl; base nearly flat, concentrically lirate, the lirae more or less tuberculate, five or six in number, their interstices regularly striate; aperture subtriangular, white within, the lower margin fluted; columella arcuate, broad, excavated at position of the umbilicus, and terminating in a tooth-like prominence below. Alt., 45; diam., 55–62 mm. (Tryon and Pilsbry, *Manual of Conchology*.)

TYPE in Martyn Cabinet. Type locality ?

RANGE. Vancouver Island, British Columbia, to San Pedro, California.

## Astraea inaequalis pacifica Dall, 1919

### Plate 108, figs. 3, 4

*Proceedings of the United States National Museum,* **56**:356.

Shell of moderate size, of a reddish color, trochiform, carinate, of more than five flattish whorls, the apex defective; the carina overhangs a somewhat obscure suture; axial sculpture of numerous (on the last whorl about 70) obliquely protractive, close-set, rounded riblets, which are cut anteriorly into segments by three rather obscure grooves; these riblets, bunched in pairs and swollen, undulate the carina, giving it a somewhat stellate profile; on the base from the carina to the umbilical callus are six deep channeled grooves separated by wider interspaces, which are crossed by numerous minute laminae; the squarish interspaces obliquely nodulous; aperture diamond-shaped, simple, body with a glaze of enamel; pillar concavely arched, white with a callous deposit on the umbilical area behind it. Height, 25; maximum diameter, 42 mm. (Dall.)

TYPE in United States National Museum, No. 222320. Type locality, Pacific Beach, San Diego, California.

RANGE. Known only from type locality.

## Astraea inaequalis montereyensis n. subsp.

### Plate 108, figs. 5, 6

Shell thick, imperforate; whorls 8, sculptured with corrugated, obliquely descending folds; periphery more rounded than the typical species; base with five coarse squamose lirae. Aperture subtriangular, pearly; columella slightly arcuate. A thin callus covers the umbilical region and part of the base. Shell much taller and heavier than the typical species.

TYPE in Oldroyd Collection, Stanford University, No. 620–1. Type locality, Monterey, California.

Altitude, 60; diameter, 50 mm., of type; a larger specimen showed altitude, 80; diameter, 60 mm.

RANGE. Monterey, California. This shell has been called *Astraea inaequalis,* by collectors, but differs enough to be given a subspecific name. The range of the typical species is Vancouver Island, British Columbia, to San Pedro, California. That of *Avar. pacifica* Dall, Pacific Beach, San Diego County, California.

## Astraea barbarensis Dall, 1919

*Proceedings of the United States National Museum,* **56**:357.

Shell of moderate size, trochiform, white, covered by a reddish-brown periostracum, with six or more flattish whorls, the nucleus defective; suture rather obscure, not appressed; axial sculpture of (on the last whorl about 30) rounded, protractively oblique ribs, with equal or wider interspaces, reaching from the suture to the periphery which they undulate; halfway from the suture to the periphery these are cut into segments by three deeply incised grooves; there may be, near the periphery, a few intercallary short ribs; there are also fine, close, inconspicuous, incremental lines; on the base inside the undulate border are five strong regularly beaded cords with narrower channeled interspaces; aperture subquadrate, the upper part of the outer lip much produced; the body glazed, the pillar concavely arcuate, ending in a blunt projection, behind it a porcellanous white semilunar callus around the imperforate umbilical depression. Height of shell, 26; of aperture, 12; diameter, 37 mm. (Dall.)

TYPE in United States National Museum, No. 223819. Type locality, U.S. Fish Commission Station 2945, off Santa Cruz Island, California.

RANGE. Known only from type locality.

## Genus LEPTOTHYRA (Carpenter) Pease, 1869

Shell small or minute, globose-depressed, solid, compact; umbilicate or imperforate, whorls three to seven, spirally sculptured, the last generally somewhat deflexed at the aperture; aperture subcircular, white and nacreous within; columella generally but not always bluntly denticulate near the base. Operculum subcircular, nearly flat or concavo-convex, inside with a very thin corneous layer, slightly convex, with many gradually increasing whorls, the nucleus subcentral; outside calcareous, subspiral, with a slightly convex concentric elevation or ridge around the margin, most prominent at its termination, the middle portion concave and more or less rugose. (Tryon and Pilsbry, *Manual of Conchology.*)

TYPE. *Leptothyra carpenteri* Pilsbry.

DISTRIBUTION. In nearly all tropical and subtropical seas, but most numerous in the Pacific.

## Leptothyra carpenteri Pilsbry, 1888

Tryon and Pilsbry, *Manual of Conchology,* 10:247; Pl. 39, *a,* figs. 26–29.

Shell small, globose, very solid, imperforate, spire conic, more or less depressed; suture moderately impressed; whorls five, slightly convex, the last decidedly deflected toward the aperture, encircled by about fifteen subequal spiral lirae, separated by interstices about as wide as the ridges; incremental striae generally strongly developed, causing the lirae to appear nodose or somewhat irregular, and the interstices to appear pitted; aperture oblique, pearly white within, about half the length of shell; columella arcuate, base obsoletely uni-, bi-, or tri-dentate; color red, ashen, or purple. Alt., 8; diam., 8–9 mm. (Pilsbry.)

TYPE in Philadelphia Academy of Sciences. Type locality, California.

RANGE. Sitka, Alaska, to San Diego, California, to Cape San Lucas.

This is the shell that has been called *Leptothyra sanguineus.*

## Leptothyra lurida Dall, 1885

*Proceedings of the United States National Museum,* 8:542. Tryon and Pilsbry, *Manual of Conchology,* 10: Pl. 39, *a,* figs. 27–29.

This form has a dull olivaceous cast, sometimes mottled with whitish. The sculpture is more compact and closer, the nacre less brilliant than that of Monterey specimens; they are also smaller on the average. (Dall.)

TYPE in United States National Museum. Type locality, Puget Sound.

RANGE. Puget Sound to Cape San Lucas, Lower California.

## Leptothyra bacula Carpenter, 1864

*Proceedings of the California Academy of Sciences,* 3:177. Tryon and Pilsbry, *Manual of Conchology,* 10.

L. t. *L. sanguineo jun.* sed rufocinerea, sculptura obsoleta; anfr. iv. planatis, suturis vix distinctis, marginibus spirae valde excurvatis; lirulis obsoletis latioribus, et circa basim striis crebris, vix sculpta; apertura rotundata, declivi; columella vix callosa. Long., 0.08; long. spir., 0.06; lat., 0.14. (Carpenter.)

Shell small, depressed-globose, solid, imperforate, rufus ashy; whorls four, slightly convex, rapidly increasing, obsoletely but regularly spirally striate; aperture large, oblique, deflexed above. Alt., 4; diam., 5 mm. (Tryon and Pilsbry, *Manual of Conchology*.)

TYPE in Cooper Collection, No. 1056. Type locality, Catalina Island. RANGE. Puget Sound to San Martin Island, Lower California.

## Leptothyra paucicostata Dall, 1871

*American Journal of Conchology,* 7:131.

Shell small, depressed, rounded, of three whorls. Sculpture consisting of seven revolving strong ridges, with the interstices deeply channeled, and marked by evident lines of growth. In a few specimens these are thickened at regular intervals, forming faint transverse riblets, which do not reach as high as, or pass over, the revolving ribs. Suture deep, channeled. Apical whorl smooth, translucent, the others solid and strong. Aperture rounded, peristome thickened, continuous with an internal groove for the operculum. Columella ending in a toothlike knob or callosity, which is especially sharp and prominent in young shells. Throat pearly white. Operculum of a few whorls, smooth, externally concave with thickened edges; circular, calcareous. Color of shell usually of a rusty brown, on the ribs alternating darker and lighter in spots, or reddish, in a similar way, or pure white. Interspaces usually light, but sometimes of the same color as the ribs. Alt., .12; lat., .2 in. (Dall.)

TYPE in United States National Museum. Type locality, Monterey, California.

RANGE. Monterey to Coronado Islands.

## Leptothyra paucicostata fenestrata Bartsch, 1919

*Proceedings of the United States National Museum,* 56:358.

Shell in which, under the strong spiral sculpture, there is visible in the interspaces a certain number of raised radial threads forming more or less evident reticulation. (Bartsch.)

TYPE in United States National Museum. Type locality, Tia Juana, Lower California.

RANGE. Monterey to Coronado Islands.

## Leptothyra juanensis Dall, 1919

*Proceedings of the United States National Museum,* 56:358.

Shell solid, turbinate, very dark olive, with a very minute smooth nucleus and about five well-rounded whorls; suture obscure, not appressed; spiral sculpture of (on the penultimate whorl five, on the last whorl about a dozen) strong, undulate or bearded cords with wider rather deep inter-spaces; axial sculpture of close, oblique, rather prominent incremental lines; aperture oblique, internally nacreous, the outer lip thin, the body with a glaze of nacre, the pillar lip concave, white, nacreous, with two rather formless nodulations anteriorly. Height, 8; maximum diameter, 9 mm. (Dall.)

TYPE in United States National Museum, No. 186070. Type locality, Tia Juana, Lower California.

RANGE. San Diego, California, to Tia Juana, Lower California.

## Leptothyra grippi Dall, 1911

*Nautilus,* **25**: No. 3, 25.

Shell small, solid, of about five whorls; the apex slightly flattened, nepionic whorls one and a half, small, nearly smooth, whitish; sculpture on the spire, on the second whorl three, increasing to five on the last whorl, strong, prominent, squarish spiral cords, articulated in the type with crimson and white, the interspaces at first smooth on the later whorls with one to three intercalary much smaller spiral threads; on the last whorl between the periphery cord and the next posterior cord five uniform fine threads, though this feature is probably variable; base flattened, trans-lucent white, with one articulated crimson and white color band around the umbilical region, which is also white; pillar broad, white, with one prominent knob of callus in the middle of it; throat brilliantly pearly; the whorl is laid slightly above the peripheral cord, which is covered by the advancing whorl; body color crimson; faint spiral striae on the flattened base; axial sculpture only of incremental lines. Height of shell, 5.25; of last whorl, 4.0; max. diam., 6.0 mm. (Dall.)

TYPE in United States National Museum. Type locality, off entrance to San Diego Harbor, in 100 to 150 fathoms.

RANGE. Known only from type locality.

## Family LIOTIIDAE

## Genus **LIOTIA** Gray, 1847

Shell turbinated or depressed, varicose, perforated or umbilicated; whorls ribbed or cancellated; aperture rounded, pearly within; peristome thick, callously margined. (Tryon, *Structural and Systematic Con-chology.*)

TYPE. *Liotia scalaroides* Reeve.

DISTRIBUTION. Tropical and subtropical. Fossil, Jurassic. . . . .

## Liotia acuticostata Carpenter, 1864

*Supplementary Report,* British Association for the Advancement of Science, 652.
*Proceedings of the California Academy of Sciences,* 3:159. Tryon and Pilsbry,
*Manual of Conchology,* 10: Pl. 36, fig. 1.

L. t. parva, subglobosa, alba; anfr. nucl. ii. laevibus, apice satis extante;
anfr. normalibus iii., carina in spira maxime extantibus ii., anfr. ult. vi.;
suturis subrectangulatis; apertura circulari; labio extus parum contracto;
labio conspicuo; umbilico haud magno. (Carpenter.)

Turbinate, with revolving riblets, which are more or less nodose above,
imperforate. Diameter, 4 mm. (Tryon and Pilsbry, *Manual of
Conchology.*)

TYPE in State Collection, Species 519a. Type locality, Catalina Island,
California.

RANGE. Catalina Island, California, to Magdalena Bay, Lower California.

## Liotia acuticostata radiata Dall, 1919

*Proceedings of the Biological Society of Washington,* 31:8.

Shell resembling the type but with numerous radiating riblets visible
in the interspaces between the revolving costae. (Dall.)

TYPE in United States National Museum, No. 223291. Type locality,
off South Coronado Island.

RANGE. San Diego to Coronado Islands.

## Liotia fenestrata Carpenter, 1864

*Supplementary Report,* British Association for the Advancement of Science, 652.
*Proceedings of the California Academy of Sciences,* 3:158.

L. t. parva, primum subdiscoidea, postea variante, albido-cinerea; anfr.
nucl. laevibus, planatis, apice depresso; anfr. norm. ii. et dimidio, convexis;
clathris validis distantibus circ. xv. radiantibus, et vii. spiralibus, sub-
aequantibus, conspicue fenestrata; apertura circulari, saepius plus minusve
declivi, parieti vix attingente; umbilico maximo, anfractus monstrante;
labio, regione umbilicari, sinuato. (Carpenter.)

Depressed, clathrate by equidistant spiral and radiating riblets, with
deep interstices, sculpture terminating with a spiral ridge surrounding the
rather wide, deep umbilicus. Diameter, 4.5 mm. (Tryon and Pilsbry,
*Manual of Conchology.*)

Type in State Collection, Species 1006. Type locality, Catalina Island, California.

Range. Monterey to Catalina Island, California.

## Subgenus Arene H. and A. Adams, 1854

### Liotia cookeana Dall, 1918

### Plate 91, figs. 12, 13, 15

*Proceedings of the Biological Society of Washington*, 31:8.

Shell small, white depressed, with a narrow crenulated suture and three rapidly enlarging stellate whorls; axial sculpture of (on the last whorl 13) pronounced barrow ribs with wider interspaces, acutely produced at the periphery, and extending nearly to the crenate margin of the funiculate umbilicus; the whole surface in addition is covered with fine, close-set, sharp, almost microscopic, axial threads; beyond the periphery are four or five spiral cords which do not appear on the ribs; aperture circular, the margin crenated by the sculpture. Breadth of shell, 3; height, 1 mm. (Dall.)

Type in United States National Museum, No. 223290. Type locality, off South Coronado Island.

Range. South Coronado Island to Gulf of California.

## Genus MÖLLERIA Jeffreys, 1865

Shell remarkably solid, with strong and partly dichotomous transverse ribs; peristome continuous. Operculum calcareous, multispiral. (Tryon and Pilsbry, *Manual of Conchology*.)

Type. *Mölleria costulata Möller.*

Distribution. Europe and west coast of North America.

### Mölleria quadrae Dall, 1897

### Plate 91, figs. 11, 16

*Bulletin,* Natural History Society of British Columbia, No. 2, 14; Pl. 1, fig. 14.

Shell small, solid, straw-colored or brownish, with three, rapidly enlarging, well-rounded whorls; spire depressed, suture very distinct, umbilicus narrow and deep; surface microscopically spirally striate, the sculpture coarser on the base, but visible with a lens on the upper surface also; aperture circular, peritreme continuous, simple; operculum multispiral, calcareous, centrally slightly concave on the outer side. Diameter of shell, 1.8; alt., 1.0 mm. (Dall.)

TYPE in ? Type locality, Cumshewa Inlet, British Columbia.

RANGE. Amchitka Island, Aleutians, to Queen Charlotte Islands, British Columbia.

## Mölleria drusiana Dall, 1919

*Proceedings of the United States National Museum,* **56**:358.

Shell minute, of two and a half rapidly enlarging whorls, whitish, covered with an olivaceous periostracum; suture distinct, rather deep; whorl section circular, surface smooth except for microscopic incremental lines; base convex with a rather wide umbilicus; aperture circular, simple, sharp-edged; operculum calcareous, multispiral, centrally depressed on the outer side. Height, 1; larger diameter, 1.5 mm. (Dall.)

TYPE in United States National Museum, No. 31117. Type locality, Constantine Harbor, Amchitka Island, Aleutian.

RANGE. Amchitka Island to Glacier Bay, Alaska.

## Family TROCHIDAE

## Genus **NORRISIA** Bayle, 1880

Shell thick, conoidal, orbicular, covered by an epidermis, smooth; widely umbilicated, umbilicus surrounded by the callous extension of the columella; outer lip not thickened or sculptured. (Arnold.)

TYPE. *Norrisia norrisii* Sowerby.

DISTRIBUTION. Monterey, California, to Cerros Island, Lower California.

Described as *Trochiscus norrisii* Sowerby.

## Norrisia norrisii Sowerby, 1838

*Annals and Magazine of Natural History,* new ser. **2**:96; text fig., p. 97. Tryon and Pilsbry, *Manual of Conchology,* **11**:276; Pl. 61, fig. 30.

In its general form it approaches very nearly to *Rotella,* but differs from that genus in having a rather large and deep open umbilicus. The shell is nearly smooth; its lines of growth are, however, distinct, but not prominent. Its outer coat is of chestnut-brown color, but the inside is of a brilliant pearly luster. The border of the outer lip is nearly black, that of the inner is green, and the inside of the umbilicus is nearly colorless. Width, 1.7; axis, 1.3 in. (Sowerby.)

TYPE in British Museum. Type locality ?

RANGE. California to Cerros Island, Lower California.

## Genus **HALISTYLUS** Dall, 1889

Shell small, cylindrical, holostomate, polychromatic; operculum multispiral, coriaceous; dental formula $\dfrac{1}{x+4.4+x}$

TYPE. *Halistylus columna* Dall (*Proceedings of the United States National Museum*, **12**:341.)

DISTRIBUTION. Queen Charlotte Islands, British Columbia, to Panama Bay on the west coast of North America.

Described as a subgenus of *Cantharidus* Montfort.

### **Halistylus subpupoideus** Tryon, 1887

Tryon, *Manual of Conchology*, **9**:394; Pl. 60, fig. 77.

Whorls about 6, convex, closely spirally striate, thin, light olivaceous or brownish, with a series of chestnut spots under the suture; lip simple, sharp. Length, 6 mm. (Tryon.)

TYPE in Academy of Natural Sciences, Philadelphia. Type locality, Monterey, California.

RANGE. Queen Charlotte Islands, British Columbia, to Panama Bay.

Described as *Fenella pupoidea* Carpenter, which was preoccupied.

## Genus **TEGULA** Lesson, 1832

Shell conical, spire pointed, with revolving granulated ribs; columella spirally twisted, terminating anteriorly in a large obtuse tubercle. (Tryon, *Structural and Systematic Conchology*.)

TYPE. *T. pellis-serpentis* Wood.

DISTRIBUTION. Western tropical America.

### Subgenus CHLOROSTOMA Swainson, 1840

### **Tegula funebrale** A. Adams, 1854

### Plate 98, figs. 8, 9

*Proceedings*, Zoölogical Society of London, 316. Tryon and Pilsbry, *Manual of Conchology*, **11**:170; Pl. 28, figs. 42–44.

C. testa turbinata, imperforata, nigra, glabra, longitudinaliter oblique striata, ad suturas crenulata, anfractibus convexiusculis, ultimo rotundato, basi planiusculo, regione umbilicali valde impressa, callo albo obtecta; columella superne sinuata, antice bituberculata, tuberculo supremo prominente; labro nigro marginato. (A. Adams.)

This species is similar to *C. gallina* in form and character of the aperture. It is lusterless, purple or black, the apex usually eroded, orange-colored; the teeth of the columella are white; and there is never a yellowish streak at the base, as in the variety *tincta* of the last species. The whorls are spirally lirate, sometimes smooth except on the base, sometimes strongly lirate above. The suture is margined below by an impressed line, and by elevated, foliaceous incremental lamellae. This last feature may almost always be detected, although sometimes but very slightly developed. Alt., 35; diam., 32; alt., 25; diam., 26 mill. (Tryon and Pilsbry, *Manual of Conchology.*)

TYPE in Museum Cuming. Type locality, California.

RANGE. Vancouver Island to Cerros Island, Lower California.

Described as *Chlorostoma funebrale*. A. Adams.

## Tegula funebrale subaperta Carpenter, 1864

*Supplementary Report,* British Association for the Advancement of Science, 252.

With umbilical pit. (Carpenter.)

TYPE in ? Type locality, Vancouver Island?

RANGE. Neah Bay, Washington, and southward. Fossil: Pleistocene, San Pedro.

Described as *Chlorostoma funebrale* var. *subapertum.*

## Tegula gallina Forbes, 1850

*Proceedings,* Zoölogical Society of London, 271; Pl. XI, fig. 8. Tryon and Pilsbry, *Manual of Conchology,* 11:169; Pl. 20, fig. 5; Pl. 28, figs. 52, 53.

T. testa obtuse pyramidali, crassa (adultus ponderosus), spira magna, anfractibus 5, glabris, obsolete oblique striatis, convexiusculis, albidis, fasciis angustis numerosis purpureis ornatis, anfractu ultimo prope suturam subcanaliculato, basi lateribus rotundatis, umbilico albo, imperforato, apertura subquadrata, labro externo subpatulo, margine acuto, laevi, nigrescente, labro columellari bidentato, albo, faucibus margaritaceo-albis, operculo circulari, corneo, fusco, spiris numerosissimis, confertis. Testa junior spira depressiuscula. Alt., $1\frac{1}{10}$; lat. max., $1\frac{2}{10}$ alt. apert., $\frac{6}{10}$ unc. (Forbes.)

Shell imperforate, heavy, solid, thick, conoidal, dark purplish or blackish, longitudinally striped or speckled with whitish, the stripes occupying the interstices between close, narrow, superficial folds of the surface, which may be well-marked or obsolete, continuous or cut into granules by equally close spiral furrows, the latter sometimes predominating; spire conoidal, the apex usually blunt, eroded, and yellow; body-whorl rounded at the periphery; base convex, more or less eroded in front of the aperture; whorls 5 to 6; aperture oblique; outer lip black-edged, smooth and pearly within; columella short, arcuate, strongly bidentate near the base; place of the umbilicus marked by a pit. Alt., 28–40; diam., 26–34 mm. (Tryon and Pilsbry, *Manual of Conchology*.)

TYPE in British Museum? Type locality ?

RANGE. San Francisco Bay to Gulf of California. Fossil: Pleistocene, San Pedro and San Diego, California.

## Tegula gallina umbilicata Dall, 1919

*Proceedings of the United States National Museum,* 56:359.

Most of the specimens of *Chlorostoma gallina* are imperforate, some have the umbilical callus concave, forming a sort of pit, and a rare variety has an entirely open, narrow, deep umbilicus. The types of the latter are United States National Museum No. 152998, and while in other characters similar, are smaller than the average adult, measuring height 22 and diameter 25 mm. (Dall.)

TYPE in United States National Museum, No. 152998. Type locality, San Quentin Bay, Lower California.

RANGE. San Diego, California, to San Quentin Bay, Lower California.

## Tegula gallina tincta (Hemphill) Pilsbry, 1893

Tryon and Pilsbry, *Manual of Conchology,* 11:169.

In this form the longitudinal markings and sculpture are obsolete and the spiral grooves generally scarcely visible above; the color is yellow-greenish or blackish, the apex eroded, yellow; and there is a streak of yellow on the base just below the columellar teeth.

TYPE in Hemphill Collection. Type locality not known.

RANGE. Halfmoon Bay, California, to San Diego and Socorro Island. Described as *Trochus (Monodonta) gallina* var. *tincta.*

## Tegula gallina multifilosa Stearns, 1893

### Plate 98, figs. 3, 6

*Nautilus* 6:86 and *Proceedings of the United States National Museum,* 16:351; Pl. 50, figs. 8, 9.

Shell imperforate, large, solid, turbinate, globosely conical, elevated; whorls five and a half to six and a half, rounded; suture simple not channeled; apex obtusely pointed, eroded and yellowish; color nearly black; sculpture consisting of numerous, spiral or revolving, closely set, narrow, rounded ridges or costae, alternating with fine, incised, whitish lines or grooves. Aperture rounded oblique, interior pearly, outer edge rimmed with black and finely crenulated by the projecting ends of the modified whitish grooving. Columella arcuated with two blunt tubercles near the base and a shallow umbilical pit above; base convex. Dimensions: Altitude, 36; diameter maximum, 34 mm. (Stearns.)

TYPE in United States National Museum, No. 125315. Type locality, Guadalupe Island, off Lower California.

RANGE. San Pedro, California, to Guadalupe Island, Lower California. Described as *Chlorostoma gallina* var. *multifilosa* Stearns.

## Tegula brunnea Philippi, 1848
### Plate 98, fig. 2

*Zeitschrift für Malakozoologie*, 189. Tryon and Pilsbry, *Manual of Conchology*, 11:170; Pl. 27, figs. 36–38.

Testa imperforata, conoidea, laevi, fusca; anfractibus parum convexis, ultimo ad peripheriam subangulata, ad suturam impresso; apertura subrhombea; columella perobliqua, bidentata; foveola umbilici regionem occupante. Satura, apertura, columella, fovea in regione umbilici, callus lubialis illam ex parte cingens exacte ut in *T. argyrostomo;* color fere exacte ut in *T. pulligine.* Alt. aoeos fict., 11'''; diam. 15½'''. (Philippi.)

Shell imperforate, conical, solid, russet-yellow, brown, orange-colored, or deep crimson; spire conic; sutures deeply impressed; whorls about seven, convex, smooth, obliquely lightly striate, the last sometimes obsoletely undulated or plicate below the suture; base depressed, deeply concave in the center; aperture very oblique; columella one- or two-toothed near the base; umbilical callus white; place of the umbilicus deeply excavated. Alt., 32; diam., 36; alt., 38; diam., 35 mm. (Tryon and Pilsbry, *Manual of Conchology.*)

TYPE in Philippi Collection. Type locality, California.

RANGE. Mendocino County, California, to Santa Barbara Islands. Fossil: Pleistocene, Santa Barbara and San Pedro; Pliocene, San Pedro. Described as *Trochus (Chlorostoma) brunneus.*

## Tegula brunnea fluctuosa Dall, 1919

*Proceedings of the United States National Museum,* **56**:359.

The common form of *brunnea* has a nearly smooth surface or a surface affected by incremental rugosities, but there is another form rather widely spread which exhibits well-marked obliquely protractive ribs to the number of 18 or 20 on the last whorl, reaching from the suture nearly to the periphery with narrower interspaces. It is rather less elevated than the average of the typical *brunnea* and has a more depressed suture. (Dall.)

TYPE in United States National Museum, No. 60055. Type locality, Monterey, California.

RANGE. Purisima to San Nicolas Island, California.

### Tegula montereyi Kiener, 1850

*Species Général des Coquilles Vivantes, Trochus,* 104; Pl. 33, fig. 1. Tryon and Pilsbry, *Manual of Conchology,* 11:171; Pl. 27, figs. 27, 28, 29.

Testa umbilicata, tenuis, conica, sublaevigata, spiraliter et obsolete striata, pallide cenerea, epidermide cornea induta; anfractus 8 planulati, sutura vix impressa discreti; anfractus ultimus carinatus, infra planulatus et concentrice striatus; apertura ampla; margina dextro valde obliquo; area umbilicari alba, lata, funiculo spirali munita; columella dedio denticulata, et infra dentem emarginata. Operculum typicum, anfractus 12–13 gerens. (Kiener.)

Shell umbilicate, strictly conical, rather thin, light olivaceous or pale corneous; spire conical, with nearly straight outlines; apex acute; sutures linear; whorls seven, flattened, encircled by numerous fine lirae, which become obsolete on the lower whorl, which shows usually very ill-defined obliquely descending small folds, at right angles to the incremental striae; body-whorl acutely angular at the periphery; base flat, spirally, subobsoletely lirate; aperture subhorizontal; outer lip thin, margined with brown or corneous; columella subhorizontal, curved, toothed below the middle, receding above, not spreading around the umbilicus as in the other species; umbilicus funnel-shaped, rapidly becoming very narrow, white within, its edge defined by an angle. Alt., 28–29; diam., 34–42 mm. (Tryon and Pilsbry, *Manual of Conchology.*)

TYPE in Museum, Paris. Type locality, California coast.

RANGE. Bolinas Bay to Santa Barbara Islands, California. Fossil: Pleistocene, Santa Barbara to San Diego, California; Pliocene, San Pedro. Described as *Trochus montereyi* Kiener.

### Tegula rugosa A. Adams, 1851

*Proceedings,* Zoölogical Society of London, 182. Tryon and Pilsbry, *Manual of Conchology,* 11:173; Pl. 26, fig. 26.

C. testa turbinato-conoidali, profunde umbilicata, luteo-fusca, nigro variegata, longitudinaliter nodoso-plicata, transversim sulcata; anfractu ultimo rotundato, infra suturam angustato; columella incurvata, antice bituberculata, tuberculo supremo magno, prominente; labro fusco marginato. (A. Adams.)

Shell narrowly umbilicate, conoidal, solid, heavy, dull cinereous, more or less variegated by brown, blackish or red streaks; spire conoidal, generally eroded and white or yellow at the apex; whorls, about five, obliquely striate, radiately coarsely and irregularly plicate and rugose above, sometimes nearly smooth; periphery rounded; base convex, concentrically lirate; aperture oblique; columella strongly dentate in the middle or below it, with a second small tooth at the base; edge of the columella rather deeply curved above the tooth, but spreading at its junction with the whorl, bounding and somewhat narrowing the umbilicus by a white callus, which does not extend to the upper margin of the aperture; umbilicus deep, white within. Alt., 26; diam., 27. (Tryon and Pilsbry, *Manual of Conchology*.)

TYPE in British Museum? Type locality ?

RANGE. San Diego, California, to Acapulco, Mexico.

Described as *Chlorostoma rugosum* A. Adams.

## Tegula regina Stearns, 1893

### Plate 98, figs. 5, 7

*Nautilus,* **6**:85. *Proceedings of the United States National Museum,* **16**:350; Pl. 50, figs. 6, 7.

Shell conical, imperforate, black or purplish-black; whorls six to seven, concave, longitudinally somewhat obliquely plicated, the plicae more or less projecting at the suture, and on the edge of the basal whorl, producing an undulating or crenulated effect. Otherwise sculptured by incremental striae which traverse the surface and cross the plicae at right angles. Base concave, radiately, closely and prominently striated, more conspicuous, flattened, coalescing, and sinuously curving at the edge. Commencing at the point where the outer lip joins the body whorl, a shallow groove follows parallel to the periphery and extends toward the aperture, without interrupting the basal sculpture. Aperture obliquely subangulate, black-rimmed and crenulated on the thin edge of the outer lip; nacreous, silvery white toward the edge, bright lustrous golden yellow within and around the umbilical region, which latter though deeply pitted is not open. Columella white, calloused, arcuated with a moderately developed rib bounding the umbilical depression, and terminating in a single tubercle. This rib is paralleled by a shallow furrow terminating in a notch just below the tubercle, and by an exterior or outer ridge, part of the way double, of a brilliant orange color; this orange-colored rib is also exteriorly bounded by a shallow furrow which becomes obsolete toward the aperture. The base of the shell otherwise exhibits faint revolving sculpture. Dimensions: Altitude, 36 mm.; diameter maximum, 34 mm. (Stearns.)

TYPE in United States National Museum, No. 125314. Type locality, San Clemente Island, California.

RANGE. Catalina and San Clemente Islands, California, to Madalena Bay, Lower California, and Guadalupe Island.

Described as *Uvanilla regina* Stearns.

## Subgenus PROMARTYNIA Dall, 1909

### Tegula pulligo Martyn, 1784

#### Plate 91, figs. 1, 4

*Universal Conchology*, Table 2, Pl. 76. Tryon and Pilsbry, *Manual of Conchology*, 11:171; Pl. 26, figs. 23, 24, 25.

Martyn figured and named but did not describe.

Shell deeply and widely umbilicate, conical, solid, dull purplish or brown, when worn often orange, obliquely streaked with white or unicolored; spira elevated; whorls seven, flattened, the upper ones finely spirally striate and sometimes very obsoletely plicate; the remainder smooth, obliquely finely striate; base flattened, slightly convex, obliquely streaked, concave and white around the umbilicus; body-whorl bluntly angled at the periphery; aperture very oblique, smaller than usual in *Chlorostoma;* columella thin, obtusely dentate, ending above in a white callus which partly covers the umbilicus. Alt., 35; diam., 32; alt., 22; diam., 22–27 mm. (Tryon and Pilsbry, *Manual of Conchology*.)

TYPE in ? Type locality, not known to writer.

RANGE. Sitka, Alaska, to Santa Barbara Islands, California. Fossil: Pleistocene, San Pedro and San Diego, California.

Described as *Trochus pulligo* Martyn.

### Tegula pulligo taylori I. Oldroyd, 1922

#### Plate 91, figs. 3, 6

*Marine Shells of Puget Sound and Vicinity*, Puget Sound Biological Station, 4:171.

Shell very like the typical form, but the periphery is rounded instead of angular; the umbilicus larger; blunt dentition on the inner lip as in the typical form; color of shell reddish, and the white patch around the umbilicus larger, with the umbilical area very much deeper; the base convex and the edge of the outer lip thin and red, with a dull strip between that and the pearly interior. Shell nearly twice the size of the typical, and more elevated. Height of type, 40; of the base, 36 mm. (Oldroyd.)

TYPE in collection of Rev. G. W. Taylor, Stanford University. Type locality, Hope Island at the north end of Vancouver Island.

RANGE. Known only from the type locality.

## Subgenus OMPHALIUS Philippi, 1847

### Tegula aureotincta Forbes, 1850

*Proceedings,* Zoölogical Society of London, 271; Pl. 11, fig. 7. Tryon and Pilsbry, *Manual of Conchology,* 11:172; Pl. 27, figs. 31, 32, 33.

T. testa obtuse pyramidali, crassa, spira mediocri, anfractibus 4 vel 5, convexiusculis, obtuse, angulatis, subcanaliculatis, spiraliter 1–2 late sulcatis, striis spiralibus minutis, longitudinalibus minutissimis sculptis, colore nigro obscure minutissimeque griseo-lineato, ultimo anfractu basi subplanato 4–5 sulcis profundis spiralibus sculpto, margine obtuse subangulato, umbilico profunde perforato, laete aurantio, apertura subrotunda, labro externo tenui, nigro marginato, labro columellari albo 1–2 dentato, dentibus inaequalibus munitis, dente inferiore minimo, fauce albo-margaritaceo. Alt., $7/10$; lat. max., 1; alt. apert., $4/10$ unc. (Forbes.)

Shell umbilicate, conoidal, thick, solid, black or cinereous; spire conical, apex generally eroded; sutures impressed; whorls about five, convex, spirally coarsely ridged below, radiately plicate above; the revolving ridges five in number, the folds of the upper surface disappearing with age; base rounded, concave round the umbilicus; periphery rounded; aperture oblique; columella with a blunt tubercle in the middle and a smaller one below it, slightly reflexed above, joined to the upper margin by a heavy white callus extending across the parietal wall; umbilicus wide, deep, bright orange-colored within. Alt., 38; diam., 35; alt., 22; diam., 26 mm. (Tryon and Pilsbry, *Manual of Conchology.*)

TYPE in? Type locality?

RANGE. Santa Barbara Islands, California, to Magdalena Bay, Lower California. Fossil: Santa Barbara and San Pedro.

### Tegula impressa Jonas, 1848

*Conchylien Cabinet,* 2d ed., "Trochus," 318; Pl. 45, fig. 6. Tryon and Pilsbry, *Manual of Conchology,* 11:185; Pl. 63, figs. 8, 9.

Tr. testa conica, perforata, nigra; anfractibus planiusculis, in parte superiore undatoplicatis, et, praesertim in interstitiis transverse sulcatis, ultimo rotundato-undulato; basi tenuiter et confertim concentrice striata; apertura suborbiculari; columella arcuata, basi dentibus duobus terminata, superius in lobum brevem, partem umbilici cingentem, contenuata.

Shell narrowly and profoundly perforate, rather thick, conoid, dull cinereous, ornamented with castaneous radiating flammules; whorls six, rather convex, spirally finely lirate, the lirae seven to eight on the penultimate whorl, five on the preceding; last whorl rounded, compressed below the suture above, somewhat convex beneath, and provided with about ten concentric lirae; aperture slightly dilated, ovate, the lip plicatulate within; columella thin in the middle, arcuate, concave, bearing two or three tubercles below; columellar callus thick, green, slightly impinging upon the umbilicus. Alt., 13; diam., 13 mm. (Fischer.) (Tryon and Pilsbry, *Manual of Conchology.*)

TYPE in Museum, Hamburg. Type locality, not known to writer.

RANGE. San Diego, California, to Tres Marias Island.

Described as *Trochus impressus* Jonas.

### Tegula ligulatus Menke, 1850

*Zeitschrift für Malakozoologie,* 173. Tryon and Pilsbry, *Manual of Conchology,* 11:177; Pl. 29, figs. 58–60.

Testa convexo-oblique conoidea, obtusiuscula, anfractibus quenque vel sex, convexuisculis, liris confertis subaequalibus argutis, granulosis cincta; granulis oblongis; squalide cinerea, nigro nebulosa; anfractu ultimo superius radiatim obsolete plicato; pagina infera convexo-planuiuscula; umbilico aperto, spiraliter: sulcato; columella basi sinuato-truncata: sinu medio denticulo oblongo conspicuo instructo. Alt., 5.4; diam., 7 lin. (Menke.)

This is an extremely variable form. The shell may be either very much depressed or as high as broad. It may be spirally sculptured with numerous narrow, unequal lirae, or as strongly cingulate as the preceding form. The best development of this variety is shown by the specimens before me from San Diego. They are elevated, turbinated, strongly granose-lirate; the base is deeply eroded in front of the aperture; the color is brownish-yellow, with numerous, close, narrow, longitudinal, purplish-brown stripes, but the whole surface is so dingy that it appears uni-colored; the spiral lirae are subequal, the grains low and elongated in the direction of the lirae. The whorls are rounder than in *C. viridulum,* and the aperture decidedly smaller, and lacking green tinge on the columella. Alt., 22; diam., 22; alt., 14; diam., 18 mm. (Tryon.)

TYPE in ? Type locality, Mazatlan.

RANGE. Monterey, California, to Acapulco, Mexico. Fossil, Pleistocene, San Pedro.

Called *Trochus viridulus* var. *ligulatum* by Tryon.

## Tegula reticulata Wood, 1828

*Index Testaceologicus,* Supplement. Tryon and Pilsbry, *Manual of Conchology,* 11:176; Pl. 29, figs. 63, 64, 68, 69.

This form is more depressed than *C. viridulum* and smaller; the base is flatter; the periphery carinate; the lirae of the upper surface more regular, more distinctly beaded, the interliral interstices with fewer spiral striae, often with minute beaded lirulae. Coloration as in *C. viridulum,* consisting of radiating stripes. Aperture and columella tinged with green or white. Alt., 15; diam., 18–19 mm. (Tryon and Pilsbry, *Manual of Conchology.*)

TYPE in ? Type locality, Gulf of California.

RANGE. San Pedro, California, to Panama.

## Tegula mariana Dall, 1919

*Proceedings of the United States National Museum,* **56**:359. Tryon and Pilsbry, *Manual of Conchology,* **11**; Pl. 24, figs. 80–83.

This is *O. coronatus* Pilsbry, 1889, not of C. B. Adams, 1852, and *O. turbinatus* Pease, 1869, not of A. Adams, 1851. A new name for the species being needed, the above is proposed. (Dall.)

TYPE in United States National Museum, No. 4033. Type locality ?

RANGE. Santa Barbara Island to Payta, Peru.

Described as *Omphalius.*

## Genus **CALLIOSTOMA** Swainson, 1840

Shell imperforate or rarely umbilicate, conical, rather thin; whorls smooth, spirally ridged or granular, the last angulated at the periphery; aperture quadrangular; columella simple, usually ending anteriorly in a slight tooth. (Tryon and Pilsbry, *Manual of Conchology.*)

TYPE. *C. ziziphinum* L.

RANGE. Universally distributed.

## Calliostoma splendens Carpenter, 1864

### Plate 98, fig. 1

*Supplementary Report,* British Association for the Advancement of Science, 653. Tryon and Pilsbry, *Manual of Conchology,* **11**. *Proceedings of the California Academy of Sciences,* 3:156.

Orange-chestnut, with fleshy nacre; small, rather flattened, base glossy. (Carpenter.)

C. t. parva, latiore, tenuiore; exquisite rufo-castaneo et purpureo, interdum intensiorbus, et livido, varie nebulosa et punctata; anfractu primo nucleoso diaphano, granuloso, apice mamillato; dein iv. normalibus, subtabulatis; primo costibus spiralibus ii. acutis, valde expressis, altera parva suturali; anfr. penult. costis iii. quarum media extantior, superior subgranulosa; anfr. ult. aliis intercalantibus, supra peripheriam v. quarum tertia magis extans; interstitiis a lineis incrementi vix decussatis; costa circa peripheriam angulatam conspicua; basi costulis rotundatis, haud extantibus, peripheriam et axim versus conspicuis, medio saepe obsoletis; basi nitida, subplanata; apertura subquadrata, intus carneo-nacrea, valde splendente: operculo tenuissimo, levissimo, pallido, diaphano, concavo; anfr. circ. x. crebris, parum definitis. Long., 0.23; long. spir., 0.15; lat., 0.24 poll. (Carpenter.)

TYPE in State Collection, Species 530a. Type locality, Monterey, California.

RANGE. Monterey to San Diego, California.

## Calliostoma costatum Martyn, 1784

### Plate 97, figs. 8, 10

*Universal Conchology,* Table 1, Pl. 34. Tryon and Pilsbry, *Manual of Conchology,* 11:362; Pl. 16, figs. 6, 9; Pl. 18, fig. 16.

Martyn figured and named but did not describe.

Shell conical, rounded at periphery, base flattened; imperforate; solid; dark chestnut colored, the spiral riblets lighter, apex dark, usually purple. Surface encircled by numerous spiral smooth riblets, their interstices closely finely obliquely striate; riblets usually seven to nine on the penultimate whorl, about nine on the base. Spire conic; apex acute; sutures impressed. Whorls about seven, convex, the last rounded (or a trifle angled) around the lower part, slightly convex beneath; aperture rounded, oblique, outer lip fluted within, with a beveled opaque white submargin; throat pearly, iridescent; columella simple, arcuate. Alt., 20; diam., 18 mm.

TYPE in? Type locality?

RANGE. Sitka, Alaska, to San Diego, California. Fossil: Pleistocene, San Pedro to San Diego; Pliocene, San Fernando, Los Angeles County, California.

Described as *Trochus costatus* Martyn.

## Calliostoma costatum caeruleum Dall, 1919

*Proceedings of the United States National Museum,* 56:359.

Shell resembling the ordinary typical form except that the apex and part of the whorls in front of the suture are colored with bands of a brilliant mazarin blue when fresh, but which unfortunately fades after a few years in the cabinet. Fresh specimens have a very different aspect from the common shells carrying reddish spiral lines on a yellowish ground. (Dall.)

Typical specimens in United States National Museum, No. 59808. Type locality, Monterey, California.

RANGE. Neah Bay, Washington, to Monterey, California.

### Calliostoma costatum pictum Dall, 1919

*Proceedings of the United States National Museum,* **56**:359.

These shells resemble the type except that on the periphery of the whorl, and sometimes on the whorl between the periphery and the preceding suture, the shell is adorned with alternating light and dark patches or clouds of color. (Dall.)

Typical specimens in United States National Museum, No. 12612a. Type locality, Neah Bay, Washington.

RANGE. Neah Bay, Washington, to Monterey, California.

### Calliostoma annulatum Martyn, 1784
### Plate 97, figs. 1, 3

*Universal Conchology,* Table 1, Pl. 32. Tryon and Pilsbry, *Manual of Conchology,* 11:363; Pl. 67, fig. 43.

Shell elevated-conic, imperforate, rather thin; light yellow, dotted with brown on the spiral rows of grains, the periphery or lower edge of each whorl encircled by a zone of violet or magenta, the axis surrounded by a tract of the same. Surface with numerous granose lirae, about seven on the penultimate whorl, nine or ten on the base. Spire conical, apex acute, minute, reddish; sutures slightly impressed. Whorls about nine, slightly convex, the last angular at periphery, flattened beneath; aperture rhomboidal, oblique, fluted within. (Tryon.)

TYPE in? Type locality?

RANGE. Forrester Island, Alaska, to Catalina Island, California. Fossil: Pleistocene, San Pedro to San Diego, California; Pliocene, San Diego well.

Described as *Trochus annulatus* Martyn.

### Calliostoma canaliculatum Martyn, 1784
### Plate 97, figs. 4, 6

*Figures of Nondescript Shells,* Table 1, Pl. 32. Tryon and Pilsbry, *Manual of Conchology,* 11:361.

This is, perhaps, less variable than any other Californian species. The ribs have a slight tendency to become beaded, and always are more or less so on the apical whorls. The nucleus is heliciform, smooth and whitish-hyaline, comprising a whorl and a half. The ribs of the first three whorls are more or less beaded. Five radiating brown spots are to be seen on the apical whorls, and obsolete but similar markings may be seen on the others. The apex when eroded is light blue; the principal variations in color are in the darker or lighter brown of the interspaces. Northern specimens are usually darker and smaller. There is usually a small blue patch on the umbilical callus. The shell is distinguished from the sharp-ribbed specimens of *costatum* by the more pointed apex, stronger carina, and less rounded aperture. (Dall.)

TYPE in? Type locality?

RANGE. Sitka, Alaska, to San Diego, California.

### Calliostoma canaliculatum nebulosum Dall, 1919

*Proceedings of the United States National Museum,* **56**:360.

Shell resembling the typical form except that the whorls are painted with small brown nebulous patches in a radial fashion. (Dall.)

Typical specimens in United States National Museum, No. 159251. Type locality, San Diego, California.

RANGE. Known only from type locality.

### Calliostoma canaliculatum transliratum Dall, 1919

*Proceedings of the United States National Museum,* **56**:360.

Shell in which the channels between the yellowish spiral cords are of a dark reddish-brown, giving the shell a somber appearance, much in contrast with the light straw-colored type. (Dall.)

TYPE in United States National Museum, No. 160558. Type locality, Biorka Island, Sitka, Alaska.

RANGE. Known only from type locality.

### Calliostoma variegatum Carpenter, 1865

Plate 100, fig. 10

*Proceedings,* Philadelphia Academy of Natural Sciences, p. 61. Dall, *Proceedings of the United States National Museum,* **24**:552; Pl. 39, fig. 10.

C. t. parva, conica, variegata; nucleo rosaceo; anfr. vi. planatis, suturis haud impressis; costulis in spira iii. regularibus, nodulosis; nodulis albidis, subdistantibus; interstitiis elegantissime rosaceis; lirulis basalibus viii. haud nodulosis, rosaceo maculatis. Long., .24; long. spir., 13; lat., .21; div., 50°. (Carpenter.)

This species was described from a young shell dredged by Dr. Kennerley, and only a quarter of an inch (6 mm.) in height. It has long remained unique, but of late years the U.S. Fish Commission and the Seaside Laboratory of the University of California at San Pedro have obtained adult specimens which reach a height of 28 mm. and a width of 26 mm. The adolescent shell has the apical whorls rose color, the rest yellowish-white, with the alternate spirals stronger, and articulated with madder brown both on the spire and the base. As the shell gets fully adult the color becomes less lively and the articulation less distinct, so that the general tone of the shell appears to be of a yellowish pink with indications of the nacre shining through. (Dall.)

TYPE in United States National Museum, No. 122567. Type locality, Puget Sound, Washington.

RANGE. Puget Sound to Cerros Island.

Described as *C. (annulatum* var.) *variegatum* Carpenter.


## Calliostoma platinum Dall, 1889

*Proceedings of the United States National Museum,* **12**:343; Pl. 7, fig. 2.

Shell large, thin, polished, iridescent white, with seven whorls beside the nucleus; nucleus minute, lost; subsequent whorls slightly flattened behind the periphery, full and rounded on the base; longitudinal sculpture of obscure spiral lines behind the periphery and somewhat stronger flattish threads, separated by shallow grooves, on the base; at the periphery is a single prominent thread, immediately in front of which is the suture, the succeeding whorl being appressed against the thread; the single specimen obtained has a second prominent thread about two millimeters behind the peripheral one on the last whorl, but it is probable that the development of this thread was stimulated by an injury of which traces are plainly visible just before the second thread begins; base full and rounded; aperture rounded quadrate; the outer lip thin and sharp, its plane oblique and slightly flexuous; body with a very faint wash of callus; pillar slender, pearly, slightly arched, very little reflected, simple; interior polished, iridescent, without lirae; the external sculpture faintly perceptible owing to the tenuity of the shell. Max. longitude of shell, 32; maximum latitude, 29 mm.

TYPE in United States National Museum. Type locality, U.S.S. "Albatross" Station 2839, near the Santa Barbara Islands, California, in 414 fathoms, sand.

RANGE. Farallon Islands to San Diego, California, in deep water.

## Calliostoma gloriosum Dall, 1871

*American Journal of Conchology,* 7:127.

Shell six-whorled, acute, whorls gently rounded, with fine, revolving, thread-like ribs; four or five ribs near the suture granulated. Last whorl, roundly carinated, base flattened, with about twenty-five revolving striae. Columella thick, not reflected, but base somewhat grooved or depressed behind it. Aperture about one-third of the length of the whole shell, rhomboidal, pearly, smooth. Shell of a beautiful light salmon color, ornamented near the suture and carina with alternate patches of light yellow and chestnut brown. Alt., 1.1; maj. diam., .9 in. (Dall.)

Type in United States National Museum. Type locality, Soquel, north side of Monterey Bay, California.

Range. South of San Francisco to San Diego, California.

## Calliostoma antonii Koch, 1843

Philippi, *Abbildungen und Beschreibungen neue oder wenig gekannte Conchylien,* "Trochus," Pl. 1, fig. 4. Tryon and Pilsbry, *Manual of Conchology,* 11:365; Pl. 67, figs. 46–48.

Testa conica, solida, imperforata, soride carnea; anfractibus planis, seriebus granulorum 8 inaequalibus, secunda maxima, cinctis; basi cingulis granulorum frequentibus aspera. Alt., 9‴; diam., 11½‴. (Koch.)

Shell conical, solid, imperforate, dull flesh-colored, granulate; whorls flat, encircled by eight unequal series of granules, the second largest; base roughened by numerous granose cinguli. (Phil.)

The shell is thick, quite conical, and consists of seven to eight whorls, difficult to distinguish in the neighborhood of the apex. These are flat and apparently margined, for one is likely to take the second series of granules of the following whorl for a margin. The whorls show about eight rows of very pretty granules of three sizes; the upper, third, fifth, and seventh rows have the smallest granules, the second the largest; the fourth and sixth have middle-sized granules. The periphery consists of several closely crowded rows of the smallest size, and is rounded on the lower whorls. The base is slightly convex, with a multitude of granulose series, the granules becoming larger near the center, which is a semicircle, its cord being the columella, formed of inferior, gray nacre. The aperture is rhomboidal; the outer lip has a little distance within, a brown streak. The color is dirty flesh color, with a few very pale brown clouds and fewer dark brown points. Alt., 18; diam., 23 mm. (Phil.) (Tryon and Pilsbry, *Manual of Conchology.*)

Type in ? Type locality, not known to writer.

RANGE. Santa Barbara, California, to Panama.

Described and figured as *Trochus antonii* in Philippi, *Abbildungen und Beschreibungen neue oder wenig gekannte Conchylien,* "Trochus," Table I, fig. 4a.

## Calliostoma eximium Reeve, 1842

*Proceedings,* Zoölogical Society of London, 185. Tryon and Pilsbry, *Manual of Conchology,* 11:366; Pl. 63, figs. 84–86.

Troch. testa conico-pyramidali, pallide carneola, lineis transversis interruptis sparsim ornata; anfractibus superne concavis, in medio depressoplanis, bifuniculatis, funiculis albonigroque tessellatis; umbilico tecto. Alt., ⅞; diam., ¾ poll. (Reeve.)

Shell conical, imperforate, rather solid and strong, light yellowish or grayish, with irregular, bluish-black, longitudinal maculations and streaks, the base dotted or with small maculations; sometimes without dark flames, their place taken by obscure brownish clouding, the larger spiral cords both above and below articulated with deep red. There are about eight whorls, each one more or less markedly biangular at the circumference, the lower angle obtuse, concealed by the suture on the spire, the upper one acute, continuing nearly to the apex; whorls concave above, slightly excavated around the periphery, a little convex beneath; encircled by numerous unequal spiral threads, the larger ones beaded, the smaller irregularly crenulated by rather decided incremental striae. Base radiately striate, with about eight to twelve smooth spirals, their interstices without secondary riblets. Aperture oblique, rhomboidal; columella heavy, smooth, its face concave, obtusely subdentate at base. Alt., 24; diam., 23 mm. (Tryon and Pilsbry, *Manual of Conchology.*)

TYPE in Museum Cuming. Type locality, Payanam, in 10 fathoms.

RANGE. Catalina Island, California, to Mazatlan, Mexico.

## Calliostoma tricolor Gabb, 1865

*Proceedings of the California Academy of Sciences,* 3:186. Williamson, *Proceedings of the United States National Museum,* 15:201; Pl. 19, fig. 8.

C. t. conica, spira vix elevata; anfr. I nucleari, laevi; anfr. VI ad marginem truncatis, supra declivibus, infra planulatis; tota superficie confertim costata; costata; costis minute granulatis; colore fulvo. lineis purpureis albo maculatis, spiraliter fasciato; apertura subquadrata, intus margaritacea; labio crasso, labro acuto; umbilico albo. Long., .45; lat., .5; long. aper., .19, lat. aper., .24.

Shell conical, spire somewhat elevated, nuclear whorls smooth; other whorls sloping above, truncated on the margin, nearly flat below; surface marked by numerous, finely granulated, revolving ribs; color yellowish-brown, banded by a variable number of spiral purple lines, interrupted by white spots; aperture subquadrate, nacreous within; inner lip heavy, outer lip and base acute, umbilicus white. (Gabb.)

TYPE in California Geological Survey Cabinet, Mollusca, No. 602. Type locality, San Pedro, California.

RANGE. Santa Cruz to San Diego, California. Fossil: Pleistocene, San Pedro to San Diego, California; Pliocene, San Pedro.

## Calliostoma gemmulatum Carpenter, 1864

*Supplementary Report,* British Association for the Advancement of Science, 653. Tryon and Pilsbry, *Manual of Conchology,* 11:371; Pl. 67, fig. 54.

Very swollen; painted like *eximium;* with 2 principal and 2 smaller rows of granules. (Carpenter.)

Shell conic-elevated, solid but rather thin, imperforate, greenish-olive, with narrow irregular longitudinal blackish-olive stripes. Whorls about seven, of a rounded form, separated by deep sutures, encircled by three principal granulose carinae, the base and interstices with smaller lirulae and regular incremental striae; whorls of the spire with two strong carinae. Base rather flattened, with about 10 concentric lirae, dotted with brown. Aperture rounded-quadrangular, iridescent; columella pearly, iridescent, not truncate below, bounded outside by a whitish-yellow streak. Alt., 17; diam., 14 mm. (Tryon and Pilsbry, *Manual of Conchology.*)

TYPE in United States National Museum. Type locality, California.

RANGE. San Pedro, California, to Gulf of California.

## Calliostoma supragranosum Carpenter, 1864

*Supplementary Report,* British Association for the Advancement of Science, 653. Tryon and Pilsbry, *Manual of Conchology,* 11:369; Pl. 67, fig. 71.

Swollen, with sharp ribs; posterior 1–4 granular. (Carpenter.)

Shell small, conical, rather thin, imperforate, light chestnut-brown with a few short subsutural white flames and a peripheral circle of alternating chestnut and white spots, the ribs of the base minutely articulated with chestnut white. The spire is conical, short, composed of five convex whorls, the apical one very minute, smooth, whitish; the next two whorls are encircled by two strong, articulated ribs; on the next whorl these become beaded, and smaller beaded riblets appear above them; the last whorl has four (or five) strong, elevated ribs around the middle, above them two or three beaded ribs; the base has nine fine, distinct, smooth, concentric lirae. The last whorl is somewhat biangular at periphery, slightly convex beneath. Aperture nearly round, oblique; peristome thin, a trifle crenulated inside; the columella has a slight excavation, and is very bluntly nodulous near the base; its inner face is dark, pearly. Alt., 5; diam., 5½; alt., 7½; diam., 7½ mm. (Tryon and Pilsbry, *Manual of Conchology*.)

TYPE in ? Type locality, San Pedro, California.

RANGE. San Pedro to San Diego, California.

### Calliostoma turbinum Dall, 1895

#### Plate 100, fig. 1

*Proceedings of the United States National Museum,* **18**:8; also **24**; Pl. 39, fig. 1.

Shell small, margaritae-form, with six and a half rounded whorls; nucleus minute, white, smooth, of one whorl, followed by strongly sculptured, rather inflated whorls separated by an inconspicuous suture; sculpture on the spire of rather elevated, narrow, spiral ridges, of which the most posterior is always beaded, though the beading on the others fails on the apical whorls; in front of this ridge is a smaller one, then three, or on the last whorl five, subequal, larger ones, the third forming the periphery of the whorl, the suture being laid against the most anterior ridge; the base has about twelve, subequal, more crowded, spiral threads, faintly or not at all beaded, larger toward the axis; the body of the shell is of a nacreous waxen tint, with transverse flammules of dark brown, which articulate the spirals, are much fainter on the interspaces, but do not reach the base, on which the spirals are more or less articulate with reddish brown; the base is somewhat flattened, the periphery not keeled, the pillar short, white, with a minute umbilical chink; aperture subquadrate, nacreous, sulcate by the external sculpture; there is no projection at the distal end of the pillar. Height, 12; major diameter, 12.5 mm. (Dall.)

TYPE in United States National Museum, No. 122578. Type locality, U.S. Fish Commission Stations 2902 and 2972, among the Santa Barbara Islands.

RANGE. Point Conception to San Diego and the islands.

### Genus TURCICULA Dall, 1881

Shell globosely conical, white, thin; umbilicus reduced to a chink under the thin callus of the upper part of the pillar lip; mouth rounded rectangular, margins all thin; columella concave; outer surface with tuberculose ridges. (Tryon and Pilsbry, *Manual of Conchology*.)

TYPE. *Margarita* (*Turcicula*) *imperialis* Dall, 1881.

DISTRIBUTION. Atlantic, eastern Pacific, and Japanese seas. Fossil: Tertiary of the Pacific coast of North America.

## Turcicula bairdii Dall, 1889

*Bulletin,* Museum of Comparative Zoölogy, **18**:377. Dall, *Proceedings of the United States National Museum,* **12**:346; Pl. 7, fig. 3.

Dall named this species and described the soft parts in *Bulletin,* Museum of Comparative Zoölogy, **18**; the shell was described in the *Report on the Voyage of the "Albatross."*

Shell large, turbinate, elevated, thin, inflated, with four and a half or five whorls, of which the last is much the largest; surface apt to be eroded, but where perfect covered with an extremely thin, dense, vernicose, pale apple-green epidermis; whorls inflated; suture deep, not channeled; apex moderately pointed; spiral sculpture of (1) numerous fine, faint, rather irregular scratches or impressed lines; (2) sparse, slightly elevated, revolving bands which are usually more or less nodulous, the nodules when prominent being sharp and laterally flattened as if pinched up; of these there are on the upper whorls usually three series between the sutures, of which one at the periphery is the most prominent and persistent, the next one behind it, halfway between the periphery and the suture, being the least marked; on the base the cinguli are six or seven in number, becoming narrower toward the axis, smaller than those behind the suture, with smaller, less prominent, rounder, and more numerous nodules; there is some variation in number and strength of all the cinguli, but that on the periphery is the most prominent and constant; the whorls are particularly round and inflated above and below, so that the outline of the aperture is often nearly circular; interior of the aperture brilliantly pearly, a thin wash of callus on the body; the outer lip very slightly thickened and distinctly reflected in the adult; pillar thin, simple, arching roundly into the curve of the base without any interruption, angle, or tooth; axis imperforate; the external sculpture showing through the thin shell. Altitude, 50; maximum diameter, 42 mm. Maximum diameter of operculum, 18 mm., with about 12 whorls. The operculum is externally polished, smooth, and deeply concave; the inner side presents a minute, central, rounded, elevated point; the margin is very thin but entire.

TYPE in United States National Museum. Type locality, U.S.S. "Albatross" Station 2839, off San Clemente Island, California, in 414 fathoms, sand; bottom temperature not registered.

## Genus TURCICA A. Adams, 1854

Shell conoidal, thin, subdiaphanous, imperforate; whorls with transverse series of granules, the last rounded on the periphery; columella thick, spirally twisted posteriorly, ending anteriorly in an obtuse, prominent point; outer lip thin, simple, acute. (H. and A. Adams.) (Tryon and Pilsbry, *Manual of Conchology.*)

TYPE. *T. monilifera* A. Adams.

DISTRIBUTION. Australia, Japan, China, California, Philippines, Lord Hoods Island, Sea of Korea.

### Turcica caffea Gabb, 1865

*Proceedings of the California Academy of Sciences, 3:187. Paleontology of California, 2:16; Pl. 3, fig. 27.*

T. conoidea, tenui; spira elevata; anfr. 1 nucleari, laevi, anfr. V ad marginem subangulatis, supra recte declivis, ultimo infra sub-planato; sutura valde impressa; tota superficie concentrice tuberculato costata; costis circitur XVIII ad ult. anfr.; epidermide flavida; apertura oblique sub-quadrata, intus albida. (Gabb.)

Shell conical, thin; spire, elevated; one nuclear whorl, smooth, five perfect whorls subangulated at the margin, obliquely flat above; body whorl very slightly convex at the base; suture strongly impressed; surface marked by revolving tubercular ribs, about eighteen on the body whorl, one broad one on the margin, and about eight or nine on the under side; these ribs are less numerous on the preceding volutions, only four can be counted on the upper side of the penultimate whorl; the tubercles are arranged so as to present an irregular quincunix; epidermis a rich coffee brown, darker between the tubercles than on their summits; aperture obliquely quadrangular, outer lip and base acute, internally pearly white, columella bearing two strong folds, the lower one of which borders the truncated end and terminates in a faint tubercular enlargement; behind the columella is a slight groove running from the base to the end of the upper columellar fold. Long., 55; lat., 58; long. aper., 26; lat. aper., 28 poll. (Gabb.)

TYPE in Mollusca, Survey Collection, No. 355. Type locality, California.

RANGE. San Pedro, California, to Cape San Lucas, Lower California.

## Genus CIDARINA Dall, 1902

Shell large, whitish, unicolor, with strong spiral sculpture sharply nodose, the umbilicus closed by a reflected layer of callus, the suture channeled. (Dall.)

TYPE. *Margarites cidaris* A. Adams.

DISTRIBUTION. Deep waters of the west coast of North America.

## Cidarina cidaris A. Adams, 1864

Plate 91, fig. 7

*Annals and Magazine of Natural History*, ser. 3, **14**:426. Dall, *Proceedings of the United States National Museum*, **12**; Pl. 2, fig. 6.

M. testa magna, conica, Turcicoidea, tenui; albido-cinerea, nacreo-argentato; anfr. nucleosis? . . . . (decollatis), norm. vii., subplanatis; suturis alte insculptis; superficie spirae tota valde tuberculosa, seriebus tribus, alteris postea intercalantibus; peripheria et basi rotundatis, carinatis; carinatis; carinis circ. viii., haud acutis, irregularibus, scabris, haud tuberculosis; lacuna umbilicali vix conspicua; apertura subrotundata; labro tenuissimo; labio obsoleto; columella arcuata. Long., 1.1; long. spir., .65; lat., .75; div., 60.

Shell conical; spire elevated, subacute; whorls six to seven; four upper whorls only slightly convex; lower whorls very convex; surface ornamented with spiral rows of nodes which grade into nodose ribs on the lower side of the body-whorl; suture deeply impressed; aperture circular; outer lip thin; inner lip and columella enameled, the incrustation completely obscuring the small umbilicus in most specimens. Alt., 40; lat., 33; body-whorl, 26.5; aperture, 13 mm.; defl., 66°. (Arnold.)

TYPE in United States National Museum. Type locality, vicinity of Neah Bay, Washington.

RANGE. Kasa-an Bay, Alaska, to Cape San Quentin, Lower California. Fossil: Pleistocene, San Marcial; Pliocene, San Pedro.

Described as *Margarites cidaris*.

## SECTION SOLARICIDA Dall, 1902

### Cidarina carlotta Dall, 1902

*Proceedings of the United States National Museum*, **24**:553.

Shell rather depressed, pearly white, covered with a dense, rather fibrous, olive-gray periostracum; nuclear whorls eroded, but the shell exhibits about four and a half whorls; sculpture of, on the base eight minutely distantly nodulous spiral threads stronger and more distant as one proceeds from the verge of the umbilicus to the periphery; peripheral spiral separated from another above it by an excavated channel; these two are the strongest on the shell, and between the upper one and the suture is another much feebler thread; the upper two are all that show on the spire, as the outer lip runs just above the peripheral thread; the radial sculpture comprises incremental lines, and on the last whorl about twenty low narrow somewhat oblique riblets about a millimeter apart, extending from the suture to the first peripheral keel, but not beyond; these riblets nodulate the weak spiral, but are only about half as numerous as the nodules on the peripheral spirals; suture distinct, not channeled; base rounded; the umbilicus funicular, of moderate size, bounded by an inconspicuous keel, above which the walls are vertically striated; margins of the aperture simple, sharp, the upper lip advancing where it joins the body; pillar lip thin, slightly excavated, the distal angle not prominent. Alt., 9.0; max. diam., 13.5 mm. (Dall.)

TYPE in United States National Museum, No. 109020. Type locality, U.S.S. "Albatross" Station 3342, off the Queen Charlotte Islands, in 1,588 fathoms, ooze, bottom temperature 36°3 F.

RANGE. Known only from type locality.

Described as *Solariella carlotta* Dall.

### Cidarina ceratophora Dall, 1895

*Proceedings of the United States National Museum,* **18**:9. *Bulletin,* Museum of Comparative Zoölogy, **43**:350; Pl. 3, fig. 2.

Shell thin, with a pale olive, silky epidermis, and six whorls besides the (decollate) nucleus; early whorls smooth, gradually taking on two rows of projecting points or sharp nodules, which are, on the later whorls, connected by a slender spiral thread; periphery with a slender granular thread, on which the suture is laid; base with five similar threads, closer as they approach the umbilicus; umbilicus small, vertically striate; aperture rounded, slightly angulated by the sculpture; the outer lip thin, sharp; inner reflected over part of the umbilicus. Height, 28; diameter, 24 mm. (Dall.)

TYPE in United States National Museum, No. 122960. Type locality, U.S. Fish Commission Station 3432, in 1,421 fathoms, off La Paz, Gulf of California.

RANGE. San Diego, California, to Mazatlan, Mexico, in deep water.

Described as *Solariella ceratophora.*

### Cidarina equatorialis Dall, 1908

*Bulletin,* Museum of Comparative Zoölogy, **43**:351; Pl. 5, fig. 11.

Shell thin, pearly, the nacre shining through the translucent outer coating, and a pale yellowish, axially striated, silky periostracum; whorls six, exclusive of the (lost) nucleus; suture distinct, not channeled; whorl in front of it horizontal with a fine spiral thread at a short distance, giving a somewhat tabulate effect; somewhat more distant is a second stronger thread at the shoulder, and a similar one at the periphery, while a fourth forms the margin of the base against which the suture is laid; on the base are four more similar threads, becoming gradually more adjacent and feebly beaded or nearly simple, except the fourth, which is distinctly, minutely beaded, while a fifth, forming the brink of the large, wide, and deep umbilicus, is even more strongly beaded; axial sculpture of minute, feeble radial wrinkles which at their intersection with the posterior thread crenulate it, and on the second and third threads produce, at intervals of about a millimeter, sharp, triangular, subspinose nodules; the entire shell is covered with axial, fine, retractive, silky striation; whorls full, base rounded; umbilicus very deep, funicular, the walls axially striated; aperture rounded-quadrate, the margin thin; pillar oblique, slightly excavated, not callous, slightly expanded; body and throat pearly. Alt., 21; of last whorl, 14; of aperture, 8.5; maximum diameter of shell, 19.5; of umbilicus, 5.5 mm. (Dall.)

TYPE in United States National Museum, No. 125964. Type locality, U.S.S. "Albatross" Station 3376 off the Ecuador coast, in 1,132 fathoms.

RANGE. San Diego, California, to Ecuador.

Described as *Solariella equatorialis.*

## Genus **SOLARIELLA** S. Wood, 1842

Shell umbilicated, conical; whorls with spiral granose lirae; umbilicus with carinated margin. (Tryon and Pilsbry, *Manual of Conchology.*)

TYPE. *S. maculatum* Wood.

DISTRIBUTION. Arctic to California, circumboreal.

### **Solariella peramabilis** Carpenter, 1864
Plate 91, fig. 8; Plate 101, fig. 7

*Supplementary Report,* British Association for the Advancement of Science, 653. Tryon and Pilsbry, *Manual of Conchology,* 11:312; Pl. 67, figs. 59–61.

Shell very thin, very elegantly sculptured, livid, spotted with pale rufous-brown; nuclear whorls 2, very tumid, smooth, apex mamillated; following whorls 4, tabulated, sutures nearly rectangular; upon the spire there are two or three carinae, and intercalated carinulae; the entire surface is most elegantly and densely radiately costate, costae very acute, subgranulose upon the carinae, interstices on the first whorl fenestrated, posteriorly decussated; base deeply rounded; sculptured with about 5 lirulae, anteriorly granulose; and radiating costulations continued from the base. Aperture iridescent inside, nacreous; operculum very thin, multi-spiral, with about 10 elegantly radiately rugulose whorls. (Carpenter.) Alt., 8; diam., 8 mm. (Tryon and Pilsbry, *Manual of Conchology.*)

TYPE in ? Type locality, California.

RANGE. Forrester Island, Alaska, to San Diego and the Coronado Islands. Also Japan. Fossil: Pliocene, San Pedro.

## Solariella triplostephanus Dall, 1910

*Nautilus,* **24**:34. Dall, *Bulletin 112,* United States National Museum; Pl. 18, figs. 1, 2.

Shell trochiform, with six tabulate whorls; nucleus very minute, glassy, slightly tilted; subsequent whorls flat above, with closely appressed suture; three strong spirals girdle the whorls; one at the shoulder strongly beaded; one at the middle of the whorl minutely undulate, and the third at the suture, simple, and obscured on all the whorls but the last by the suture being laid against it; on the last whorl there may be a few microscopic spiral threadlets between the shoulder and the median spiral; between the anterior spiral and the edge of the umbilicus on the base are six or eight fine-channeled spiral grooves; the chord bordering the funicular umbilicus is coarsely beaded; within the umbilicus are three or more similar but smaller beaded threads; axial sculpture consisting of fine, sharp, uniform, and closely set elevated lines corresponding with the lines of growth, but frequently more or less obsolete; aperture nearly circular, oblique, with simple edges, hardly interrupted on the body; throat pearly. Height of shell, 5.25; of aperture, 2.5; maximum diameter of shell, 7.0 mm. (Dall.)

TYPE in United States National Museum, No. 97001. Type locality, near La Paz, Lower California, in 12 fathoms, sand.

RANGE. Off San Diego, California, and south to Panama, in deep water.

## Solariella rhyssa Dall, 1919

*Proceedings of the United States National Museum,* **56**:360.

Shell small, trochiform, translucent, with four whorls, including a smooth nucleus of a whorl and a half; suture distinct; whorl-section circular; axial sculpture of (on the last whorl 35) narrow threadlike ribs, regularly spaced, uniform, the interspaces wider; extending from the suture to the verge of the umbilicus; spiral sculpture of minute spiral threads showing in the interspaces under a lens, but not modifying the axial ribs; umbilicus wide, funicular; aperture circular, simple, sharp-edged; operculum brown, multispiral, horny; height, 1.7; longer diameter, 2 mm. (Dall.)

TYPE in United States National Museum, No. 173803. Type locality, Catalina Channel, California.

RANGE. Catalina Channel, California.

## Solariella nuda Dall, 1895

### Plate 91, figs. 2, 5

*Proceedings of the United States National Museum,* 18:9. *Bulletin,* Museum of Comparative Zoölogy, 43: Pl. 3, figs. 5, 7.

Shell turbinate, recalling *Margarita,* smooth, polished, except for obscure spiral markings which do not interrupt the surface, of about four whorls; color white, with a pink or blue nacre glowing through; whorls rounded, flattened in front of the suture; base rounded; umbilical margin keeled; umbilicus wide, funicular; aperture rounded, oblique, hardly angulate by the umbilical rib, and with a very short interruption between the inner and outer lips; operculum light brown, thin, with about ten whorls. Height, 15; major diameter, 19; minor diameter, 15.5 mm. (Dall.)

TYPE in United States National Museum, No. 122580. Type locality, U.S. Fish Commission Stations 2928, 3187, and 3348, in 298 to 455 fathoms off Lower California.

RANGE. Monterey, California, to Clarion Island.

## Subgenus MACHAEROPLAX Fride, 1877

### Solariella varicosa Mighels and Adams, 1842

*Boston Journal of Natural History,* 4:46; Pl. 4, fig. 14.

M. testa parva, tenui, conica; anfractibus quatuor, convexis; longitudinaliter costulatis, transverse striatis; sutura subcanaliculata; umbilico magno, profundo.

Shell small, thin, low, conical, of a dingy white or drab color; whorls four, convex, covered with numerous longitudinal, oblique ribs, intersected by a great number of transverse, revolving striae, which are most conspicuous on the lower part and base of the lower whorl. The striae on the upper part of the whorls can only be seen with a magnifier. Suture distinct, subcanaliculate; umbilicus rather large and deep, bounded by two rather rugged varices, intersected by the ribs which are continued to the verge of the umbilicus; aperture circular; labrum simple, sharp; within perlaceous. Height, .25 inch; diameter of base equal to the height; divergence, 90°. (Mighels and Adams.)

TYPE in cabinet of J. W. Mighels. Type locality, Bay Chaleur.
RANGE. Arctic Ocean to San Diego, California. Also Atlantic.
Described as *Margarita*.

## Solariella obscura Couthouy, 1838

*Boston Journal of Natural History,* 2:100; Pl. 3, fig. 2.

T. testa subconica, perforata, fusco-rubente; anfractibus convexis,
superne leviter angulatis; basi convexa, apertura rotunda, labro tenui,
laevi, intus iridescente; operculo corneo. (Couthouy.)

Diameter of axis, $\frac{5}{20}$; of basis, $\frac{7}{20}$ in.

Shell sub-conical, thin, color an obscure reddish-brown, whorls five in
number, convex, traversed by numerous, very minute, revolving striae,
intersected by almost imperceptible longitudinal ones; a single slightly
elevated line or rib revolves a short distance below the sutures, which are
tolerably well defined; base convex, concentric basal striae scarcely perceptible, transverse striae very apparent; umbilicus moderately wide and
extending nearly to the apex; aperture circular, lip sharp, smooth internally, and slightly reflected upon the umbilicus. Interior iridescent; operculum thin, horny, concentrically spiral. (Couthouy.)

TYPE in Couthouy Collection. Type locality, Massachusetts Bay.
RANGE. Arctic Ocean to Straits of Juan de Fuca. Circumboreal.
Described as *Turbo obscurus* Couthouy.

## Solariella paupercula Dall, 1919

*Proceedings of the United States National Museum,* 56:363.

Shell coarse, usually more or less eroded, of a yellowish white over a
brilliant nacre, with five or more moderately rounded whorls, the nucleus
eroded; suture distinct, not appressed; axial sculpture of numerous somewhat irregular narrow close-set wrinkles, extending over the whorl from
the suture to the verge of the funicular umbilicus; spiral sculpture of a
few spiral lines near the umbilicus; aperture subcircular, oblique, produced
above, the lips joined over the body by a layer of enamel; the operculum
dark brown, multispiral, with 8 or 10 turns; height of shell, 9.5; of
aperture, 4.5; maximum diameter, 11 mm. (Dall.)

TYPE in United States National Museum, No. 109457. Type locality,
Arctic Ocean, north of Bering Straits.
RANGE. Arctic Ocean to the Aleutian Islands.
Described as *Margarites pauperculus* Dall.

## Genus **MARGARITES** Leach, 1847

Shell thin, globular-conical, umbilicated; whorls rounded, smooth; aperture rounded, pearly; lip sharp, smooth. (Tryon, *Structural and Systematic Conchology.*)

TYPE. *M. helicina Fabr.*

DISTRIBUTION. Boreal.

Described as *Margarita.*

### Margarites parcipicta Carpenter, 1864

### Plate 101, fig. 4

*Supplementary Report,* British Association for the Advancement of Science, 653. *Proceedings of the California Academy of Sciences,* 3:157. Dall, *Bulletin 112,* United States National Museum; Pl. 17, fig. 3.

Like strong growth of *Margarites lirulata* var. Carpenter.

G. testa solidiore, parva, conica, pallida, purpureo-fusco varie nebulosa et maculata; anfr. v., rotundatis; carinis ii. validis in spira se monstrantibus, minore intercalante; interstitiis subsuturalibus, sublaevibus, inter carinas obtuse decussatis; lira peripherica definita, saepe in spira se monstrante; basi valde rotundata; lirulis basalibus circ. v. rotundatis, subdistantibus; apertura subcirculari; columella arcuata; umbilico majore, infundibuliformi, haud angulato. Long., .14; long. spir., .07; lat., .13 poll.

With rather solid shell, small, conical, pale, nebulous variety mottled with tawny purple and spotted; with five rotund whorls; with two carinae showing strongly on the spire, with a smaller one intercalating; with subsutural interstices nearly smooth, obtusely crossed between the carinae; the peripheral ridge is defined, often showing itself on the spire; with base strongly rotund, with about five rounded sub-distant basal lirulae; aperture subcircular; columella arcuate; umbilicus rather large, funnel-shaped, not angulate. (Translation.)

TYPE in United States National Museum. Type locality, Neah Bay, Washington.

RANGE. Sitka, Alaska, to San Diego, California.

Described as *Gibbula.*

### Margarites funiculata Carpenter, 1864

*Supplementary Report,* British Association for the Advancement of Science, 653. *Annals and Magazine of Natural History,* ser. 3, 14:425. Dall, *Bulletin 112,* United States National Museum; Pl. 18, fig. 10.

Shaped like *Montagui:* with rounded spiral riblets. (Carpenter.)

G. testa parva, elevata, compacta, fusca; marginibus spirae excurvatis;

anfr. vi., haud tumidis, suturis parum impressis; lirulis crebris rotundatis undique cincta, quarum v. in spira monstrantur; interstitiis parvis; basi rotundata, haud angulata; umbilico parvo, haud carinato; apertura suborbiculari, parum declivi; columella vix arcuata. Long., .24; long. spir., .11; lat., .2; div., 70°. (Carpenter.)

Shell small, elevated, compact, tawny; with excavate margins spiral, with six whorls, not thick, with sutures little impressed; girdled with frequent rounded lirulae, of which five show on the spire; with small interstices; with rounded base, not angulate; with small umbilicus, not crenate; with suborbicular aperture; little sloping; columella scarcely arcuate. (Translation.)

Type in United States National Museum. Type locality, Neah Bay, Washington.

Range. Known only from type locality.

Described as *Gibbula* Carpenter.

## Subgenus Pupillaria Dall, 1909

### Margarites pupilla Gould, 1849

#### Plate 101, figs. 2, 3

*Proceedings of the Boston Society of Natural History,* 3:91. *Mollusca and Shells of the United States Exploring Expedition,* 12:186, fig. 208.

Testa parva, elevato-conica, margaritacea extrinsecus incana, filis virido-fuscis ubique cincta, ad intervallos minutissime clathrata: spira anfr. 6 convexis: basis planiuscula, fissura umbilicali perforata: apertura circularis columella arcuata: faux fulgida, minutissime punctata. Lat., ¼; alt., ³⁄₁₀ poll. (Gould.)

Shell small, ovate-conic, rather solid, perforate, ash-colored, with darker greenish on the ribs, sub-surface brilliant silvery; surface with small, flattened, nearly equal and equidistant ribs, about five on the upper whorls separated by interspaces of the same width, and with fine, crowded, lamellar lines of growth, by which the interspaces are distinctly barred: spire of six convex whorls, the last obtusely angular, flattened at base, and with much finer and more crowded ribs and grooves. Aperture circular; columella somewhat arcuate, with a minute, groove-like umbilicus at its side; lip sharp; interior pale and opaque near the lip; minutely punctured, and with crimson iridescence within. (Gould.)

Type in United States National Museum. Type locality, New Zealand. Gould in his description of this species gives New Zealand as the type locality. It must have been a ballast shell.

RANGE. Nunivak Island, Bering Sea, to San Pedro, California; San Diego, in deep water.

Described as *Trochus pupillus* Gould.

## Margarites cinerea Couthouy, 1838

*Journal,* Boston Society of Natural History, 2:99; Pl. 3, fig. 9.

T. testa pyramidali, tenui, cinerea; anfractibus convexiusculis, costellis numerosis cinctis, longitudinaliter tenuissime striatis; basi subconvexa, perforata; labro tenui, crenulato; intus margaritacea; operculo corneo.

Diameter of axis, nine-twentieths; of basis, eight-twentieths of an inch. Shell conical or pyramidal, thin, uniform ash color, whorls slightly convex, varying from five to seven; sutures distinctly marked; the lower whorl has four and sometimes five or six revolving, elevated striae or small ribs, which diminish in number as they approach the apex; the most central of these is the largest, and gives rather a carinated aspect to the middle portion of the whorls. The shell is longitudinally traversed by closely set, delicate, sub-laminar, oblique striae, not interrupted by the costae, and giving two or three of the apical whorls something of a nodulous or gemmulated appearance. Base slightly convex, with transverse and concentric striae, like the whorls, and perforated about half the length of the shell. Aperture circular, slightly angulated by the junction of the outer lip, which is crenulated by the termination of the striae and slightly reflected in its columellar portion, so as to conceal a small part of the umbilicus, in adult specimens. Interior pearly; operculum horny, transparent and multispiral. (Couthouy.)

TYPE in Couthouy Collection. Type locality, Massachusetts Bay.

RANGE. Bering Strait to Port Etches, Alaska. Circumboreal.

Described as *Turbo cinereus* Couthouy.

## Margarites vorticifera Dall, 1873
### Plate 100, figs. 7, 8

*Proceedings of the California Academy of Sciences,* 5:59; Pl. 2, fig. 4.

Shell depressed, with three flattened, rapidly expanding whorls, which have a tendency, in old individuals, to overhang the suture anterior to them. The upper surface is traversed by numerous slender, slightly elevated, revolving threads, which are crossed by faint lines of growth. Outer edge of whorls subcarinate. The basal surface is less flattened, but similarly sculptured, except that the very wide and funnel-shaped umbilicus is destitute of revolving striae, and the lines of growth are here a little stronger. Aperture excessively oblique, with the anterior angle much produced; lips hardly thickened, and but slightly interrupted at the junction with the body whorl. Nacre, salmon-color; external surface pinkish-white, brilliantly pearly where eroded. Lat. of largest specimen, 0.85 in.; alt., 0.5 in.; defl., 88°. (Dall.)

TYPE in ? Type locality, Iliuliuk Harbor, Captain's Bay, Unalaska, Alaska.

RANGE. Bering Strait and Sea, south and east to Unalaska, Alaska.

### Margarites vorticifera sharpii Pilsbry, 1898

*Proceedings,* Philadelphia Academy of Natural Sciences, 486.

Shell thin, of low-conoid form, with extremely broad funnel-shaped umbilicus and very rapidly expanding whorls. Color, dull salmon or brick-pink, becoming ashy on the spire and within the umbilicus. Sculpture, numerous spiral cords and threads, which on the spire are strong, alternately smaller, then with a tertiary series intercalated, the whole becoming less pronounced on the last whorl, where by further intercalation of threads the spirals become very numerous in some individuals, and in others mostly obsolete; the base with close, strong spiral cords outside the edge of the umbilicus; the whole surface with fine, crowded, and somewhat lamellar growth-striae, the spire with some spaced coarser radial riblets. Whorls 4½, very rapidly expanding, the last at the aperture about three times the width of the preceding (seen from above); gently convex; periphery angular; base convex, the umbilical region broadly excavated nearly as large as the aperture. Aperture large, very oblique, salmon colored within, with brilliant green reflections, but having a wide border within the lip appearing dull whitish from in front, but showing red and white reflections seen from below. Peristome thin, deeply excised in the umbilical region, above the excision produced forward as a low wall curved around the umbilical edge, continuing as far as the posterior termination of the outer lip. Alt., 7.5; greater diam., 14; lesser, 11 mm. (Pilsbry.)

TYPE in Academy of Natural Sciences, Philadelphia, No. 70554. Type locality, Dutch Harbor, Unalaska.

RANGE. Unalaska to Port Althorp, Alaska.

Described as *Margarita sharpii* Pilsbry.

### Margarites vorticifera ecarinata Dall, 1919

*Scientific Results of the Canadian Arctic Expedition,* 8:22A; Pl. 2, fig. 5.

Shell pinkish-gray, depressed, with about five rapidly enlarging whorls; nucleus minute, glassy; subsequent whorls moderately inflated, separated by deep but not channeled suture, having a rounded periphery, a wide, completely pervious umbilicus and a large, very oblique, iridescent aperture. Axial sculpture of very fine silky incremental lines; spiral sculpture of low flattish threads separated by narrower interspaces sometimes carrying a finer intercalary thread; this sculpture is carried over the base but is absent from the walls of the wide umbilicus; aperture rounded, very oblique, the margins sharp, hardly meeting over the body except by a thin layer of enamel. Operculum brown, thin, multispiral. Height of shell, 8; max. diameter, 15 mm. (Dall.)

TYPE in Museum at Ottawa, No. 4071. Type locality, Station 22, in N. Latitude 69° 35 ′ and W. Longitude 163° 27′, in 11–12 fathoms.

RANGE. Southwest of Point Barrow, Arctic Ocean, and south and east to the Aleutians.

Described as *Margarites ecarinatus* Dall.

## Margarites salmonea Carpenter, 1864

*Supplementary Report,* British Association for the Advancement of Science, 653. *Proceedings of the California Academy of Sciences,* 3:158. Dall, *Bulletin 112,* United States National Museum; Pl. 18, figs. 6, 9. Tryon and Pilsbry, *Manual of Conchology,* 11:295.

Between pupilla and undulata: salmon-tinted, sculpture fine, not decussated; sutures not waved. 6–40 fm. (Carpenter.)

M. t. inter *M. undulatae* et *M. pupillae* intermedia; minore, spira satis elevata; anfr. nucl. iii. purpureis; dein iv. normalibus, colore salmoneo; liris spiralibus in spira viii., quarum ii. suturales, minimae; suturis haud undulatis; interstitiis a lineis incrementi creberrimis, haud elevatis, sculptis; basi lirulis creberrimis, aequalibus, circ. xviii. ornata; apertura subquadrata; umbilico minore, angulato; operculo tenuissimo, diaphano, anfr. circ. x. vix definitis. Long., 0.22; long. spir., 0.14; lat., 0.22; div. 80°. (Carpenter.)

Si ·ll intermediate between *M. undulata* and *M. pupilla;* smaller, with spire well elevated; with three purplish nuclear whorls; then four normal ones, of salmon color; with eight spiral lirae on the spire, of which two are very small, on the suture; sutures not undulate; with very numerous interstices on the lines of growth, not raised, incised; basis ornamented with numerous (about 18) equal lirulae; aperture subquadrate; umbilicus rather small, angulate; with very thin translucent operculum, with about 10 scarcely defined whorls. (Translation.)

TYPE in State Collection, Species 352. Type locality, Monterey, California.

RANGE. Known only from the type locality.

## Margarites rhodia Dall, 1920

Plate 101, fig. 5

*Proceedings,* Philadelphia Academy of Natural Sciences, **62.**

New name for *Margarita inflata* Carpenter, 1865.

M. t. tumida, tenui, albida, narceo pallide aureo; anfr. vi. valde inflatis, suturis ad angulum fere rectum impressis; tota superficie tenuissime spiraliter lirulata; lirulis acutis, haud elevatis; in spira circ. viii., minoribus saepe intercalantibus; interstitiis a lineis incrementi extantibus creberrimis tenuissime decussatis; basi obtuse subangulata, striis creberrimis circ. xx. ornata; apertura subquadrata; columella arcuata; umbilico infundibulifori, laevi, angulato: operculo tenui, planato, suturis distinctis. Long., .44; long. spir., .22; lat., .45; div., 85°. (Carpenter's description of *M. inflata.*)

TYPE in United States National Museum. Type locality, Puget Sound.

RANGE. Port Althorp, Alaska, to San Diego, California.

## Margarites rudis Dall, 1919

*Proceedings of the United States National Museum,* **56**:364. Dall, *Bulletin 112,* United States National Museum; Pl. 18, figs. 13, 14.

Shell of moderate size, white, with a pale olivaceous periostracum, a smooth nucleus of about one whorl and five subsequent whorls; spiral sculpture of two strong cords with wider interspaces and a third on which the suture is laid and which forms the margin of the base; there is also a small thread between the suture and the posterior cord and on the last whorl a similar thread in the interspaces; on the base there are six or seven smaller closer cords separated by obscurely channeled interspaces between the verge of a narrow umbilicus and the basal margin; axial sculpture of (on the penultimate whorl about 20) retractive riblets extending from suture to periphery, with wider interspaces, slightly nodulous at the intersections with the spiral cords; there are also close, obvious, incremental, regular lines over the whole surface; aperture rounded quadrate, simple, a glaze on the body, the pillar lip slightly thickened; operculum multispiral. Height, 12; diameter, 12.5 mm. (Dall.)

TYPE in United States National Museum, No. 213951. Type locality, Coal Harbor, Shumagins, Alaska, in 8 fathoms.

RANGE. Bering Sea to Cook's Inlet, Alaska; also Kamchatka.

## Margarites healyi Dall, 1919

*Proceedings of the United States National Museum,* **56**:363.

Shell large, livid yellowish white, with a thin very pale periostracum and about six whorls including a smooth white turbinate nucleus of two moderately rounded whorls; suture distinct, not appressed, the whorls between not inflated; axial sculpture of oblique quite incremental lines, occasionally developed into minute wrinkles; spiral sculpture of rather irregularly spaced low threads, about a dozen on the penultimate whorl, with wider interspaces; especially near the periphery on the spire; on the last whorl, including the base, they are more numerous and closer; base slightly flattened with a narrow, perforate unbilicus; aperture oblique; simple, nacreous; body with a coat of enamel; pillar straight, slightly callous, with no projection at the base; height of shell, 20.5; of last whorl, 15; maximum diameter, 20 mm. (Dall.)

TYPE in United States National Museum, No. 223801. Type locality, Arctic Ocean, north of Bering Strait, Station 10 of U.S.S. "Corwin."

RANGE. Arctic Ocean north of Bering Strait.

### Margarites simbla Dall, 1913

*Proceedings of the United States National Museum,* **45**:592. Dall, *Bulletin 112,* United States National Museum; Pl. 18, fig. 3.

Shell pale gray, beehive-shaped, with a blunt apex and five and a half rapidly enlarging convex whorls; nucleus minute; subsequent whorls polished, finely spirally striate, crossed by very fine flexuous striae corresponding to the lines of growth, which more or less microscopically crenulate the interspaces between the spirals; suture not impressed; base with an obscure angulation peripherally, the sculpture similar to the rest of the shell but more pronounced; umbilicus narrow, deep; aperture subquadrate, oblique; the pillar thin, white; the throat pearly. Height of shell, 13; of last whorl, 10; maximum diameter of base, 14 mm. (Dall.)

TYPE in United States National Museum, No. 267172. Type locality, deep water off Santa Barbara Channel, California.

RANGE. Off Santa Barbara Islands, California.

### Subgenus LIRULARIA Dall, 1909
### Margarites succincta Carpenter, 1864

*Supplementary Report,* British Association for the Advancement of Science, 653. *Annals and Magazine of Natural History,* ser. 3, **14**:424. Dall, *Bulletin 112,* United States National Museum; Pl. 17, fig. 9.

Small, scarcely sculptured, with spiral brown pencillings. (Carpenter.)

G. testa parva, subelevata, solidiore; livida, testa jun. strigis angustis, creberrimis, fusco-purpureis penicillata, testa adulta maculis quoque magnis nebulosa; anfr. v., subquadratis; liris obtusis medianis et striis subobsoletis cincta, suturis valde impressis; basi rotundata, obtuse angulata, striis saepe evanidis spiralibus ornata, testa adulta circa umbilicum magnum, infundibuliformem, vix angulatum, saepe tumidore, medio obtuse impressa; apertura subquadrata, parum declivi; columella subarcuata. Long., .16; long. spir., .07; lat., .16; div., 70°. (Carpenter.)

Shell small, subelevate, rather solid; pale with young shell ornamented with narrow very frequent tawny-purplish bands, with adult shell also nebulous with large spots; with five subquadrate whorls; girdled with obtuse median lirae and subobsolete striae, with strongly impressed sutures; with rotund basis, obtusely angulated, ornamented often with evanescent spirals, adult shell often more swollen around the large infundibuliform umbilicus; aperture subquadrate, little sloping; columella subarcuate. (Translation.)

Type in United States National Museum. Type locality, Neah Bay, Washington.

RANGE. Sitka, Alaska, to San Diego, California.

Described as *Gibbula succinta* Carpenter.

### Margarites lacunata Carpenter, 1864

*Supplementary Report,* British Association for the Advancement of Science, 653. *Annals and Magazine of Natural History,* ser. 3, **14**:425.

Very small, nearly smooth; umbilicus hemmed-in by swelling of columella. (Carpenter.)

G. testa parva, fusco-purpurea, solidiore; marginibus spirae valde excurvatis; anfractibus nucleosis normalibus, postea iv. subplanatis, suturis distinctis, apice mamillato; sublaevi, circa basin vix angulatam striolata, striolis spiralibus distantibus; apertura suborbiculari, parum declivi; labio juxta umbilicum constrictum, quasi lacunatum, lobato; columella callositate parva umbilicum constringente. Long., .11; long. spir., .05; lat., .11; div., 80°. (Carpenter.)

Shell small, tawny purplish, rather solid; with margins of spire strongly excurved; with nuclear whorls normal, afterward four in number subplanate, with distinct sutures; apex mammillate, rather smooth, striate around the scarcely angulate base, with distinct spiral striae; with aperture suborbicular, little sloping; with inner lip lobate next to the constricted and somewhat hollowed-out umbilicus; columella with small callosity constricting the umbilicus. (Translation.)

Type in United States National Museum. Type locality, Neah Bay, Washington.

RANGE. Neah Bay, Washington, to San Diego, California.

Described as *Gibbula lacunata.*

## Margarites inflatula Dall, 1920

*Supplementary Report,* British Association for the Advancement of Science, 653, 1864.

New name for *Margarita inflata* Carpenter.

Thin, whorls very swollen; sculpture very fine; spiral hollow inside keeled umbilicus. (Carpenter.)

TYPE in United States National Museum. Type locality?

RANGE. Vancouver Island, British Columbia, to Puget Sound.

## Margarites lirulata Carpenter, 1864

### Plate 101, fig. 1

*Supplementary Report,* British Association for the Advancement of Science, 653. *Proceedings,* Philadelphia Academy of Natural Sciences, 61:1864. Tryon and Pilsbry, *Manual of Conchology,* 11:296; Pl. 65, figs. 81, 82, 87.

Small; operculum smooth; 2 sharp principal riblets on spire; outline variable. (Carpenter.)

M. t. parva, cineracea, tenus, tumentiore, nacreo rosaceo; anfr. v. plerumque subdepressis, suturis distinctis; interdum purpureo-fusco pallide maculata; lirulis acutis spiralibus haud elevatis, supra valde distantibus, in spira ii., circa basim rotundatum circ. viii.; apertura subquadrata; umbilico magno, infundibuliformi, angulato; interstitiis lirularum laevibus, seu ab incrementis epidermidis decussatis: operculo tenuissimo, pallido, subplanato, suturis distinctis. Long., .18; long. spir., .07; lat., .2; div. 80°.

Shell umbilicate, globose-conical, solid, lusterless or slightly shining, purplish, unicolored, or with large radiating white patches above, or around the periphery, or spiral darker lines, or spiral articulated lines. Surface either with (1st) a few (2–4) strong lirae above, their interspaces smooth, the base with about eight concentric lirulae, or (2d) more numerous narrow irregular lirulae above, those of the base still smaller, or (3d) the spiral sculpture obsolete, surface smooth or nearly so above and beneath. The spire is more or less elevated; apex obtuse; suture impressed, sometimes subcanaliculate; body-whorl convex beneath; aperture oblique, oval-rhomboidal, very brilliantly iridescent within, but the acute peristome has a rather broad marginal band of opaque white; columella simple; umbilicus tubular, with incremental striae within. (Tryon and Pilsbry, *Manual of Conchology.*)

TYPE in United States National Museum. Type locality, Puget Sound.

RANGE. Port Etches, Alaska, to San Diego, California.

## Margarites lirulata subelevata Carpenter, 1864

*Supplementary Report,* British Association for the Advancement of Science, 653. *Proceedings,* Philadelphia Academy of Natural Sciences, **61**:1865.

Raised, livid. (Carpenter.)

T. elatiore; colore livido, intensiore; lirulis vix acutis. (Carpenter.)

Shell somewhat elevated, with livid very intense color; lirulae scarcely acute. (Translation.)

TYPE in United States National Museum. Type locality, Neah Bay, Washington.

RANGE. Known only from type locality.

## Margarites lirulata obsoleta Carpenter, 1864

*Supplementary Report,* British Association for the Advancement of Science, 653. *Proceedings,* Philadelphia Academy of Natural Sciences, **61**:1865.

Sculpture evanescent. (Carpenter.)

T. ut in var. *subelevata;* lirulis evanescentibus; operculo planato, tenuissimo, suturis indistinctis. (Carpenter.)

Shell as in var. *subelevata;* with evanescent lirulae; operculum flattened, very thin, with indistinct sutures. (Translation.)

TYPE in United States National Museum. Type locality, Neah Bay, Washington.

RANGE. Known only from type locality.

## Margarites acuticostata Carpenter, 1864

*Supplementary Report,* British Association for the Advancement of Science, 653. *Proceedings of the California Academy of Sciences,* **3**:157. Dall, *Bulletin 112,* United States National Museum; Pl. 18, fig. 5.

Small, painting clouded: 3 sharp ribs on spire. 8-20 fm. (Carpenter.)

M. t. *M. lirulatae* simili; parva, tenui, albido-cinerea, olivaceo-fusco varie maculata, seu punctulata; anfr. nucleosis ii. laevibus, tumidis, fuscis, apice mamillato; anfr. norm. iii. tumidis, tabulatis, suturis rectangulatis; carinis acutis in spira iii., quarta peripheriali, aequidistantibus; interstitiis spiraliter striatis; in spira et circa basim radiatim creberrime striulatis; basi subrotundata, lirulis distantibus circ. ix. ornata; umbilico magno, infundibuliformi, vix angulato, intus interdum striis spiralibus paucis sculpto; apertura subrotundata, pariete parum attingente: operculo anfr. paucioribus, circ. vi. suturis subelevatis. Long., 0.18; long. spir., 0.12; lat., 0.19; div., 87°. (Carpenter.)

Shell like *M. lirulata;* small, thin, whitish-ashy, spotted irregularly with tawny-olive, or punctulate; with nuclear whorls two, smooth, thick, tawny, with mammillate apex; with normal whorls three, thick tabulate with rectangular sutures; with three acute, equidistant carinae on the spire; with a fourth peripheral; with interstices spirally striate; on the spire and around the base radially thickly striate; with base subrotund, arcuate with about nine distant lirulae; with large umbilicus, funnel-shaped, within sometimes sculptured with few spiral striae; with aperture subrounded, with wall hardly touching; operculum with about six small whorls, with subelevate sutures. (Translation.)

TYPE in United States National Museum. Type locality, Santa Barbara, California.

RANGE. Bodega Bay to Coronado Islands.

## Margarites smithi Bartsch, 1927

*Proceedings of the United States National Museum, 70:32.*

Shell minute, rather elevatedly helicoid, white. Nuclear whorls one and a half, well-rounded with a carina about one-third of the distance between the summit and the suture, anterior to the summit. Postnuclear whorls well-rounded, marked by the continuation of the nuclear carina, which forms a rather strong cord and a slender cord about midway between this and the summit, and two strong cords which divide the space between the carina and the suture into equal spaces. Between these three cords a lesser one is present. From the third cord, which almost marks the periphery on the last whorl, the base curves gently to the rather open umbilicus. The base is marked by spiral cords, which increase steadily in strength from the periphery to the umbilical angle; the last five are very strongly developed. The umbilical wall appears to be free of sculpture, excepting incremental lines. The spire and base of the shell are marked by strong incremental lines, which in crossing the base form slight riblets between the spiral cords. Aperture, oval; posterior angle, obtuse; outer lip, thin; peristome complete; operculum multispiral, horny. (Bartsch.)

TYPE in United States National Museum, No. 340814. Type locality, China Point, Monterey, California. It measures, altitude, 1.6 mm.; greater diameter, 1.7 mm.

## Margarites althorpensis Dall, 1919

*Proceedings of the United States National Museum, 56:365.*

Shell small, solid, trochiform, nacreous white, with a minute subglobular smooth nucleus and five subsequent whorls; spiral sculpture of five uniform prominent threads on the upper half of the last whorl and between the sutures of the spire, with wider interspaces; on the base the threads are more numerous, smaller, and with subequal interspaces, extending from near the periphery to the umbilicus, which is perforate and not internally sculptured; axial sculpture of fine regular incremental lines not modifying the spirals; aperture rounded-quadrate, simple, sharp-edged, the lips connected by a glaze on the body, not anywhere reflected; height, 3; larger diameter, 3.2 mm. (Dall.)

TYPE in United States National Museum, No. 208559. Type locality, Granite Cove, Port Althorp, Alaska, 14 fathoms.

RANGE. Known only from type locality.

Subgenus MARGARITES Leach, s.s.

## Margarites helicina Phipps, 1773

*Voyage to the North Pole,* Appendix, 198. Binney's Gould, *Report on the Invertebrata of Massachusetts,* 281, fig. 542.

Testa umbilicata convexa obtusa: anfractibus quatuor laevibus. (Phipps.)

Shell small, orbicular, depressed, thin, and translucent, smooth and shining, of a light yellowish horn color or light olive; whorls four or five, very convex, the last very large and tumid, a little flattened above; minutely wrinkled by the lines of growth, and at its base marked with very fine spiral lines; suture well-impressed; aperture large, circular, somewhat expanded; edge sharp and simple, a little reflected at the umbilicus, which is large and profound, not bounded by an angular ridge; operculum horny, multispiral. Length, one-fifth of an inch; breadth, nearly three-tenths of an inch. (Binney and Gould.)

TYPE in ? Type locality, north side of Spitzbergen.

RANGE. Bering Strait, all coasts of Bering Sea, to Catalina Island, California.

## Margarites helicina excavata Dall, 1919

*Proceedings of the United States National Museum,* **56**:366.

Shell small, depressed, thin, polished, lurid flesh color with a darker globular glassy nucleus, and about three subsequent whorls, on which a few spiral lines of obscurely lighter color are sometimes apparent; suture distinct, not appressed; surface with faint incremental lines as the only sculpture; base rounded with, in the adult, a widely excavated funicular umbilicus, aperture rounded, simple, more or less patulous when mature; inner lip with a moderate coat of enamel continued on to the pillar lip and slightly reflected there; larger diameter, 9; shorter diameter, 5.5; height, 3.5 mm. (Dall.)

TYPE in United States National Museum, No. 219144. Type locality, Constantine Harbor, Amchitka, Aleutian Islands.

RANGE. Amchitka and Middleton Islands, Alaska.

## Margarites helicina elevata Dall, 1919

*Proceedings of the United States National Museum,* **56**:366.

Shell small, trochiform, polished, purple-brown, with a hard glassy nucleus and about five subsequent well-rounded whorls; suture distinct, rather deep; sculpture of evident incremental lines without any spiral striae; base rounded with a small umbilical chink; aperture simple rounded, slightly angular at the suture, the body with a well-marked glaze uniting the outer lips with a rather wide, white, slightly reflected pillar lip; operculum brownish with about 10 turns; larger diameter, 9; shorter diameter, 7; height, 6.5 mm. (Dall.)

TYPE in United States National Museum, No. 205833. Type locality, Bear Bay, Baranoff Island, Alaska.

RANGE. Known only from type locality.

## Margarites beringensis Smith, 1899

### Plate 99, figs. 7, 8

*Proceedings of the Malocological Society of London,* 3:206, fig. 1. Dall, *Bulletin 112,* United States National Museum; Pl. 16, figs. 5 and 6.

Testa anguste umbilicata, depressa, olivaceo-lilacea vel pallide rufo-lilacea, nitida, lineis incrementali obliquis curvatis sculpta; spira brevis ad apicem nigrescens; anfractus 5, celeriter accrescentes, perconvexi, sutura profunda sejuncti, ultimus magnus, paulo dilatatus, antice leviter descendens; apertura pulcherrime iridescens, subrotundata; peristoma haud continuum, margine externo tenui, columellari incrassato, albo reflexo. Diam. maj., 11; min., 8.5; alt., 8 mm. (E. A. Smith.)

Shell narrowly umbilicate, depressed, olive-lilac or reddish-lilac, polished, sculptured with oblique curved lines of growth; spire short, turning black toward the apex; five whorls, increasing rapidly, very convex, divided by a profound suture, the last whorl large, in the rear dilated, in front gently sloping; aperture beautifully iridescent, subrotund; peristome not continuous, with outer margin thin, with thickened columella, white, reflexed. (Translation.)

TYPE in British Museum. Type locality, Arctic.

RANGE. Arctic Ocean, Plover Bay, Commander Islands, and Petrel Bank, Bering Sea.

Described as *Valvatella*.

## Margarites albolineatus E. A. Smith, 1899

### Plate 99, figs. 4, 5

*Proceedings of the Malocological Society of London,* 3:206; fig. 2. Dall, *Bulletin 112,* United States National Museum; Pl. 16, figs. 3, 4.

Testa depressa, suborbicula, imperforata, rosacea, linis albis filiformibus volventibus numerosis picta, tenuis, nitida; spira brevis; anfractus 5, celeriter accrescentes, convexi, lineis incrementi obliquis tenuissimis indistincte striati; apertura magna, rotundata, intus pulcherrime iridescens; peristoma tenue, margine columellari albo incrassato reflexo, umbilicum quasi obtegente. Diam. maj., 8; min. 6; alt., 4.5 mm. (E. A. Smith.)

Shell depressed, suborbicular, imperforate, rosy and ornamented with numerous revolving filiform white lines, thin, polished; spire short; five whorls, increasing rapidly, convex, striated indistinctly with very fine oblique lines of growth; aperture large, rotund, inside very beautifully iridescent; peristome thin, with white columellar margin thickened and reflexed, almost touching the umbilicus. (Translation.)

TYPE in British Museum? Type locality, Bering Sea.

RANGE. All coasts of Bering Sea.

Described as *Valvatella*.

## Margarites pribiloffensis Dall, 1919

*Proceedings of the United States National Museum,* 56:366.

Shell small, solid, trochiform, pale straw color, with a small glassy nucleus and about five and a half subsequent well-rounded whorls; suture distinct, slightly appressed; surface dull, with fine incremental lines crossed by extremely fine spiral striae; base well-rounded with a deep, not funicular umbilicus, aperture rounded, simple, the outer lip produced at the suture and united with the pillar by a thin glaze of enamel over the body, the pillar lip a little thickened, not reflected; operculum brownish with eight or more turns; larger diameter, 8.5; shorter diameter, 7; height, 8 mm. (Dall.)

Type in United States National Museum, No. 210130. Type locality, U.S. Fish Commission Station 3504, near the Pribilof Islands, Bering Sea, in 34 fathoms.

Range. Arctic Ocean to off Pribilof Islands, Bering Sea.

## Margarites frigidus Dall, 1919

*Proceedings of the United States National Museum, 56:357.*

Shell small, polished, conic, pale flesh color, of six whorls, including a minute subglobular nucleus; suture distinct, not appressed, whorls only moderately rounded; axial sculpture of faint incremental lines, spiral sculpture of a few very faint lines near the umbilical region; base rounded, imperforate; operculum pale brown with about 8 turns; aperture rounded, slightly angular above, outer lip simple sharp, body with a thin nacreous glaze, pillar lip rounded, broader than the rest; height of shell, 9; of last whorl, 6.5; of aperture, 3; diameter, 6 mm. (Dall.)

Type in United States National Museum, No. 223423. Type locality, Arctic Ocean north of Bering Strait.

Range. Arctic Ocean and south to Nunivak Island, Bering Sea, and Windfall Harbor, Admiralty Island, Alaska.

## Margarites marginatus Dall, 1919
### Plate 99, figs. 1, 2

*Proceedings of the United States National Museum, 56:367.*

Shell small, thin, trochiform, pale gray or pink, with a minute glassy nucleus and about five subsequent whorls; suture distinct, rather deep, in front of it the last whorl is marginated by a series of eight or more slightly arcuate broad convex waves with narrow interspaces, extending about halfway to the periphery which is somewhat angular though not distinctly keeled; other spiral sculpture of minute almost microscopic striae over the whole surface; base moderately convex with a narrow umbilicus; aperture rounded, simple, the margin not expanded, the body with a thin layer of enamel uniting the lips, the pillar lip not reflected, slightly thickened; the operculum pale with six or more turns; larger diameter, 6.5; shorter diameter, 5; height, 5 mm. (Dall.)

Type in United States National Museum, No. 109464. Type locality, Adakh Island, Aleutians.

Range. Arctic Ocean and Bering Sea, south to Oregon; also Atlantic.

## Margarites hypolispus Dall, 1919

*Proceedings of the United States National Museum, 56:367.*

Shell small, solid, turbinate, pale flesh color, polished, smooth, with five well-rounded whorls, including a minute subglobular nucleus; suture very distinct, not appressed; base rounded with a narrow deep perforate umbilicus; aperture subcircular, simple, the pillar lip hardly thickened, the body with a thin coat of enamel; height, 3.5; diameter, 4.5 mm. (Dall.)

TYPE in United States National Museum, No. 274122. Type locality, Arctic Ocean north of Bering Strait.

RANGE. Arctic Ocean north of Bering Strait.

## Margarites (vahlii var?) tenuisculptus Carpenter, 1865

*Proceedings,* Philadelphia Academy of Natural Sciences, 61.

M. t. *M. Vahlii* forma, colore, et operculo simillima; sed striulis spiralibus, plus minusve obsoletis cincta, quarum iv.–vi. in spira monstrantur. Long., .22; long. spir., .11; lat., .13; div. 70°. (Carpenter.)

Shell like *M. vahlii* in form and color and operculum; but girdle with spiral small striae more or less obsolete, of which four to six show on the spire. (Translation.)

TYPE in ? Type locality, Neah Bay, Washington.

RANGE. Neah Bay and Puget Sound. (Plover Bay, Krause?)

## Margarites umbilicalis Broderip and Sowerby, 1829

*Malacological and Conchological Magazine,* 1:26. *Conchological Illustrations,* "Margarita," fig. 5. *Zoölogical Journal,* 4:371.

M. testa obtuse conica, obliqua, anfractibus sensim majoribus, longitudinaliter striatis; umbilico maximo. Long., 6/10; lat., 8/10; poll. (Broderip and Sowerby.)

Only one specimen of this interesting species has occurred. The outer lip is much broken. (Broderip and Sowerby.)

TYPE in British Museum. Type locality, Oceano Boreali.

RANGE. Arctic coast. Circumboreal.

## Margarites olivaceus Brown, 1827

*Illustrations of the Conchology of Great Britain and Ireland;* Pl. 46, figs. 30, 31.

Shell thin, olive-colored, pellucid, smooth, subglobose; body large, inflated; spire small, short, with three depressed volutions, terminating in a moderately pointed apex; aperture large, circular, standing out from the body; outer lip thin, continuous with the inner lip above, which is narrow, and a small circular umbilicus behind. Length, two-tenths of an inch. (Brown.)

TYPE in ? Type locality, Greenock.

RANGE. St. Lawrence Island; Atka Island, Aleutians. Also Atlantic. Described as *Trochus olivaceus.*

## Genus **MARGARITOPSIS** Thiele, 1906

### **Margaritopsis frielei** Krause, 1886

*Archiv für Naturgeschichte,* **51:** heft 1, 263; Pl. 16, figs. 2, *a–c. Nachr. Mal. Ges.* **15:**1906.

Testa tenuis, alba intus margaritacea, depresso-conica, aspira obtusata, anfractibus 4, aequiliter rotundatis, ultimo valde dilatato; axis sutura profunde impressa; apertura rotundata, labro externo et interno aequiliter arcuatis, umbilico lato a basi rotundata omnino non definito. Superficies tota striis spiralibus undulatis subtilissimis obducta. (Krause.) Diam. bas., 10; alt., 6 mm.

Shell thin, pearly whitish within, depressed-conical, spire obtuse, with four whorls equally rotund, the last one strongly dilate, deeply impressed at the suture; aperture rotund with six externally equally arcuate, with broad umbilicus always undefined at the rounded base. Whole surface most finely covered with undulate striae. (Translation.)

TYPE in British Museum. Type locality, St. Lorenzbau, one example. (Krause.)

RANGE. Bering Strait region.

## Genus **GIBBULA** Risso, 1826

Shell conoidal, umbilicated; umbilicus cylindrical or infundibuliform; whorls frequently tuberculated above and with channeled suture; columella sometimes terminating in a tubercular tooth. (Tryon, *Systematic and Structural Conchology.*)

TYPE. *G. magus* Linn.

DISTRIBUTION. European and Australian seas, Indian Ocean, Red Sea, west coast of North America.

### **Gibbula adriatica canfieldi** Dall, 1871

#### Plate 100, fig. 2

*American Journal of Conchology,* **7:**129. *Proceedings of the United States National Museum,* **24:** Pl. 39, fig. 2.

Shell of seven whorls, the last comprising more than half the shell. Above, sutures small but deeply channeled; whorls smooth, with three revolving ribs close to the suture, also three or four on the lower part of the whorl. Color pearly, with bronze yellow pencillings obliquely to the suture. Surface of the whorls rather flattened, semicarinated, convex. Shell umbilicated with nine basal revolving ribs. Umbilicus strongly carinate internally, smooth, narrow and small. Aperture rhomboidal, pearly, with grooves answering to the exterior ribs. Columella straight, with a slight callosity, but not reflected. Min. diam., 3; maj. diam. 4; alt., 4 inch.

TYPE in United States National Museum. Type locality, Monterey, California.

RANGE. Monterey to San Diego, California. Fossil: Pleistocene.

## Family VITRINELLIDAE

## Genus **VITRINELLA** C. B. Adams, 1850

Shell turbiniform, vitreous, minute, with a large, orbicular, aperture; either umbilicated, or with the umbilical region deeply and widely indented. (C. B. Adams.)

TYPE ?

DISTRIBUTION. West Indies, Panama, and west coast North America.

### Vitrinella williamsoni Dall, 1892

### Plate 106, figs. 4, 6

*Proceedings of the United States National Museum,* **15**:202; Pl. 21, figs. 2, 3.

Shell small, white, depressed, with two and a half whorls; spire flattened; suture appressed with a shallow channel or excavation outside of the appressed margin of the whorl, outside of which the convexity of the whorl rises higher than the suture. Base slightly more rounded than the upper side, with a wide and flaring umbilicus; periphery rounded; aperture rounded, oblique; surface polished, finely striate here and there by the incremental lines which are most prominent above. Maximum diameter of shell, 5.5; minimum diameter, 4.5; altitude, 1.25 mm. (Dall.)

TYPE in United States National Museum, No. 106856. Type locality, San Pedro, California.

RANGE. Known only from type locality. Fossil: Pleistocene, San Pedro, California.

### Vitrinella oldroydi Bartsch, 1907

### Plate 107, figs. 1, 2, 3

*Proceedings of the United States National Museum,* **32**:167, figs. 2, 3.

Shell small, sublenticular, semitransparent, a little more convex above than below. Nepionic whorls not differentiated from the rest, the entire upper surface smooth and shining, marked only by irregularly distributed incremental lines. The upper sides of the whorls are moderately and evenly rounded. Sutures well-marked. Periphery of the last whorl well-rounded. Base moderately well-rounded, openly umbilicated to the very apex. Columellar wall of the base well-rounded (not concaved). Aperture decidedly oblique, broadly oval; outer lip thin; parietal wall covered by a rather strong callus, which partly fills the posterior angle. Greater diameter, 2.1; lesser diameter, 1.6; altitude, 0.8 mm. (Bartsch.)

TYPE in United States National Museum, No. 158777. Type locality, Point Loma, California.

RANGE. San Pedro, California, to Point Abreojos, Lower California.

### Vitrinella eshnauri Bartsch, 1907

#### Plate 106, figs. 5, 7, 9

*Proceedings of the United States National Museum*, 32:168; fig. 2.

Shell moderately elevated, subglobose, thin, almost transparent, glassy. Nepionic whorls 1½, scarcely differentiated from those which follow, well-rounded, smooth. Succeeding whorls well-rounded, somewhat inflated, marked only by exceedingly fine lines of growth. Sutures well-impressed. Periphery of the last whorl well-rounded. Base moderately rounded, marked only by incremental lines, with narrow but open umbilicus, which is obsoletely angled at the outer edge. Columellar wall of umbilicus vertical from the outer edge to within a short distance of the parietal wall, where it bends outward to join the preceding turn. Aperture decidedly oblique, almost circular; outer lip thin and translucent; columella quite strong and decidedly curved; parietal wall covered by a moderate callus, which forms an acute angle with the posterior margin of the lip. Greater diameter, 2.3; lesser diameter, 1.9; altitude, 1.3 mm. (Bartsch.)

TYPE in United States National Museum, No. 127557. Type locality, San Pedro, California.

RANGE. Known only from type locality.

### Vitrinella alaskensis Bartsch, 1907

#### Plate 106, figs. 8, 10, 12

*Proceedings of the United States National Museum*, 32:168; fig. 3.

Shell small, subglobose, semitransparent. Nepionic whorls 1⅓, well-rounded, smooth. Succeeding turns somewhat inflated, well-rounded, separated by strongly impressed sutures, marked only by incremental lines. Periphery and base of the last whorl well-rounded. The latter narrowly and openly umbilicated to the very apex. Columellar wall of the umbilicus not flattened nor angulated at the outer edge, but evenly rounded with the rest of the base. Aperture forming a broad semioval, of which the columellar side forms the short diameter; outer lip thin and semitransparent; columella slender and curved; parietal wall covered by a thin callus. Greater diameter, 1.6; lesser diameter, 1.2; altitude, 1.2 mm. (Bartsch.)

TYPE in United States National Museum, No. 109470. Type locality, Unalaska, Alaska.

RANGE. Known only from type locality.

Subgenus DOCOMPHALA Bartsch, 1907

**Vitrinella stearnsii** Bartsch, 1907

Plate 106, figs. 11, 13, 14

*Proceedings of the United States National Museum,* **32**:169; fig. 4.

Shell depressed, lenticular, a little more convex above than below. The nepionic portion of the shell consists of the first one and a half turns, which are small, slightly convex, and smooth. The turns which succeed the nepionic part of the shell are strongly, obliquely, transversely ribbed on the upper side, but these ribs gradually grow weaker as the shell increases in size and disappear entirely after one and one-half turns, the remaining portion being marked by mere lines of growth on the upper surface. Periphery of the last whorl well-rounded. Base very gently rounded, crossed by rather strong incremental lines. Umbilicus wide and open to the very apex, decidedly angulated at the outer margin. The columellar wall is strongly concaved from the outer angulation to the junction with the preceding whorl, the inner half of it bears a series of strong ribs behind the aperture. Aperture decidedly oblique; outer lip acute, forming a regular semioval of which the parietal wall and columella form the short diameter; columella short, stout, concave; parietal wall covered by a thick callus, which renders the peristome continuous and forms an acute angle with the outer lip posteriorly. The parietal callus and columella form a strongly sigmoid curve. Greater diameter, 3.8; lesser diameter, 3; altitude, 1.5 mm. (Bartsch.)

TYPE in United States National Museum, No. 74011. Type locality, Monterey, California.

RANGE. Known only from type locality.

**Vitrinella berryi** Bartsch, 1907

Plate 107, figs. 4, 5, 6

*Proceedings of the United States National Museum,* **32**:170; fig. 5.

Shell small, semitransparent, lenticular, with the upper part only slightly more convex than the base. Nepionic whorls forming a little more than one and two-thirds turns, smooth, and moderately convex. The portion following the nepionic part is crossed on the upper surface by quite regularly spaced, sublamellar riblets, which become weaker as the shell increases in size, and disappear completely after one and one-half turns; the remaining part of the upper surface being marked by weak incremental lines only. Sutures well-marked. Periphery of the last whorl well-rounded, the first half showing continuations of the ribs, the rest being smooth. Base moderately rounded with a fairly strong spiral keel which is situated about one-third of the way toward the umbilical angle from the periphery. Umbilicus much narrower than *V. stearnsi,* its outer edge terminating in a blunt angle. Columellar wall decidedly concaved, the inner half marked by a series of strong riblike nodules as in *V. stearnsi.* Aperture decidedly oblique, subcircular; outer lip thin; columella very thick, concaved, provided with a moderately strong callus, which bends back into the umbilicus. Greater diameter, 2.2; lesser diameter, 1.7; altitude, 1 mm. (Bartsch.)

TYPE in United States National Museum, No. 192686. Type locality, off Del Monte, Monterey Bay, California.

RANGE. Monterey to San Diego, California.

## Vitrinella columbiana Bartsch, 1921

*Proceedings,* Biological Society of Washington, **34**:39.

Shell moderately large, depressed, helicoid, semitranslucent, bluish-white. Nuclear whorls decollated. Postnuclear whorls gently rounded, almost appressed at the summit, marked by rather strong incremental lines which extend over both the upper and lower surface; the lower surface is a little more convex than the upper; the umbilical wall is marked by strong notches. Aperture decidedly oblique, almost circular; parietal wall marked by a thin callus, which renders the peristome almost complete. Altitude, 1.5; greater diameter, 3.1 mm. (Bartsch.)

TYPE in United States National Museum, No. 340848. Type locality, Departure Bay, British Columbia.

RANGE. Departure Bay, British Columbia.

## Vitrinella smithi Bartsch, 1927

*Proceedings of the United States National Museum,* **70**: Art. 11, **33**: Pl. 4, figs. 6, 8, 9.

Shell minute, pale brown. Nuclear whorls one and a half, smooth. Postnuclear whorls two, well-rounded, with well-impressed suture, marked by strongly curved lines of growth only. Under surface widely, openly umbilicated, marked by lines of growth only. The last whorl strongly curved. Aperture slightly oblique, subcircular; peristome thin, rendered complete by the thick callus on the parietal wall. Altitude, 0.6; diameter, 1.2 mm. (Bartsch.)

TYPE in United States National Museum, No. 340813. Type locality, Whites Point, California.

RANGE. California.

## Genus CYCLOSTREMA Marryatt, 1818

Shell orbicular, depressed, widely umbilicated, spire short; whorls transversely striated or cancellated; aperture round, not nacreous; peristome continuous, simple. (Tryon, *Structural and Systematic Conchology.*)

TYPE, *C. cancellata,* Marryatt.

DISTRIBUTION. Cape, India, Philippines, Australia, Peru, California, Washington.

### Cyclostrema diegensis Bartsch, 1907
### Plate 107, figs. 7, 8, 9

*Proceedings of the United States National Museum,* **32**:172; fig. 7.

Shell exceedingly small, thin planorboid, with a prominent, compressed peripheral keel, translucent, yellow horn-colored. Nepionic whorls 1½, moderately rounded, not elevated, smooth. The succeeding turns have their highest elevation at about one-third of the distance from the suture to the periphery, at which place they are raised into a broad, well-rounded ridge from which they slope abruptly, convexly rounded, to the suture and more gently concavely to the angulated periphery. On the upper surface the whorls are ornamented with slender, regularly spaced, oblique riblets, which are best developed on the elevated ridge, where they are about one-fourth as wide as the spaces that separate them. Sutures strongly marked. Periphery with a compressed, obtuse angle. Base moderately rounded, broadly, openly umbilicated, with a slender thread bounding the outer edge of the umbilicus and a slender spiral cord situated about halfway between the umbilical thread and the periphery. The riblets seen on the upper surface extend feebly beyond the peripheral keel to the first basal cord but are reduced to simple incremental lines between this and the umbilical thread; columellar wall well-rounded, marked by incremental lines only. Aperture oblique, irregularly pentagonal, one blunt angle being formed by the elevated part of the whorl, another equally obtuse one by the periphery, the third by the basal keel, the fourth by the umbilical angle, and the fifth by the junction of the columella with the parietal wall. Outer lip thin, showing the sculpture of the shell within. Columella straight and very obliquely placed. Parietal wall covered by a thin callus. Greater diameter, 1.0; lesser diameter, 0.8; altitude, 0.25 mm. (Bartsch.)

Type in United States National Museum, No. 105488. Type locality, San Diego, California.

Range. Known only from type locality.

## Cyclostrema miranda Bartsch, 1911

Plate 105, figs. 1, 2, 3

*Proceedings of the United States National Museum,* **39**:230; Pl. 39, figs. 1–3.

Shell small, subdiaphanous, depressed. Nuclear whorls two, depressed helicoid, smooth. Post-nuclear turns one and a third, appressed at the summit, marked above with three strong, spiral cords, one of which is at the periphery, while the other two divide the space between the periphery and the summit into three equal areas. The space between the summit and the first spiral cord is decidedly concave, while the space between the first spiral cord and the median one is very slightly concave, that between the median and the peripheral cord being well-rounded. Axially, the upper surface of the shell is marked by about sixty-two slender, well-developed, equally spaced riblets, which are about half as wide as the spaces that separate them. These riblets are decidedly retractively curved between the summit and the first keel, less so between the first and the median keel, while between the median and the peripheral one they are practically vertical. The junctions of riblets and keels are not nodulose. Periphery of the last whorl strongly angulated; base marked by two spiral cords, one of which bounds the broad, open, funnel-shaped umbilicus and is a little weaker than the other, which appears as a very strong cord halfway between it and the periphery. In addition to these spiral cords, the base is marked by the undiminished continuations of the axial riblets which become bifurcated here and extend deep within the umbilicus. Aperture very large, ovate, very oblique, the columellar border being considerably behind the outer lip; the posterior and anterior angles are acutely angulated; outer lip thin, showing the external sculpture within; columella slender, decidedly curved, the free edge continuing to the posterior angle of the aperture, rendering the peritreme complete. (Bartsch.)

Type in United States National Museum, No. 211108. Type locality, San Pedro, California.

Range. Known only from type locality.

## Cyclostrema baldridgae Bartsch, 1911

Plate 105, figs. 7, 8, 9

*Proceedings of the United States National Museum,* **39**:229; Pl. 39, figs. 7–9.

Shell rather large, bluish-white, subdiaphanous. Nuclear whorls two and a third, smooth, forming a decidedly depressed spire. Post-nuclear whorls with a strong, broad, rounded keel at the periphery and another almost as strong about one-third of the distance between the periphery and the summit, where it forms a decided shoulder. The space between the appressed summit and the shoulder is marked by twelve subequal and subequally spaced, spiral cords, the spaces between which are crossed by very slender, retractive, axial threads; the latter are about one-fourth as wide as the spaces which separate them, while the spaces between them are only about one-half the width of the spiral cords. The space between the peripheral keel and the strong shoulder is crossed by nine subequal and subequally spaced spiral cords, and the continuations of the axial threads. Here the spiral sculpture is not quite as strong as on the upper surface and the spaces inclosed between the axial riblets and the spiral threads are more or less quadrangular pits. Base well-rounded, marked like the upper surface, with the sculpture a little less pronounced. Umbilicus open; parietal wall showing ten equal and equally spaced spiral cords, which are as wide as the spaces that separate them, the latter being crossed by feeble continuations of the axial riblets. Aperture subcircular; outer lip rendered somewhat angular by the two keels; columella evenly curved. (Bartsch.)

Type in United States National Museum, No. 214100. Type locality, Gulf of California.

Range. San Pedro, California, to Gulf of California.

## Genus CIRCULUS Jeffreys, 1865

Coquille subdiscoïdale, à peine nacrée; ombilic très large; ouverture subquadrangulaire; opercule normal. (Fischer.)

Type. *Circulus striatus*, Philippi.

Distribution. Europe, west coast of America.

### Circulus rossellinus Dall, 1919

*Proceedings of the United States National Museum,* **56**:368.

Shell minute, white, solid, depressed turbinate, of two and a half whorls, including the smooth nucleus; suture distinct; sculpture of numerous close-set spiral threads rather large for the size of the shell, crossed by microscopic incremental lines; base rounded, with a narrow deep umbilicus; aperture subcircular, the outer lip sharp, much produced above, pillar lip thickened, not reflected; height, 0.75; longer diameter, 2 mm. (Dall.)

Type in United States National Museum, No. 223286. Type locality, off South Coronado Island, near San Diego, California, 3 fathoms.

Range. San Diego, California.

## Circulus cosmius Bartsch, 1907

*Proceedings of the United States National Museum,* 32:173; fig. 8.

Shell decidedly depressed, planorboid, creamy white, shining. Nepionic whorls 2, well-rounded, helicoid, polished. Succeeding turns marked by a low, rather broad, spiral thread at the summit and a strong, acute, lamellar ridge at the periphery, and another equally strong halfway between the periphery and the summit. The last has the free edge pointing outward, forming an angle of 45° with the peripheral lamella. The spaces between these keels are gently rounded and marked by incremental lines only. The middle keel forms the most elevated portion on each whorl, the summit of the whorl at the suture being considerably lower. Base very broadly umbilicated, marked by a spiral, lamellar carina, which is as strong as the peripheral one and is situated halfway between this and the umbilical angle. In addition to this carina the entire base shows fine incremental lines. Umbilicus limited by an obtuse angle; columellar wall almost vertical, marked by three, slender, equally spaced, spiral lirations. Aperture decidedly oblique, pentagonal, the angles being formed by the three carinae, the posterior angle of the aperture, and the umbilical angle; outer lip thin; columella decidedly curved; parietal wall covered by a strong callus, which renders the peritreme almost continuous. Greater diameter, 2.5; lesser diameter, 2.1; altitude, 1 mm. (Bartsch.)

TYPE in United States National Museum, No. 192708. Type locality, U.S.S. "Albatross" Station 2799, near Atacames, Ecuador.

RANGE. Catalina Island, California, to Ecuador.

## Genus CYCLOSTREMELLA Bush, 1897

Shell minute, thin, semitransparent when fresh, planorbiform, of few convex whorls, nearly symmetrically coiled, forming a concavely depressed spire and large umbilical cavity. Epidermis thin, nearly colorless. Nuclear whorl relatively large, smooth, turned downward, seen only in a basal view, leaving a small pit above. Suture deep and channeled. Aperture triangular-ovate, expanded below, angulated above, with a relatively wide deep sinus just below the suture. Peritreme thin, simple, continuous, not modified, slightly attached. (Bush.)

TYPE. *Cyclostremella humilis* Bush.

DISTRIBUTION. Atlantic and California.

## Cyclostremella californica Bartsch, 1907
### Plate 106, figs. 1, 2, 3

*Proceedings of the United States National Museum,* 32:174; fig. 10.

Shell small, planorboid, semitransparent, closely spirally striated. Nepionic whorls 1¼, moderately rounded, smooth, and shining. Succeeding turns increasing regularly in size like *Planorbis,* rendering the apex considerably lower than any of the succeeding turns, the last being the most elevated. Whorls well-rounded, separated by strongly impressed sutures and marked by many equally strong and equally spaced, somewhat wavy, incised spiral lines and the fine incremental lines. At more or less regular intervals there appear slight constrictions which coincide with the lines of growth. Periphery of the last whorl well-rounded. Base well-rounded, very broadly and openly umbilicate to the very apex, marked like the upper surface. Aperture oblique, suboval; outer lip thin; columella short, forming almost a straight line with the faint callus of the parietal wall. The type has a little more than three and a half whorls and measures: greater diameter, 2.3; lesser diameter, 1.8; altitude, 0.8 mm. (Bartsch.)

TYPE in United States National Museum, No. 125537. Type locality, Long Beach, California.

RANGE. San Pedro to San Diego, California. Fossil: Pleistocene.

## Cyclostremella concordia Bartsch, 1920

*Journal of the Washington Academy of Sciences,* **10**:1920.

Shell very small, planorboid, hyaline, semitransparent. Early whorls eroded in all the specimens seen. The last two whorls curve suddenly to the deeply channeled suture on the upper surface; the rest gradually, evenly rounded. Periphery of the last whorl well-rounded. Base openly umbilicated. The entire surface of spire and base is marked by rather strong, irregularly developed incremental lines and more or less equal and equally spaced fine spiral lirations. The intersections of these two sculptured elements give to the surface of the shell the characteristic beaded sculpture of the genus. Aperture very broadly ovate, almost subcircular, the narrower position being at the posterior angle; peristome thin, not reflected; parietal wall covered by a thin callus. Operculum thin, corneous, paucispiral. Altitude, 1; diameter, 2 mm. (Bartsch.)

TYPE in United States National Museum, No. 340862. Type locality, Olga, Orcus Island, Washington.

RANGE. Known only from type locality.

## Genus SCISSILABRA Bartsch, 1907

*Vitrinella*-like shells with the middle of the outer lip deeply and broadly notched, the center of the notch coinciding with the periphery of the shell. (Bartsch.)

TYPE. *Scissilabra dalli.*

DISTRIBUTION. California to Gulf of California.

## Scissilabra dalli Bartsch, 1907

### Plate 104, figs. 10, 11, 12; Plate 107, figs. 10, 11, 12

*Proceedings of the United States National Museum,* **32**:176; fig. 11.

Shell small, depressed, lenticular, with acutely angulated periphery, having 3½ transparent, vitreous whorls which are separated by well-marked sutures. The nepionic portion consists of the first 1⅔ turns and is scarcely differentiated from the rest of the shell. The upper surface is evenly and gently rounded from the summit to the periphery, which is strongly and sharply carinated. Under side openly umbilicated, much less convex than the upper. The umbilical edge is marked by an acute carina from which the columellar wall in the last whorl extends almost vertically to where it joins the preceding turn. This carina and vertical umbilical wall are characteristic of the last turn only; in all the others which are visible in the umbilicus it appears evenly rounded. Aperture very large, decidedly oblique; outer lip very broadly and strongly notched, the blunt angle of the notch coinciding with the periphery of the shell; the portion of the lip posterior to the sinus and its basal part somewhat sinuous; columella vertical and slightly concave; parietal wall covered by a thick callus which renders the peritreme almost continuous. Greater diameter, 2; lesser diameter, 1.5; altitude, 0.75 mm. (Bartsch.)

TYPE in United States National Museum, No. 192712. Type locality, San Diego, California.

RANGE. Monterey, California, to Gulf of California.

## Genus **LEPTOGYRA** Bush, 1897

Shell minute, semitransparent, dull, dirty white or faintly brown shells covered with a thin, rather tough, delicate straw-colored epidermis, consisting of a few convex whorls forming an elevated spire with relatively large, smooth, slightly twisted nuclear whorl and large body-whorl. Suture deep, somewhat channeled. Umbilicus relatively large, round, deep, showing some of the whorls, with well-rounded walls. Aperture very oblique, somewhat ovate. Periostome simple, continuous, modified on the body-whorl into a thin glaze, sometimes in the adult having a free edge; strongly sinuate along the umbilical region and anteriorly, slightly angulated below, at the junction of the two lips; above arching well upward, forward, then backward, from the body-whorl, forming a distinct sutural notch. Interior of the aperture smooth and very lustrous, with the conspicuous, exterior transverse lines showing through by transparency. There is no opaque internal line; in all the specimens the operculum is drawn well into the shell. Operculum is very thin, circular, of a delicate horn-color, with central nucleus of about 7 whorls defined by a fine spiral line. (Bush.)

TYPE. *Leptogyra verrilli* Bush.

DISTRIBUTION. Alaska, Cooks Inlet, circumboreal.

## Leptogyra alaskana Bartsch, 1910

### Plate 104, figs. 4, 5, 6

*Nautilus,* 23, No. 11, 136; Pl. 11, figs. 4–6.

Shell minute, depressed helicoid. Nuclear whorls one and one-half, light yellow horn color, marked by faint incremental lines. A single postnuclear turn follows, which is bluish-white, rather broad, and gently, almost evenly, curved from the well-impressed suture to the periphery. This whorl is marked by about twelve fine, incised, spiral lines between the suture and the periphery, which are stronger toward the periphery than at the suture. Periphery of the last whorl rounded. Base broadly and deeply umbilicated, strongly arched, with a slender cord at the junction of the basal and parietal wall, surface of the base marked by incised lines which are equal in strength and number to those occurring upon the upper surface. Wall of the umbilicus almost flat, marked by faint spiral lines. Aperture very large, subcircular, posterior angle obtuse; outer lip thin; columella curved, somewhat expanded and thickened basally; parietal wall covered with a thin callus. Operculum thin, horny. . . . . the type measures: greater diameter, 0.85; lesser diameter, 0.7; altitude, 0.4 mm. (Bartsch.)

TYPE in United States National Museum, No. 208433. Type locality, Port Graham, Alaska.

RANGE. Port Graham, Cooks Inlet, Alaska.

## Genus TEINOSTOMA A. Adams, 1854

Shell orbicular, depressed, subspiral, polished or spirally striated, last whorl rounded, or angulated at the periphery; umbilical region covered with a large, flat callosity; aperture transverse; inner lip smooth, callous; outer lip thin, simple, not margined or reflected. (Tryon, *Structural and Systematic Conchology.*)

TYPE. *T. politum* A. Adams.

DISTRIBUTION. Philippines, Japan, Mazatlan.

## Teinostoma salvania Dall, 1919

*Proceedings of the United States National Museum,* 56:369.

Shell small, translucent white, moderately depressed, smooth, of about three and a half whorls, with a minute inflated nucleus; the only sculpture is a few faint incremental lines; base rounded, convex, imperforate, with a minute umbilical dimple, behind which is a small callus; aperture ovate, simple, the outer lip arcuate, simple, hardly produced except near the suture, with a retractive wave near the periphery; pillar thick with a small callus behind it; height, 1.2; larger diameter, 1.8 mm. (Dall.)

TYPE in United States National Museum, No. 225190. Type locality, off South Coronado Island, near San Diego, California.

RANGE. Known only from type locality.

## Teinostoma bibbiana Dall, 1919

*Proceedings of the United States National Museum*, 56:369.

Shell minute, very similar to *T. sapiella*, but smaller, more depressed relatively, and more transparent and thin; aperture circular, with a thinner callus on the body, and only a small linguiform pad behind the pillar lip. Height, 0.75; longer diameter, 2 mm. (Dall.)

TYPE in United States National Museum, No. 274123. Type locality, San Diego, California.

RANGE. Known only from type locality.

## Teinostoma sapiella Dall, 1919

*Proceedings of the United States National Museum*, 56:369.

Shell minute, white, translucent, solid, of about two whorls, including a minute dark brown subglobular nucleus; suture distinct; surface glossy, smooth; last whorl only moderately enlarged and little produced at the aperture; base rounded, depressed in the center, with a pad of enamel leaving a slight chink but no perforation in the umbilical region; aperture subcircular, the outer lip thin, a thick callus on the body and a large ovate pad behind the pillar lip; height, 1; longer diameter, 2 mm. (Dall.)

TYPE in United States National Museum, No. 127560. Type locality, San Pedro, California.

RANGE. San Pedro to San Diego, California.

## Teinostoma invallata Carpenter, 1864

*Supplementary Report*, British Association for the Advancement of Science, 652. *Proceedings of the California Academy of Sciences*, 3:215.

Without keel. (Carpenter.)

E. t. E. *supravallatae*, aliter exacte simili; sed vallo spirali omnino carente; basi angulata, haud carinata. (Carpenter.)

Shell otherwise exactly like that of *E. supravallata,* but always lacking the spiral furrow; with angulate base, not carinate. (Translation.)

TYPE in United States National Museum. Type locality, San Diego, California.

RANGE. Monterey to Gulf of California.

Described as *Ethalia* var. *invallata* Carpenter.

## Family SCISSURELLIDAE

## Genus SCISSURELLA Orbigny, 1828

Shell minute, thin, not pearly; body-whorl large; spire small; surface striated; aperture rounded, with a slit in the margin of the outer lip; operculate. The young have no slit. (Tryon, *Structural and Systematic Conchology.*)

TYPE. *Scissurella laevigata* Orbigny.

DISTRIBUTION. Europe and west coast of North America. Fossil: Tertiary.

### Scissurella kelseyi Dall, 1905

*Nautilus,* **18**:124.

Shell large for the genus, trochiform, white, with about four rounded whorls, sculptured with fine (forwardly convex) arcuate threads or raised lines, which above the fasciole are spirally, microscopically striate, and on the base, with somewhat regularly spaced and stronger spirals; the fasciole is narrow, slightly above the periphery, bounded by two sharp, very thin, elevated keels; the slit extends about one-fifth of the circumference of the last whorl. The aperture is nearly circular, interrupted for a short distance by the body, the inner lip slightly reflected over a small umbilicus; the operculum is multispiral and pale yellow. Alt. of shell, 6.0; of aperture, 3.0; max. diameter, 5.5 mm. (Dall.)

TYPE in United States National Museum. Type locality, U.S. Fish Commission Station 4353, off San Diego.

RANGE. Queen Charlotte Islands, British Columbia, to South Coronado Island.

### Scissurella chiricova Dall, 1919

*Proceedings of the United States National Museum,* **56**:370.

Shell small, white, trochiform, with a minute subglobular nucleus and about four and a half subsequent whorls; spiral sculpture of two sharp narrow peripheral keels, with a narrow interspace inclosing the anal sulcus and fasciole; on the outer side of each keel is a moderate constriction of the whorl; the suture is laid just below the anterior keel; there are also very minute spiral threads over the whole surface which do not reticulate the axial sculpture; the latter is composed of retractively arcuate, uniform minute, close-set threads extending over the flattish upper surface of the whorls, and the roundly convex base; axis perforate, the umbilicus small, the aperture rounded except the rather straight, somewhat reflected, short pillar lip; operculum multispiral, whitish, subtransparent. Height, 2.5; diameter, 3 mm. (Dall.)

TYPE in United States National Museum, No. 206509. Type locality, U.S. Fish Commission Station 3340, southeast of Chirikoff Island, Alaska, in 695 fathoms.

RANGE. Known only from type locality.

## Genus **SCHISMOPE** Jeffreys, 1856

*Schismope* is a *Scissurella* in which the anal slit becomes closed in the adult, and transformed into an oblong perforation like one of the holes of a *Haliotis*. It bears much the same relation to *Scissurella* that *Trochotoma* does to *Pleurotomaria*. The species inhabits deep water; there are a number of fossil forms described. (Tryon and Pilsbry, *Manual of Conchology*.)

TYPE. *Schisomope cingulata* Costa.

DISTRIBUTION. Mediterranean, New Caledonia, West Indies, Australia, Tasmania, Cape Verde Islands, Gulf of California, and California.

### **Schismope rimuloides** Carpenter, 1865

*Proceedings,* Zoölogical Society of London, 271.

S. t. rapide augente, albida, tenuissima; apice celato; anfr. iii., radiatim liratis, liris subdistantibus, acutis, obliquis; umbilico magno; labro declivi, haud fisso, sed apertura postica, ut in *Rimula* formata, subquadrata, elongata; liris transversis gradus testae increscentis definientibus; peritremate continuo, oblique. (Carpenter.)

Shell rapidly enlarging, whitish, very thin; apex concealed; whorls three, radiately lirate, the lirae subdistant, acute, oblique; umbilicus large; lip sloping, scarcely fissured, but with an aperture formed posteriorly as in *Rimula,* subquadrate, elongate; peristome continuous, oblique. Alt., .023; diameter, .03 in. (Tryon and Pilsbry, *Manual of Conchology*.)

TYPE in Liverpool Collection. Type locality, Mazatlan, off Spondylus.

RANGE. South Coronado Islands to Mazatlan, Mexico.

## Schismope caliana Dall, 1919

*Proceedings of the United States National Museum, 56:370.*

Shell minute, translucent white, of three whorls and a minute glassy subglobular nucleus; suture distinct, last whorl with a keel halfway to the periphery, another at the periphery; half a whorl back from the aperture between these two is the oval perforation found in the genus; these interspaces are wide; a third keel is found nearer the second on the outer part of the base, and several finer ones on the base, which is funnel-shaped in the center but forms merely a pit, the axis being imperforate; aperture wide, the upper part of the outer lip protracted; the pillar lip thin, arcuate, the general form of the aperture rounded. Height, 1; maximum diameter, 1.5 mm. (Dall.)

TYPE in United States National Museum, No. 198609. Type locality, San Diego, California.

RANGE. Known only from type locality.

## Family HALIOTIDAE

## Genus **HALIOTIS** Linnaeus, 1758

The principal characters for distinguishing the species are the outline of the shell, which is either equally curved on the two sides or straighter on the right margin; the convexity of the back, which may be carinated or rounded at the row of holes; the sculpture; the position of the spire; the color of the inside; smoothness or roughness of the muscle-scar; width and slope of the columellar-plate; and, within rather wide limits, the number of open holes. It is convenient to segregate the numerous species of *Haliotis* into groups; and the following is offered as a preliminary arrangement. (Tryon and Pilsbry, *Manual of Conchology.*)

First species, *Haliotis midae.*

DISTRIBUTION. West coast of North America, Alaska, Australia, Japan.

## Haliotis cracherodii Leach, 1817

*Zoölogical Miscellany, 131. Conchologia Iconica, fig. 23.*

H. supra caerulescente-niger, umbone laterali-dorsali; interne margaritaceus iricolor.

Bluish-black above, umbo lateral-dorsal; internally pearly and iridescent. (Leach.)

Shell oval, convex, spire near the margin; surface almost smooth, but usually showing nearly obsolete spiral lirae. Perforations about eight, color greenish-black or dull purplish-black.

An oval shell with the two sides equally curved, the back regularly convex, not carinated at the row of perforations; outside covered with a thick black layer. Surface smooth, except for spiral lirae, which are sometimes wholly obsolete, and lines of growth. Spire low, near the margin. Inside smooth, silvery with red and green reflections; columellar plate not truncate below, sloping inward, its face concave; cavity of spire very small, almost concealed. Length (of average specimen), 112; width, 85; convexity, 30 mm. (Tryon and Pilsbry, *Manual of Conchology.*)

TYPE in British Museum. Type locality, California.

RANGE. Coos Bay, Oregon, to Santa Rosalia, Lower California.

### Haliotis cracherodii holzneri Hemphill, 1907

*Transactions of the San Diego Society of Natural History, 1, No. 2, 4.*

I have before me three shells that, while they possess all the characteristics of *Haliotis cracherodii*, are without perforations, and there is no evidence on these shells that they ever had any holes. Thus they have lost the most important generic character that separates the genus *Haliotis* from that of the genus *Gena*, which has smaller but similar shells in every respect except the perforations, which are absent.

The shells are unusually high and arching, as well as narrow and oblong, with the spire much nearer the posterior margin than in any other specimens of *H. cracherodii* that I have examined of the same size. Length of largest specimen, 4½; width, 3½; height, 1¾ in. (Hemphill.)

TYPE in the Frank Holzner Collection, San Diego. Type locality, Lower California.

RANGE. San Pedro, California, to Lower California.

### Haliotis cracherodii splendidula Williamson, 1892

*Proceedings of the United States National Museum, 15:198.*

A number of shells, found at one time, at Point Vincent, have brilliant blotches of color in their interior somewhat like *H. fulgens*. Some have spots of brown color. (Williamson.)

TYPE in Williamson Collection, Los Angeles Museum, Exposition Park. Type locality, Point Vincent, California.

RANGE. Known only from type locality.

### Haliotis cracherodii imperforata Dall, 1919

*Proceedings of the United States National Museum, 56:370. Bulletin 112, United States National Museum, Pl. 21.*

In *The Nautilus* for December 1910 (p. 96), I described a unique form of this species which is entirely imperforate, never having had any perforations, but appears normal in every other respect. While this can hardly be termed a variety, it seems well to give it a name in order that it may be kept in mind by those interested in teratology of mollusca. The specimen . . . . measures 100 mm. in length by 42 in height and 95 in width. (Dall.)

TYPE in United States National Museum, No. 219850. Type locality, on the coast of California not far from San Pedro.

RANGE. California to Lower California.

## Haliotis californiensis Swainson, 1822

### Plate 86, figs. 1, 2

*Zoölogical Illustrations,* 2: Pl. 80.

H. testa ovali, laevi, obscure thalassina; labio exteriore supra immarginato, interiore lato, complanato, foraminibus numerosis, minutis, orbicularibus, laevibus. (Swainson.)

Shell ovate, smooth, obscure sea green; outer lip above immarginate; inner lip broad, flat; perforations numerous, very small, orbicular and smooth. (Swainson.)

TYPE in ? Type locality, California.

RANGE. On the islands, the Farallones to Guadalupe.

## Haliotis californiensis bonita Orcutt, 1900

*West American Scientist,* 10: No. 5, 30.

A new form with 13 long narrow holes close together, without showing scars of any of the closed holes, and characterized further by the very large, rough, muscular impression (50 mm. in greatest diameter) forming a most beautiful "pearl" and showing equally well from the inside or outside in the polished type. It is evidently rare, and may be from Mexican waters. (Orcutt.)

TYPE in Orcutt Collection? Type locality, near Santa Barbara, California.

RANGE. California.

## Haliotis rufescens Swainson, 1822

### Plate 90, fig. 2; Plate 89, fig. 2

*Bligh Catalogue,* Appendix 2. *Conchologia Iconica,* Haliotis, fig. 6.

Testa ponderosa, leviter striata, aliquando nodis sparsis deformi, rufescente: spira brevi; labii exterioris extremitate antica gibba: labio interiore angusto; foraminibus levatis, remotis (3 aut 4 apertis).

Shell ponderous, obsoletely striated, sometimes deformed by scattered nodules, reddish; spire small; outer lip at the top gibbous; inner lip narrow; perforations elevated, remote, only three or four open. In proportion as the back of the shell is worn down, the color becomes a brighter red. It attains to a very large size, and when in perfection is a beautiful species. Inhabits the Galapagos and California. (Swainson.)

Shell large, heavy and solid, oval, not very convex; sculpture consisting of unequal spiral cords and threads and wide, low, radiating waves; color dull red; holes three or four. The shell is very large, sometimes attaining a length of nine inches; it is thick and heavy, covered outside with a thick, brick-red layer which projects at the edge of the lip forming a narrow, coral-red edge. The spiral cords are unequal in size, and finer than in *H. fulgens;* the waves of the surface are large and oblique. Below the row of holes there is a depression, followed by a low ridge bearing usually large, obtuse tubercles. The spire does not project above the general curve of the back. Inside the nacre is lighter than in either *H. fulgens* or *corrugata,* and the play of tints not so much broken. The colors are chiefly pink and light green, with here and there a small area of prussian blue. The muscle scar is large, peculiarly and variously striped with olive-brown, green, and blue; a portion of it is roughened by coarse raised cords which take a spiral direction. The columellar plate is rather narrow, its lower part sloping inward somewhat. Perforations large, somewhat tubular, three or four open. Length, 185; width, 150; convexity, 40 mm. (Tryon and Pilsbry, *Manual of Conchology.*)

TYPE in Bligh Collection. Type locality, California.

RANGE. Bodega Bay, California, to La Paz, Lower California.

## Haliotis fulgens Philippi, 1845

*Zeitschrift für Malakozoologie,* 150. *Conchologia Iconica,* Haliotis, fig. 9.

H. testa maxima, ovata, convexa, sulcis transversis superficialibus exarata, rubra; spira parva, laterali; foraminibus parum tuberlosis, sates magnus; margine sinistra haud prominente; margarita pulcherrima, viridi, fusco mixta. Long., 7; lat., 5⅓; alt., 2½ poll. (Philippi.)

Ab affini *H. gigantea* Chemnitzii distinguitur testa magis convexa, foraminibus parum tuberlosis et margarita pulcherrima viridi, quae margaritam *H. iris* superat, cum *H. gigantea* margaritam albam, argenteam habitata, *H. iris* margine sinistro ptominente, spira perpendiculari, sculptura et colore dorsi differt. (Philippi.)

Shell large, oval, quite convex, sculptured all over with equal rounded cords or lirae; of a reddish-brown color. Generally five holes are open. The form is oval, as in the other American *Haliotis,* the back quite convex.

It is solid, but thinner than *H. rufescens*. The outside is a uniform dull reddish-brown. It is sculptured with rounded spiral lirae, nearly equal in size, 30 to 40 in number on the upper surface. At the rows of holes there is an angle, the surface below it sloping almost perpendicularly to the columellar edge, and having an obtuse keel about midway. The spire does not project above the general curve of the back. Inside dark, mostly blue and green with dark coppery stains, pinkish within the spire; the muscle impression painted in a peculiar and brilliant pattern, like a peacock's tail. Columellar plate wide, flat, sloping inward. Cavity of the spire small, almost concealed. Perforations rather small, elevated, circular, about five in number. Length, 170; width, 130; convexity, 48 mm. (Tryon and Pilsbry, *Manual of Conchology*.)

TYPE in Philippi Collection, Chile. Type locality, California?

RANGE. Monterey Bay, California, to Lower California.

## Haliotis wallalensis Stearns, 1899

### Plate 90, fig. 1; Plate 89, fig. 1

*Proceedings of the United States National Museum*, **22**:140. *Bulletin 112*, United States National Museum, Pl. 22.

Shell of an oval form, considerably flattened, and with about two and a half whorls; color, dark brick red, with occasional mottlings of pale bluish green; holes, four in the young to six in the adult; sculpture of fine, somewhat irregular spiral threads, crossed by fine, close, slightly elevated, sharp concentric lamellae, and a few small obscure wavelets which radiate obliquely from the apex; nacre rather pale, with pink and pale-green reflections, but much less deep in color than the typical *fulgens*.

This variety differs from the type in its more elongate and flattened form, its constantly finer spiral threading, and its paler nacre. The concentric lamellation is sometimes undeveloped on the young shells. It has the same number of holes as the type. (Stearns.)

TYPE in United States National Museum, No. 98327. Type locality, in the vicinity of Walalla, Mendocino County, California.

RANGE. Mendocino County, California.

Described as *H. fulgens* var. *walallensis* Stearns.

## Haliotis kamtschatkana Jonas, 1845

### Plate 88, figs. 1, 2

*Zeitschrift für Malakozoologie*, 168. *Conchologia Iconica*, Haliotis, Pl. 3, fig. 8.

H. testa ovata, convexiuscula, tenui, rubro, viridi, et albo variegata et marmorata, striis longitudinalibus sulcisque transversis undulatis decussata, plicis obliquis costisque concentricis nodosis et tuberculosis gibberosa; lateraliter biangulata, in angulo superiore perforata: forminibus tubulosis, quique perviis; spira subterminali, prominula; intus concava, plicis multis inaequalibus iniqua, margarita splendidissime iridescente, cavitate spirali detecta; margine dextro acuto undatim flexo, sinistro subangusto, plaro. Diameter, 3" 4"".

*H*. with ovate shell, slightly convex, thin, green, high, variegated and mottled; checked with longitudinal oblique plications and concentric nodose ribs; laterally biangulate, and perforate in the upper angle; with tubular foramens, with subterminal spire, somewhat prominent; within concave, convex with many unequal plications, brilliantly iridescently pearly; ornamented with a spiral cavity; with right margin acute, undulately bent, with the left subangulate, plane. Variety with smaller shell, more plane and less rough.

TYPE in Museum Grüner. Type locality, Unalaska Island, Kamchatka Sea.

RANGE. Kamchatka Sea, Sitka, Alaska, to Redondo, California, and Sendai, Japan.

## Haliotis assimilis Dall, 1878

### Plate 87, figs. 1, 2

*Proceedings of the United States National Museum*, 1:46. Tryon and Pilsbry, *Manual of Conchology*, **12**; Pl. 22, fig. 29.

Shell solid, strong, not very thick, with a rather light pink, white and greenish nacre, usually with five open holes; spire more elevated than that of any other California species, consisting of two and a half or three whorls; aperture very oblique in adult specimens, the thickened margin of the columella narrow, somewhat concave, inclined sharply inward and upward, about three-fourths as long as the columellar side of the aperture. Between the row of openings and the columellar edge, the space is unusually broad, marked midway by an obtuse carina, separated from the row of holes by a shallow channel; surface reddish or dull greenish, with rather rough, crowded, unequal, spiral ribs and rounded, irregular, wavy, radiating undulations crossing the spiral sculpture obliquely. The muscular impression, in most specimens, is but lightly marked, and, except for occasional spot-like impressions, is smoothly nacreous, like the rest of the interior. Long., 4.5; lat., 3.0; alt. of spire, 1.5 to 2.0 in. Aperture 3 inches wide and 3.75 long, in an adult specimen. (Dall.)

TYPE in United States National Museum, No. 31267. Type locality, California.

RANGE. Farallon Islands to San Diego, California, in deep water.

## Haliotis corrugata Gray, 1828

Wood, *Index Testaceologicus,* Supplement, Pl. 8, fig. 6. *Conchologia Iconica, 3;* species 12.

Shell ovate, convex, spirally closely ridged, ridges prominently raised throughout in swollen wrinkles, sometimes obliquely crossed with waved folds, perforations few in number, large, tubiferous, three only open; exterior rayed with dark green and reddish chestnut, edged with black, very beautifully iridescent. (*Conchologia Iconica.*)

TYPE in Museum Cuming? Type locality, California.

RANGE. Monterey, California, to San Quentin Bay, Lower California.

## Family FISSURELLIDAE

## Genus **DIADORA** Gray, 1821 (*Glyphis; Fissuridea*)

Shell ovate-conical; orifice anterior to the center, elliptical. Surface cancellated by radial ribs and concentric growth lines. Internal callus around the orifice heavy and slightly truncated behind. Interior of shell smooth; margins crenulated. (I. S. Oldroyd.)

TYPE. *Fissuridea aspera* Eschscholtz.

DISTRIBUTION. Universal but mostly in warm seas.

This is equal to *Genus glyphis.*

## Diadora inaequalis Sowerby, 1834

*Proceedings,* Zoölogical Society of London, 126. *Conchologia Iconica,* 6; species 50.

Fiss. testa oblonga, tenui, subdepressa, latere antico brevi, postico longo; intus albicante, margine albo ingroque vario, crenulato; extus radiatim striata, concinne decussata, olivacea albicante subraditim variegata; apertura dorsali antica, oblonga, utrinque bidentata: long., 1.1; lat., 0.6 poll.

Shell oblong-ovate, attenuated and much inclined anteriorly, raised in the middle, slightly compressed at the sides, radiately thinly ridged, finely latticed with concentric striae, orifice rather small, oblong, a little contracted in the middle; rusty black, ash-white around the orifice. (Reeve.)

TYPE in Museum Cuming. Type locality, Galapagos Islands.

RANGE. Santa Barbara, California, to Manta, Ecuador, and the Galapagos Islands.

Described as *Fissurella inaequalis,* Sowerby.

## Diadora aspera Eschscholtz, 1833
### Plate 85, fig. 11; Plate 93, fig. 1

*Zoölogical Atlas*, Part 5, 21; Pl. 23, fig. 5.

Testa elliptica, convexo-pyramidata, albogrisea, transversim sulcata, costis prominentioribus asperas radiata, foramine ovali inclinato, margine crenato subreflexo. Lange, 1″ 8‴; Greite, 1″ 2‴; Höhe, 8‴. (Eschscholtz.)

Shell ovate, narrower in front, conical, the slopes nearly straight or a little convex behind the middle. Sculptured with numerous radiating riblets, of which 30–34 are larger, the intervals between them bearing about three smaller ones; the whole decussated by close elevated concentric lirae, which are more or less scale-like and imbricating. Color soiled whitish, with numerous wide blackish rays. Inside white, hole-callus white, very abruptly truncated behind; margin deeply and sharply crenulated. Perforation short-oval, nearly circular, in front of the middle. Length, 56; width, 40; height, 18 mm. (Tryon and Pilsbry, *Manual of Conchology.*)

TYPE in St. Petersburg Museum? Type locality, Sitka, Alaska.

RANGE. Cook's Inlet, Alaska, to Magdalena Bay, Lower California.

## Diadora aspera densiclathrata Reeve, 1850

*Conchologia Iconica*, Fissurella, fig. 64.

Fiss. testa ovata, medio elevata et antice inclinata, liris concentricis et radiantibus undique dense clathrata, orificio subrotundo; albida aut flavicante, olivaceo radiata. (Reeve.)

Shell ovate, elevated and inclined anteriorly in the middle, thickly latticed throughout with radiating and concentric ridges, orifice nearly round; whitish or yellowish, rayed with olive. (Reeve.)

TYPE in Museum Cuming. Type locality?

RANGE. Monterey, California, to San Quentin Bay, Lower California, and Socorro Island.

## Diadora murina (Carpenter MS) Dall, 1885

*Proceedings of the United States National Museum*, 8:543; also 15:197.

Two found dead, one at Catalina. (This is the *Glyphis densiclathrata* of Californian conchologists, but not of Reeve; *G. saturanlis* of Pilsbry—*Nautilus*, V., p. 105—not of Carpenter; and *G. densiclathrata* var. *murina* of Carpenter.) (Dall, *Proceedings,* 15:197.)

TYPE in United States National Museum. Type locality, Catalina Island.

RANGE. Crescent City, California, to Magdalena Bay, Lower California. Fossil: Pleistocene, San Pedro, San Joaquin Bay, Orange County, California.

Described as *Fissurella (Glyphis) murina*. Carpenter.

## Genus HEMITOMA Swainson, 1840 (Subemarginula)

Shell depressed, anterior margin slightly channeled. (Woodward, *Manual of the Mollusca*.)

TYPE. *Emarginula octoradiata*.

DISTRIBUTION. West Indies, Great Britain, Norway, Philippines, Australia, California.

Described as subgenus of *Emarginula*.

## Hemitoma bella Gabb, 1865

*Proceedings of the California Academy of Sciences,* 3:188.

E. t. alba, oblonga-ovalis, subelevata, antice parum convexa, postice excavata, parte anteriori angustiori; apice excentrico, prominente, parum recurvato; sinu mediocri; costis radiantibus circiter xix cum costis minoribus interstitialibus irregularitur alternantibus, per lineas concentricas clathratis. Long., .55; lat., .36; alt., .22.

Shell oblong oval, somewhat elevated, narrowest anteriorly; apex excentric, posterior, prominent and somewhat recurved; outline in front of the apex slightly convex, from the apex to the posterior margin slightly excavated, sides descending nearly straight; sinus moderate, variable, situated at the extremity of strong rib; surface ornamented by about nineteen large radiating ribs, with smaller ones interposed, all crossed by moderately prominent concentric ridges; color white. (Gabb.)

TYPE in California State Survey, *Mollusca,* No. 466. Type locality, Monterey, California.

RANGE. Santa Cruz to San Pedro, California.

Described as *Emarginula bella* Gabb.

## Hemitoma golischae Dall, 1916

### Plate 85, fig. 2

*Nautilus,* 30:61. *Bulletin 112,* United States National Museum, Pl. 5, fig. 2.

Shell of moderate size, radiately ribbed, concentrically zoned and radiately striped with dark rose color, the worn apex greenish, the interior whitish, the extreme edge of the slightly crenulated interior margin with the external coloration showing through. Sculpture of rather strong radial ribs, corrugated more or less by strong incremental rugosities, alternated in front and behind with a single feebler rib, on the sides there are two or three minor riblets between the major ribs; apex rather acute, somewhat anterior; marginal notch shallow, its groove distinct on the internal face of the shell; the number of major ribs in the type specimen is about twenty. Length, 20; apex behind the notch, 7; width, 13; height, 7.5 mm. (Dall.)

TYPE in United States National Museum. Type locality, San Nicolas Island, California.

RANGE. Known only from type locality.

## Hemitoma yatesii Dall, 1901

### Plate 85, figs. 1, 3

*Nautilus,* **14**:125. *Proceedings of the United States National Museum,* **24**; Pl. 38, figs. 1–3.

Shell large, coarse, strong, whitish-gray, or pale olive-green on the fresher portions especially, a very narrow margin about the base; sculptured with strong, not dichotomous, radial ribs, of which about 20 are primary, between each two of which lie from one to four secondary riblets, most numerous at the sides of the shell; besides these there is a very strong anal fasciole, higher and stronger externally than any of the ribs, extending from the apex, and ending in front at a notch about 3.5 mm. deep and rounded above and behind; the radiating sculpture is sharply and irregularly imbricated by the rude and profuse incremental sculpture, which is too close and irregular to form reticulation; apex small, pointed, not much elevated, situated three-fifths of the way from the front to the posterior margin; the fasciole descending from it swerves a little to the right of the median line of the shell; interior white, the extreme margin pale olive-green but almost linear; anal furrow deep, extending nearly to the apex, where it is lost in a very pale olive deposit of shelly matter; margins crenulated by the sculpture; muscular impressions strong, the two recurved scars unequal, the right one larger. Long. of shell, 51; lat., 36; alt., 13 mm. (Dall.)

TYPE in United States National Museum. Type locality, Monterey Bay, California.

RANGE. Forrester Island, Alaska, to Monterey, California.

## Genus **PUNCTURELLA** Lowe, 1827

Shell conical, elevated, with the apex recurved; perforation in front of the apex, with a raised border (septum) internally; surface cancellated. (Tryon, *Structural and Systematic Conchology.*)

TYPE. *Rimula noachina* Linn.

DISTRIBUTION. Greenland, boreal America, Norway, North Britain, Tierra del Fuego.

## Puncturella cooperi Carpenter, 1864

*Supplementary Report,* British Association for the Advancement of Science, 651. *Proceedings of the California Academy of Sciences,* 3:214.

Outside like *galeata,* but without props to the lamina. 30–120 fm. not r. (Carpenter.)

P. t. *P. galeatae,* fere exacte simulante; sed lamina interna solida, planata, haud antice sinuata, haud suffulta. Long., 0.30; lat., 0.21; alt., 0.24; div., 70°. (Carpenter.)

Shell almost exactly like *P. galeata;* but with internal solid flattened lamina, not sinuate in front, without props. (Translation.)

TYPE in United States National Museum. Type locality, Catalina Island, California.

RANGE. Kasa-an Bay, Alaska, to Santa Rosa Island, and San Pedro, California.

## Puncturella galeata Gould, 1846

### Plate 93, fig. 3

*Proceedings of the Boston Society of Natural History,* 2:159. *Report of the United States Exploring Expedition,* Mollusca, 369, Pl. 31, figs. 476–7.

Testa solida, cinerea, elevata, globoso-conica, apice centrali, acuto, prorsum porrecto, striis filiformibus confertis radiata, et striis minoribus concentricis decussata: fissura brevis, fusiformis: apertura rotundato-ovalis; margine expanso, acuto, minutissime crenulato; fornice costa trans-versali antice suffulto, fossam trigonalem utroque latere formante. Long., $\frac{9}{20}$; lat., $\frac{8}{20}$; alt., $\frac{8}{20}$ poll. (Gould.)

Shell solid, ashy-white, erect, elevated conic, with the outline somewhat convex; the apex central, elevated, acute, and reaching forward and upward. Surface radiated with numerous nearly equal raised lines, which are decussated by much finer concentric striae: fissure small. Aperture rounded-oval, the margin expanded and thinned to a sharp edge, which is minutely crenulated. The fornix is circularly arched over a short sulcux, and in front of it, near the summit, runs a transverse rib, leaving between it and the fornis, on either side, a triangular pit. The summit also appears to be filled with callus. Color greenish-white within. (Gould.)

Type in United States National Museum. Type locality, Puget Sound.

Range. Unalaska, Aleutian Islands, south to Santa Rosa Island, California. Fossil: Pleistocene, San Pedro, California.

Described as *Rimula galeata* Gould.

## Puncturella cucullata Gould, 1846

### Plate 93, fig. 4

*Proceedings of the Boston Society of Natural History,* **2**:159. *Report of the United States Exploring Expedition,* Mollusca, 368, Pl. 331, fig. 475.

Testa solida, cinerea, per-inequilateralis, costis ad 40 acutis, compressis, majoribus et minoribus alternantibus radiata, et striis concentricis confertis muricata; apice elevato, acuto, adunco, prorsum spectante: fissura lanceolata: apertura ovata; margine sulcis inequalibus alternantibus crenulato: facies interna porcellana; fornice tenui, simplici, arcuato. Long., ⅝; lat., ⁷⁄₁₆; alt., ⅜ poll. (Gould.)

Shell rather solid, ovate at the base, and rising to form an elevated, oblique cone, having the apex at the posterior fourth of the shell, so that the anterior slope is moderate, while the posterior slope is nearly vertical. The apex is prominent, acute; reaching forward like a curved beak. The color is dirty cinereous. The surface is radiated by sharp, prominent, compressed ribs, which are alternately larger and smaller, to the number of about 40; and these are crossed by very fine and crowded concentric lines, by which they are delicately muricated. Fissure narrow, and narrowing upward. The interior is porcelain-white; the edge is beautifully crenulated by regular furrows, alternately longer and shorter, deeper and shallower, answering to the external ribs. The fissure is rounded, and the vault enclosing it is thin, simple, semicircular, and without callus at the summit. (Gould.)

Type in United States National Museum. Type locality, Puget Sound.

Range. Kodiak Island, Alaska, to La Paz, Lower California. Fossil: Pleistocene, Santa Barbara and San Pedro, California. Pliocene: Packard's Hill, Santa Barbara County, California.

Described as *Rimula cucullata* Gould.

## Puncturella multistriata Dall, 1914

### Plate 93, fig. 2

*Nautilus,* **28**: 63.

Shell small, slightly alternated, rather close set radial threads. Length, 27; breadth, 18; height, 16 mm. (Dall.)

Type in United States National Museum. Type locality, Puget Sound.

RANGE. Atka Island, Aleutians, to San Diego, Cortez Banks, Chile.

This is the shell we have called *Puncturella cucullata* Gould. It differs from that species in having more ribs, thirty-four main ones and with fine ones in the interstices; while *Puncturella cucullata* has eighteen main ribs, with an uneven number in the interstices. (Oldroyd.)

## Puncturella major Dall, 1891

*Proceedings of the United States National Museum,* **14**:189; **17**:712; Pl. 26, fig. 4.

Shell resembling *galeata* in general and especially in the interior, but very much larger; radii alternately large and small from the beginning; shell white; anterior slope rectilinear, posterior slope slightly arched and a little longer; internal margin crenulate. Length, 57; breadth, 42; height, 27 mm. (Dall.)

TYPE in United States National Museum. Type locality, U.S. Steamer "Albatross" Station 3262, off coast of Akutan Island, Bering Sea, in 43 fathoms.

RANGE. Pribilof Islands, Bering Sea, to Dixon Entrance, Alaska.

## Puncturella longifissa Dall, 1914

*Nautilus,* **28**:63.

Shell is low, narrow, small, with an arcuate back, strong radial threads, very posterior apex, and the slit half as long as the distance from the apex to the anterior margin. (Dall.)

TYPE in United States National Museum. Type locality, Bering Island, Bering Sea.

RANGE. Known only from type locality.

## Puncturella carophylla Dall, 1914

*Nautilus,* **28**:63.

Shell minute, high, cylindro-conic, with strong, even, radial threads, resembles nothing so much as a miniature solitary coral. (Dall.)

TYPE in United States National Museum. Type locality, off San Diego, California, in 40–80 fathoms.

RANGE. Known only from type locality.

## Genus **MEGATHURA** Nuttall (in Pilsbry), 1891

Shell large, oblong-oval, embedded in the mantle, but large enough to cover most of the upper surface of the animal; apex a little in front of the middle, entirely removed by the large oval perforation; edges of shell not thickened, very regularly and finely crenulated at all stages of growth; internal callus-rim of perforation somewhat excavated posteriorly in young shells. (Tryon and Pilsbry, *Manual of Conchology.*)

TYPE. *Lucapina crenulata* Sowerby.

DISTRIBUTION. Monterey Bay, California, to Lower California.

## Megathura crenulata Sowerby, 1825

Tankerville, *Catalogue,* Appendix 6. *Conchological Illustrations,* Fissurella, fig. 18.

F. testa oblongo-ovata, depressa, alba; foramine ovato, integro; superficie striis confertis, radiantibus, decussatis; margine crenulata. Long., 3$\frac{4}{10}$; lat., 2$\frac{1}{10}$ unc. (Sowerby.)

Shell oblong-ovate, somewhat irregular, deeply convex, radiately finely and very closely ridged, ridges transversely very finely striated with obsolete scales, orifice large, nearly orbicular, basal margin crenulated; pinkish fulvous-color. (*Conchologia Iconica.*)

TYPE in Tankerville Collection. Type locality, Monterey, California.

RANGE. Monterey Bay, California, to Cerros Island, Lower California.

## Genus MEGATEBENNUS Pilsbry, 1890

The small species grouped under this name agree with *Fissurellidae* (as restricted to Orbigny's type) in having a partially internal, large-apertured shell, a mantle ample enough to cover the entire dorsal surface of the foot and head, its edges entire; they differ from *Fissurellidae* in the much greater proportional size of the shell, more elevated body, the foot (viewed ventrally) almost as extensive as the mantle, the margin of the latter not at all thickened, and the shell not white-bordered above. (Pilsbry.)

TYPE. *Megatebennus bimaculata* Dall.

DISTRIBUTION. Cape of Good Hope, West Africa, Benguela, South Africa, Natal, Australia, Tasmania, New South Wales, California.

## Megatebennus bimaculatus Dall, 1871

### Plate 85, fig. 15

*American Journal of Conchology,* 7:132; Pl. 15, fig. 7.

Shell ellipsoidal when young, subquadrangulate, and a little narrower in front than behind, when adult. Aperture the same shape as the shell, slightly encroached upon in some specimens by a point on each side. External surface furnished with radiating, round costae, not bifurcating but widening slightly toward the margin. These are crossed by evident but not very strong lines of growth, which in some individuals are rather strong. Anterior declivity of the shell concave, sides flattened, posterior declivity rounded convex. Color whitish, with numerous radiating rays of brown or slate color, usually with a broad fasciculus of darker rays in the middle of each side extending from the apex to the margin, and occasional dark dots on the ribs. Shell occasionally entirely brown or slate color, with two darker rays on the sides. Epidermis none. Interior pure white, the two dark rays sometimes showing through the shell. Extreme outer edge finely denticulate or rounded and smooth according to the stage of growth. Margin as a whole broad, smooth, differentiated from the rest of the surface by a wide, shallow groove. Margin of the aperture similarly bordered. Muscular impressions distinct, surface marked by fine radiating lines; polished. Anterior and posterior margins internally concave or emarginated, so that when laid upon a flat surface in the natural position the ends of the shell do not touch it. Adult, long., .7; lat., .5; alt., .2 in. Long. of aperture, .23; lat. of do., 12 in. (Dall.)

TYPE in United States National Museum. Type locality, Monterey, California.

RANGE. Forrester Island, Alaska, to Cape San Lucas. Fossil: Pleistocene, Santa Barbara, San Pedro.

Described as *Fissurellidaea bimaculata* Dall.

## Genus **LUCAPINELLA** Pilsbry, 1890

*Fissurellidae* with an oblong shell, not sunken in or covered by the mantle, and about as long as the foot; its apex subcentral, wholly removed by a rather large oblong perforation, which is margined within by an entire (not truncated) callus; edge of shell blunt, scarcely crenulated in adults except in front and behind; sculptured with scaly riblets; front and side-margins level, posterior margin a little elevated. Animal with a fleshy foot, much too large to be contained in the shell; mantle-edge thickened, papillose on its lower edge and having narrow processes extending up over the shell-edge. (Pilsbry.)

TYPE. *Clypidella callomarginata* Carpenter, MS.

DISTRIBUTION West coast of North and South America, West Indies, Barbadoes, Florida Keys.

### Lucapinella callomarginata (Carpenter, MS) Dall, 1871
#### Plate 85, fig. 16

*American Journal of Conchology,* 7:133; Pl. 15, fig. 8. Tryon and Pilsbry, *Manual of Conchology,* 12:196; Pl. 44, figs. 3, 4, 5; Pl. 61, figs. 1–5.

Shell oblong, a trifle narrower in front, rather depressed, the subcentral summit occupied by a rather large fissure, shaped like the shell and from one-fourth to one-fifth the shell's length. Surface having radiating riblets and concentric growth-laminae, which are elevated into imbricating scales on the ribs; color gray or white, radiated with black. The form is oblong, sides subparallel or somewhat convex. The front slope of the cone is a trifle convex, the lateral slopes straight or a little concave, the posterior slope concave. The two sides of the fissure project upward in more or less salient points. The sculpture consists of alternately larger and smaller radiating riblets, crossed by sharp, concentric laminae, elevated into imbricating scales on the ribs. The blackish rays sometimes cover most of the surface, sometimes are narrow and few. Inside bluish-white, with a rather wide callus rim around the fissure; lateral edges blunt, ends slightly crenulated. In immature specimens the edges are crenulated. Length, 19; breadth, 10; height, 4½ mm. (Tryon and Pilsbry, *Manual of Conchology.*)

TYPE in United States National Museum. Type locality, Monterey.

RANGE. Bodega Bay, California, to Magdalena Bay, Lower California. Fossil: Pleistocene, San Pedro to San Diego, California.

Described as *Clypidella callomarginata*.

## Genus **FISSURELLA** Bruguiere, 1791

Shell oval, conical, depressed, with the apex in front of the center, and perforated; surface radiated or cancellated; muscular impression with the points incurved. In very young shells the apex is entire and subspiral; but as the perforation increases in size, it encroaches on the summit, and gradually removes it. (Tryon, *Structural and Systematic Conchology.*)

TYPE. *F. picta* Gmel.

DISTRIBUTION. Universal, but mostly in warm seas.

### Fissurella volcano Reeve, 1849

*Conchologia Iconica,* Fissurella, fig. 2.

Fiss. testa ovato-conica, antice vix attenuata, radiantim costata, costis rudibus, inaequalibus, hic illic majoribus, orificio oblongo; albida, roseo radiata radiis nigropunctatis, interdum rosea, purpureo-radiata, circa orificium nigropunctata. (Reeve.)

Shell ovately conical, slightly attenuated, anteriorly rather thick, radiately ribbed, ribs rude, uneven, here and there larger, orifice oblong; whitish, rayed with rose, rays dotted with black, sometimes rose, rayed with purple, black-dotted around the orifice. (Reeve.)

TYPE in Museum Cuming. Type locality, Santa Barbara, California.

RANGE. Crescent City, California, to Panama.

## Fissurella volcano crucifera Dall, 1908

*Proceedings of the United States National Museum,* **34**:256.

A peculiar color-variety of this species has been sent to me from the Pacific coast a number of times in the hope that it was something new, and it seems worthy of a varietal name. The shell is as usual, except in color, the ground color being a brownish gray, with darker maculae, while from the apex start four broad white rays at right angles to each other, the posterior ray rapidly becoming V-shaped, the others remaining entire, each ray reaching four or five millimeters in length, and the anterior and posterior rays being in the longer axis of the shell. (Dall.)

TYPE in United States National Museum, No. 199171. Type locality, San Pedro, California.

RANGE. Monterey to Lower California.

## Family LEPIDOPLEURIDAE

## Genus **LEPIDOPLEURUS** Risso, 1826

Insertion plates absent. Girdle with minute, gravelly, smooth or striated scales, usually with a marginal fringe of longer scales. (Tryon and Pilsbry, *Manual of Conchology.*)

TYPE. *Lepidopleurus cajetanus* Poli.

DISTRIBUTION. North Atlantic and Arctic Seas, Britain, Norwegian coast, Greenland, south to Gulf of Lyons, Vigo, Spain, North Pacific, Alaska, West Indies, Mediterranean, California, Japan.

## Lepidopleurus cancellatus Sowerby, 1839

*Conchological Illustrations,* Chiton, figs. 104, 105. Tryon and Pilsbry, *Manual of Conchology,* **14**:3; Pl. 3, figs. 54–58.

Shell small, elongated, much elevated, regularly arched, not angled. Orange-ashen or whitish. Anterior valve radiately, evenly, very finely granose-lirate. Central areas of the intermediate valves having distinct longitudinal, fine, close granulous lirae, the granules being sometimes arranged in transverse lines also, giving a latticed appearance; lateral areas distinct, decidedly raised, convex, having radiating but rather irregular rows of granules. Posterior valve with central elevated apex; posterior slope concave. Interior white, the sutural plates small, triangular; jugal sinus very broad. Girdle narrow, densely beset with delicate, scarcely imbricating or striated scales. Length, 5½; breadth, 2½ mm. (Tryon and Pilsbry, *Manual of Conchology*.)

TYPE in British Museum. Type locality?

RANGE. Bering Strait to Sitka and Oregon, North Atlantic, Greenland.

## SECTION LEPTOCHITON Gray, 1847

### Lepidopleurus rugatus (Carpenter) Pilsbry, 1892

Tryon and Pilsbry, *Manual of Conchology*, 14:11; Pl. 3, figs. 67–70.

Oval, rather convex, the lateral slopes nearly straight, dorsal ridge broadly arched. Front and back valves and lateral areas of the intermediate valves sculptured with excessively fine radiating striae, which are feebly granose, and having well-marked, coarse, concentric wrinkles; central areas having an equally minute sculpture of longitudinal subgranose striae. Mucro subcentral and prominent. The lateral areas are slightly raised. Length, 13–14; breadth, 7 mm. (Tryon and Pilsbry, *Manual of Conchology*.)

TYPE in Philadelphia Academy of Natural Sciences, No. 35586. Type locality, San Tomas River, Lower California.

RANGE. Monterey, California, to Todos Santos Bay, Lower California. This is Carpenter's MS *L. internexus* var. *rugatus*.

### Lepidopleurus internexus Carpenter, 1892

Tryon and Pilsbry, *Manual of Conchology*, 12.

Shell similar to *L. nexus*, but much smaller, orange-colored. Valves regularly arcuate, without jugum. Sutural plates small, triangular; sinus larger. Girdle with more solid scales, scarcely pilose. Length, 4½; breadth, 3¼ mm. (Carpenter.) (Tryon and Pilsbry, *Manual of Conchology*.)

TYPE in? Type locality, Santa Barbara, California.

RANGE. Belkoffski, Alaska, to San Diego, California.

### Lepidopleurus ambustus Dall, 1919

*Proceedings of the United States National Museum*, 55:499.

Chiton with a rather high but rounded back, the valves reddish-brown with numerous blackish flecks, giving the interspaces a somewhat vermicular aspect; anterior valve semicircular, simple; posterior valve smaller, shorter, with a subcentral prominent mucro, behind which the surface is concavely excavated; intermediate valves subequal, dorsally rounded with hardly any trace of lateral areas; sculpture of the valves nearly uniform, fine, equal grooves, axially on the dorsal area, oblique on the lateral areas and cut by still finer transverse striae into microscopic granulations; the distal ends of the middle valves in the adult shell with concentric undulations; insertion plates short, entire; girdle with crowded arenaceous, short spinules, interspaced near the valve margins with sparse longer white spines, the body of the girdle partaking of the colors of the shell, with a silvery reflection in certain aspects; extreme edge of the eaves finely striated in accordance with the exterior sculpture; interior whitish, the jugal sinus obscure, the sutural laminae narrow, arcuate, widely separated; a conspicuous median excavation bordered behind by two wing-like, radiately grooved, narrow callosities on the internal face of the valve under the central area of the posterior valve. Length of valves, 17; width, 8; height, 3 mm. (Dall.)

TYPE in United States National Museum, No. 274120. Type locality, Santa Barbara Channel in 20–40 fathoms.

RANGE. Known only from type locality.

### Lepidopleurus lycurgus Dall, 1919

*Proceedings of the United States National Museum, 55:500.*

Chiton minute, white, with touches of brown on either side of the jugal area; dorsum rather angular with flattened sides; anterior valve less than semicircular, its edge excavated behind, simple; posterior valve axially longer, with a prominent smooth subcentral mucro over a concave posterior slope; central area axially grooved, minutely granulose; posterior area with radial granulose very minute sculpture; intermediate valves with no marked jugal area, that and the pleural tracts with oblique minute reticulations; lateral areas distinct, radially minutely striate-punctate, with some obscure concentric undulations distally; girdle spiculose with minute white spinules of uneven length; interior white. Length of valves, 4.5; width, 3; height, 1 mm. (Dall.)

TYPE in United States National Museum, No. 274119. Type locality, Catalina Island, California.

RANGE. Known only from type locality.

### Lepidopleurus oldroydi (Bartsch) Dall, 1919

*Proceedings of the United States National Museum, 55:500.*

Chiton white, with a blackish spot on either side of the jugal area, strongly sculptured; back moderately arcuate, anterior valve semicircular, sculptured with irregularly disposed, small, prominent, round pustules;

posterior valve with subcentral prominent mucro, the central area granulose, the periphery pustulose; intermediate valves with axially punctostriate jugal area, lateral areas prominent by reason of the conspicuous pustules, which are relatively large for the size of the animal; the pleural tracts coarsely axially grooved, the grooves more or less distinctly punctate; girdle with crowded minute spines of equal length, giving an arenaceous effect; interior whitish, the jugal sinus wide with a straight edge, the sutural laminae small, subtriangular. Length of dry animal, 5; width, 2; height, 0.7 mm. (Dall.)

TYPE in United States National Museum, No. 218767. Type locality, Monterey, California.

RANGE. Monterey to Catalina Island, California.

## Lepidopleurus belknapi Dall, 1878

*Proceedings of the United States National Museum,* 1:1. Tryon and Pilsbry, *Manual of Conchology,* 14:7; Pl. 1, figs. 18–22.

L. t. elongata, valde elevata, dorsualiter angulata; albida plus minusve cinereo et nigro tincta; valvis elevatis, apicibus distinctis; mucrone centrali conspicuo; sculptura ut in *L. alveolo,* sed granulis in areis dorsualis sparsim et quincuncialiter dispositis; valva postica sub apice concava, postice sinuata; zona minima spiculis tenuibus versus marginem munita. Long., 10; lat., 3 mm. Div. 90°. (Dall.)

Shell elongated, much elevated, dorsally angled; whitish, more or less tinged with ashen or black. Valves elevated, with distinct apices; mucro central, conspicuous. Sculpture as in *L. alveolus,* but the granules of the dorsal areas sparse, and disposed in quincunx. Posterior valve concave below the apex, sinuated behind. Girdle narrow, having delicate spicules toward the margin. (Dall.) (Tryon and Pilsbry, *Manual of Conchology.*)

TYPE in United States National Museum. Type locality, North Pacific Ocean, in lat. 53° 8′ N., long. 171° 19′ W., at a depth of 1,006 fathoms.

RANGE. Off Aleutian Islands; off Luzon, Philippine Islands, 1,050 fathoms.

Described as *Leptochiton belknapi* Dall.

## Lepidopleurus luridus Dall, 1902

*Proceedings of the United States National Museum,* 24:556.

Chiton small, solid, narrow, of a lurid smoky color, darker on the lateral areas; girdle densely pilose, with whitish spicules; back rounded, with the jugum defined feebly, most conspicuous as a distinct mucro, mesially, on the intermediate valves; pleural areas divided by obscure depressed lines radiating from the mucro to the inner edges of the pleural laminae; lateral areas prominent more or less concentrically rugose; anterior valve simple, normal; posterior valve with a conspicuous central mucro, behind which it is more or less concave; the whole surface is covered with minute, quincuncially arranged pustulation; on the intermediate valves the pustules on the lateral and pleural areas appear to diverge from the inner margin of the lateral areas; internally there is a wide unattached margin on the under side of the posterior edge, mesially, in the intermediate valves; the pleural laminae are short and all the valves callous internally, with the points of attachment to muscles and girdle impressed; there is no linear arrangement of the pustules on the jugum; the ctenidia only reach the seventh valve. Long. of animal about 16; lat., 6; alt., 3 mm. (Dall.)

TYPE in United States National Museum, No. 109027. Type locality, Panama Bay, "Albatross" Station 3393, in 1,020 fathoms.

RANGE. Puget Sound, 48 fathoms, to Panama, 1,270 fathoms.

### Lepidopleurus farallonis Dall, 1902

*Proceedings of the United States National Museum,* **24**:557.

Chiton small, thin, wide, with a low rounded back and yellowish-white color; girdle narrow, sparsely spiculose, with very short, fine, bristly spicules; jugum hardly defined, with no obvious mucro; lateral areas slightly elevated and feebly concentrically rugose; anterior valve simple; posterior conspicuously mucronate and, behind the mucro, concave; surface entirely covered with minute, low, close-set pustules, arranged quincuncially and to some extent concentrically from the mucronal points; pleural laminae short, subtriangular; ctenidal line reaching the fifth valve. Long. of animal about 10; lat., 5.5; alt., 2 mm. (Dall.)

TYPE in United States National Museum, No. 109025. Type locality, "Albatross" Station 3104, off Farallon Islands, California, 391 fathoms.

RANGE. Farallon Islands, California, to Panama, in deep water.

### Lepidopleurus mesogonus Dall, 1902

*Proceedings of the United States National Museum,* **24**:555.

Chiton of moderate size, yellowish or ashy white, with a narrow girdle dusted with very minute spicules; valves laterally compressed, almost keeled at the jugum, and with the sides meeting there at an angle of 69°, slightly rounded at the junction; body narrow, ctenidia about a dozen on each side, the most anterior even with the front edge of the seventh plate; anterior valve simple, without insertion plates, sculptured with moderately strong concentric resting stages and fine, low, close-set, rounded pustules; posterior plate large, similarly sculptured, with a prominent mucro nearer the posterior than the anterior edge of the plate, the anterior and posterior areas hardly defined; intermediate valves with the lateral areas more or less irregularly concentrically ridged, the pleural and jugal tracts less distinctly so, the whole covered with uniform pustulation and the inner areas defined very obscurely by faint depressions; though angular, the jugum is not beaked anteriorly. Long. of animal, about 35; lat. (with dry girdle), 9; alt., 6 mm. (Dall.)

TYPE in United States National Museum, No. 109019. Type locality, off the Queen Charlotte Islands, B.C., "Albatross" Station 3342, in 1,588 fathoms.

RANGE. Bering Sea, 688 fathoms, Queen Charlotte Islands.

## SECTION XIPHIOZONA Berry, 1919

### Lepidopleurus nexus Carpenter, 1864

*Supplementary Report,* British Association for the Advancement of Science, 650. Tryon and Pilsbry, *Manual of Conchology,* 14:11.

Like *asellus:* scarcely sculptured: mantle-margin with striated chaffy scales, like *magdalensis,* interspersed with transparent needles, 20–80 fm. (Carpenter.)

Shell small, whitish-ashen, valves gothic-arched; lateral areas scarcely defined; entire surface ornamented with series of subquadrate granules, the series longitudinal upon the central, radiating upon the lateral areas and end valves, very close, scarcely interrupted. Jugum elevated, subacute; umbones inconspicuous. Mucro conspicuous, median. Inside with strong sutural lobes and a wide plane sinus to the middle valves; insertion plates obsolete. Girdle having narrow, close, striated scales, and needle-shaped, crystalline bristles here and there and around the margin. (Carpenter.) Length, 7½; breadth 4½ mm.; div. 90°. (Tryon and Pilsbry, *Manual of Conchology.*)

TYPE in United States National Museum. Type locality, Catalina Island, California.

RANGE. Catalina Island to Gulf of California.

Described as *Leptochiton nexus* Carpenter.

### Lepidopleurus heathi Berry, 1919

*Lorquinia,* January 1919.

Shell small, strongly convex. Anterior valve with 100 or more radial series of flattened grains. Median valves similarly sculptured, the lateral area not raised and poorly defined. Tail valve with mucro much elevated and almost overhanging posteriorly, its slope strongly concave. Sutural laminae triangular. Girdle covered with minute, short, pointed spinelets, with larger scattered spines among them, particularly near the sutures as noted above. Color of shell a warm yellowish brown, copiously mottled slaty gray. (Berry.)

TYPE in Berry Collection. Type locality, off Monterey, California, in 15 fathoms.

RANGE. Known only from type locality.

## Genus **OLDROYDIA** Dall, 1894

Valves separated by narrow extensions of the girdle, reaching the jugum; jugal area prominent, sculptured differently from the pleural tracts and extending in front of them between the sutural laminae; lateral areas not differentiated; valves heavy, strongly sculptured. (Dall.)

TYPE. *Oldroydia percrassus* Dall.

DISTRIBUTION. California to Lower California.

### Oldroydia percrassa Dall, 1894

*Nautilus,* **8**:90. *Zoölogica,* **22**, Heft 56, pt. 2, 71; Pl. 7.

Shell solid, strong, small, of a pale pinkish-brown with a darker brownish girdle which appears rather narrow in the dry state; scales very minute, partly dehiscent, chaffy, with occasional slender spinules resembling hairs; scales on the base crowded, minute, sandy; an extension of the girdle is prolonged between the valves on each side as far as the jugum, the surface of these sinuses is also minutely scaly with occasional spinules; valves thick, white below, moderately arched with the prominent jugum forming a sort of keel; near the points of insertion the valves are heavily callous below; the sutural laminae are short, smooth, and separated at the median sinus by a prolongation of the jugum in advance of the anterior margins of the pleurae; sculpture of the jugum consisting of punctate fore and aft parallel grooves with some small elevated transverse ridges anteriorly; the rest of the valve has, on each side, six or eight vermicular ridges divaricating toward the posterior edge of the valve and irregularly corrugated with sharp, fine, elevated lamellae crossing the interspaces transversely but fading out on the ridges; head-valve with minutely nodulous concentric ridges; tail-valve highest at the subcentral, not very prominent mucro, in front sculptured like the intermediate valves behind the mucro like the head-valve. Length, about 14; width, 5.75; height, 2.5 mm. in the dry state. The dry girdle is about half a millimeter wide. (Dall.)

TYPE in United States National Museum. Type locality, off San Pedro, in about 75 fathoms.

RANGE. Monterey, California, to San Martin Island, Lower California.

Described as *Lepidopleurus* (*Oldroydia*) *percrassus*.

## Genus **HANLEYA** Gray, 1857

Anterior valve with an unslit insertions-plate; other valves destitute of the plates. (Tryon, *Structural and Systematic Conchology*.)

TYPE. *Chiton debilis,* Gray.

DISTRIBUTION. Mostly northern, one from deep waters of Gulf of Mexico.

### Hanleya hanleyi Bean, 1844

Thorpe, *British Marine Conchology,* 263, fig. 57. Tryon and Pilsbry, *Manual of Conchology,* **14**:17; Pl. 3, figs. 71–79.

Shell oblong oval, narrow, carinated, brownish-white, granulated, with the granulations larger toward the margin which is covered with minute spines; inside pale green. Length, .3; breadth, 1½ lines. Only two specimens of this beautiful shell have been met with at Scarborough attached to the under sides of rocks at the lowest spring tides. (Bean.)

Shell oblong, convex, the lateral slopes nearly straight, the dorsal ridge rather angular. Sculpture consisting of numerous rounded tubercles, arranged in longitudinal rows on the central areas, the series of tubercles finer and closer upon the jugum; head plate and lateral areas having coarser rather irregular tubercles. The lateral areas are not raised. The mucro is median, rather elevated. The plates of insertion of posterior and intermediate valves are obsolete, edges roughened; anterior valve having a short, acute insertion plate outwardly rugose-sulcate, the sulci indenting, but scarcely slitting the margin. Sutural plates moderate; sinus very wide, denticulated by the sculpture of the outside. Eaves very small. Girdle narrow, beset with numerous short and longer horn-colored spicules. Length, 10; breadth, 5 mm., sometimes larger. (Tryon and Pilsbry, *Manual of Conchology*.)

TYPE in British Museum? Type locality, Scarborough.

RANGE. Plover Bay, Bering Strait, to Monterey, California. Also Atlantic.

### Hanleya hanleyi spicata Berry, 1919

*Proceedings of the California Academy of Sciences,* 4th series, **9**:8; Pl. 1, fig. 3.

Shell small, oblong, rather elongate, lateral outline nearly straight. Valves sharply beaked; lateral areas distinct, but not conspicuous except by transmitted light, not ridged or grooved. Sculpture consisting of numerous rounded tubercles (transparent by transmitted light) irregularly scattered over the lateral areas, but over the central areas slightly smaller and disposed in 14–16 ill-defined, slightly oblique, longitudinal series, most crowded and irregular near the jugum. Head valve sculptured like the lateral areas. Tail valve with mucro somewhat in front of the middle, elevated. Girdle of moderate width, beset dorsally with numerous, faintly striate, glassy spicules of three main types: (1) very small, ovoid, pointed spinelets, forming a rather loose covering over the entire girdle, these being plainly striate near their apices; (2) elongate, dagger-like, marginal spinelets; and (3) a few scattered, needle-like spines, often over twice the length of the marginal spicules, some of these distributed here and there over the general surface of the girdle, but the greater proportion occurring in loose groups of 5–8 at each suture, where they extend well up between the valves; though distinct, these groups scarcely tuft-like, and unaccompanied, so far as noted, by evidence of sutural or inter-sutural pores. Ctenidia about 8 on each side. Color everywhere a pale brown or brownish-white, deepening in tone at the beaks, but without other markings. Long., 4.5; diam., 2.0 mm. (Berry.)

TYPE in Berry Collection, No. 4102. Type locality, ledge off Point Pinos, Monterey Bay, California.

RANGE. Monterey, California.

Described as *Hanleya spicata* Berry.

## Family LEPIDOCHITONIDAE

## Genus **LEPIDOCHITONA** Gray, 1821

Mantle-margin covered with minute, granule-like, round, smooth scales, not imbricate.

Shell with the valves external, broad, transverse; the hinder valve with the apex sub-central, superior. (H. and A. Adams.)

EXAMPLE. *Lepidochitona marmorea.*

DISTRIBUTION. Greenland, Britain, Mediterranean, western America.

### SECTION TONICELLA Carpenter, 1873

### Lepidochitona marmorea Fabricius, 1780

*Fauna Grönlandica,* 420. Tryon and Pilsbry, *Manual of Conchology,* 14:41; Pl. 10, figs. 8–15.

Chiton testa oxtoualui subcarinata punctata striata, corpore ochreo. Testa crassa oblongo-ovalis, glabra, punctis confertis eminentibus vix scabra, antice apulo latior, convexiuscula subcarinata, carina tamen anter extremitates definente, composita ex 8 valvulis retrorsum imbricatis (ita vi postica semper abticae subiecta sit lobis 2 acumintis lateralibus) margine antico rotundatis, postico subsinuatis, carinae apice tamen medio prominente striis minutis marginis anticae figuram describentibus striatis. Valvula postrema tamen a reliquis distert, quod tam postice quam antice rotundata sit, et strias totum circuitum describentes habeat. (Fabricus.)

Shell oblong or oval, elevated, rather acutely angular; buff, closely speckled and maculated with dark red, as in *Trachydermon ruber*. Surface densely, microscopically granulated, but apparently smooth. Lateral areas not distinct. Valves beaked, umbo of posterior valve slightly prominent, central. Interior rose tinted; anterior valve eight or nine slits. Sutural plates broad, rounded; sinus deep, angular. Girdle leathery, nude. The gill rows extend forward three-quarters the length of the foot, each containing 20–25 branchiae. Length, 40; breadth, 24 mm. Length, 27; breadth, 16 mm. (Tryon and Pilsbry, *Manual of Conchology*.)

TYPE in Zoölogy Museum, Copenhagen. Type locality, not known to writer.

RANGE. Circumboreal. Arctic and Bering Seas; Aleutian Islands south to Forrester Island, Alaska; also Japan.

### Lepidochitona submarmorea Middendorff, 1848

*Bulletin,* Imperial Academy of Sciences, St. Petersburg, **6**; No. 8, 120. *Sibirische Reise*, 178; Pl. 14, figs. 7, 10. Tryon and Pilsbry, *Manual of Conchology*, **14**:42; Pl. 10, figs. 16–24.

Shell oval, rather depressed, rather smooth and shining, the entire surface seen under a lens to be very minutely, regularly, and closely granulose. Lateral areas scarcely distinct, slightly swollen. Color rosy or yellowish-white, closely painted with spots and flames of red. Interior rose-colored; terminal valves each with five slits. Girdle smooth, shining, yellow or brown. Length, 38; breadth, 24; alt., 12–13 mm. (Tryon and Pilsbry, *Manual of Conchology*.)

TYPE in Academy, St. Petersburg. Type locality, not known to writer.

RANGE. Aleutian Islands to Puget Sound; also Okhotsk Sea and Japan.

### Lepidochitona lineata Wood, 1815

*General Conchology*, 15; Pl. 2, figs. 4, 5.

Ch. testa octovalvi laevi, valvis lineatis, limbo lato coriaceo.

Shell of eight smooth valves, valves streaked, margin broad and coriaceous. The ground color of this elegant shell is bright chestnut, with an interrupted white band running along the back; undulating white lines, edged beneath with black, pass diagonally across the marginal triangles, and concentrically on the extreme valves. The white band on the back of the shell, which is occasionally interrupted with chestnut, terminates in a white triangular spot on the posterior. The back is convex, not carinated, and is delicately striated in the direction of the margin. The inside is whitish. Its country is unknown. (Wood.)

TYPE in British Museum? Type locality, not known.

RANGE. On the west, the Okhotsk Sea and northern Japan; on the east, from the Aleutian Islands to San Diego, California.

Described as *Chiton lineatus*.

## Lepidochitona saccharina Dall, 1878

*Proceedings of the United States National Museum,* 1:2. Tryon and Pilsbry, *Manual of Conchology,* 14:44; 15:66; Pl. 15, figs. 22–24.

T. t. parva, oblonga, tota superficie saccharina, rufo et albescente picta; mucrone submediano, inconspicuo; areis lateralis inconspicue elevatis, albescentis; areis dorsualis sanguinosis, aeque quincuncialiter lente reticulatis; v. ant. 10–11, v. post. 8–10, v. centr. 1 fissatis; dent. parvis spongiosis, sinu parvo; subgrundis spongiosis, mediocris; zona coriacea ut in *Tonicellae aliis;* branchiae mediae. Long., 6.5; lat., 4 mm. (Dall.)

Shell small, oblong, the entire surface painted with lustrous red and whitish. Umbo subcentral, inconspicuous. Lateral areas indistinctly raised; dorsal area blood-colored, reticulated in quincunx. Anterior valve having 10–11, posterior 8–10, median 1 slit. Teeth small, spongy; sinus small; eaves spongy, moderate. Girdle leathery. Gills median. Length, 6½; breadth, 4 mm. (Tryon and Pilsbry, *Manual of Conchology.*)

TYPE in United States National Museum. Type locality, Aleutian Islands.

RANGE. Pribilof, Bering and Aleutian Islands, in shallow water; San Diego, California, in 101 fathoms.

Described as *Tonicella saccharina* Dall.

## Lepidochitona ruber Linnaeus, 1767

*Systema Naturae,* 12th ed., 1107. Tryon and Pilsbry, *Manual of Conchology,* 14:80; Pl. 7, figs. 50–56.

C. testa octovali arcuato-substriata; corpore rubro. (Linnaeus.)

Testa ovalis, oblongiuscula, dorso carinata, valvulis oblique subarcuato striatis. (Linnaeus.)

Shell oblong, elevated, solid, the back roundly subangular, lateral slopes somewhat convex; surface apparently smooth except for well-marked grooves or wrinkles, indicating growth-periods. Under the microscope, however, an excessively fine reticulation is visible. The color is light buff, marbled all over with orange-red in various patterns, or entirely suffused with reddish; usually having a red stripe bordered on each side with buff. The anterior valve is twice as wide as long, crescentic rather than half-round. The intermediate valves are slightly beaked, their lateral areas slightly raised and having stronger concentric wrinkles than the central areas. Posterior valve having a rather elevated but obtuse median umbo. The interior is bright pink. The sutural plates are wide, large; the jugal sinus is deep, narrow, and angular. The insertion plate of the anterior valve has 8–11 slits; intermediate valves normally 1 slit; posterior valve 7–11 slits. The insertion plates are sharp and smooth. The girdle is reddish-brown, covered with minute elongated scales. Length, 20; breadth, 12 mm. Alaskan specimens grow to the length of an inch. (Tryon and Pilsbry, *Manual of Conchology*.)

TYPE in Linnaeus Society, London. Type locality, Oceano septentrionali instar Patellae affixa.

RANGE. Arctic Ocean to Monterey, California. Circumboreal.

Described as *Chiton ruber*.

## Lepidochitona sitkensis Middendorff, 1846

*Bulletin,* Imperial Academy of Sciences, St. Petersburg, **6**:121. *Malacologia Rossica,* Pl. 13, figs. 1, 2.

Chiton minutus, testa oblongo-ovali subcarinata, exroseo flava, amculis ferruginosis irregularibus notata, subtus intense rosea. . . . . Valvae laeves, mucrone parum prominente, striis incrementi irregularibus; . . . . areis lateralibus indistinctis. . . . . Valve antica dentibus 9. . . . . Valva ultima margine postico dentibus numero 10, lateralibus majoribus. . . . . Valvarum intermediarum dentes laterales duo; mucrones apice detritu stratum testae subjacens porosum ostendunt; areae centrales inferne spectatae; antice pulchre porosae; postice vero pori in sulcos transversos extenduntur. Pallium marginale latum albidum glaberrimum. Branchiarum series ab initio tertiae quintae partis totius corporis longitudinis ad visinitatem ani usque porrigitur. Lamellulae branchiales circiter 26. . . . . Longitudo 0.11 decim. (Middendorff.)

Shell depressed, smooth, the lateral areas indistinct, under a lens seen to be sparsely granulose; reddish. Anterior valve having 8, posterior 10, central 1 slit. Length, 10; width, 6 mm. (Tryon and Pilsbry, *Manual of Conchology*.)

TYPE in Academy of Sciences, St. Petersburg. Type locality, Patria: Insula Sitka.

RANGE. Known only from type locality.

Described as *Chiton sitkensis*.

### SECTION LEPIDOCHITONA s.s.

### Lepidochitona alba Linnaeus, 1767

*Systema Naturae,* 12th ed., 1107. Tryon and Pilsbry, *Manual of Conchology,* **14**:70; Pl. 7, figs. 35–38.

C. testa octovali laevi; valvula prima postice enmarginata. Testa ovalis, laevis, alba, vix dorsata, minus caruhata. (Linnaeus.)

Shell oblong, elevated, the back keeled, lateral slopes nearly straight. Anterior valve half-circular, its posterior margin slightly concave; sculpture consisting of some scarcely perceptible, low radiating ridges, often wholly obsolete, and an excessively minute shangreening or granulation of the whole surface, the granules showing a disposition to be arranged in oblique curved lines. Intermediate valves slightly beaked, produced forward in the middle, having the same sculpture, and showing low, irregular growth wrinkles. Lateral areas scarcely raised, sculptured like the front valve. Posterior valve having the umbo slightly elevated, central, inclined backward. Interior white. Sutural plates large and wide, extending from the insertion plates nearly to the jugum. Sinus rounded. The anterior valve has 13 slits in the smooth and rather sharp insertion plate; the intermediate valves have one slit; the posterior valve has 10 slits, and the edge of the plate is decidedly roughened and irregular. Another specimen has 14 slits in the anterior valve, 12 in posterior valve, and in still another individual, some of the intermediate valves have two slits on one side. The girdle is covered with small gravelly scales, and has no marginal fringe of long scales or spines. The color is a very delicate buff tint, sometimes almost white, often shading into a light orange on the posterior part and lateral areas of each valve. Most of the specimens have more or less of a black deposit on the back part of the valves. Length, 15; breadth, 7 mm. (Tryon and Pilsbry, *Manual of Conchology.*)

TYPE in Linnaeus Society, London. Type locality, 0. Islandico Knoig.

RANGE. Arctic Ocean to San Diego, California; Greenland, Iceland, Spitzbergen, Scandinavia; British Seas, Gulf of St. Lawrence, Maine, and south to Massachusetts Bay.

### Lepidochitona sharpei Pilsbry, 1896

*Nautilus,* **10**:50.

Shell oblong, elevated, carinated, the side slopes somewhat convex. Surface to the naked eye smooth; lustreless; slightly soiled white, with some faint and ill-defined brownish spots on the lateral areas, the girdle gray. Anterior valve smooth, with some indistinct concentric grooves; the anterior slope shorter than the posterior edges; hind margin emarginate. Intermediate valves wide and short, with slightly arcuate margins at junction with girdle, hind margins emarginate. Central areas very minutely roughened by diverging wrinkles; lateral areas slightly raised, with a few arcuate faint grooves in the direction of growth-lines. Posterior valve highest at its anterior margin, the subcentral mucro but slightly projecting, the slope behind it about straight. Interior white; valve callus strong; sinus concave and shallow, not defined at the edges; sutural laminae but little projecting, broadly rounded, invading the sinus. Insertion plates hardly longer than the narrowly channelled and solid eaves, sharp and smooth. Slits in valve i, 16; valves ii to vii, 1–1 or 2–1 or 2–2, the larger number prevailing on the more anterior valves; in valve viii, 13. Posterior tooth in the median valves square and well-developed. Girdle rather unevenly covered, with convex, pebbly, coarse scales, those toward the outer margin elongated, and there is a copious marginal fringe of stout hyaline spinules. Gill-row three-fourths of length of foot, with 21 plumes on each side. Length, about 14; breadth, 8 mm. (Pilsbry.)

TYPE in Academy of Natural Sciences, Philadelphia, No. 69140. Type locality, Unalaska.

RANGE. Unalaska, Aleutian Islands.

Described as *Trachydermon sharpii* Pilsbry.


## Lepidochitona dentiens Gould, 1846

*Proceedings of the Boston Society of Natural History,* 2:145. *Report of the United States Exploring Expedition, Mollusca,* 321, fig. 433.

Testa minima, ovata, carinata, olivacea, ad dorsum albido fulminata, omnino minutissime granulata; areis vix distinctis; valvis rostratis postice subdenticulatis, dentibus albidis: margo pruinosus.

Shell minute, ovate, carinate, dark olive, with fine, white zigzag markings along the back, and dotted black and white along the posterior edge of the valves, the white portions seeming to project like minute teeth. The whole surface is minutely granulated, and faintly marked by the lines of growth. Lateral areas faintly defined, but occasionally marked by a dividing line; terminal valves like the others. Margin clothed with hoary pubescence. Length, one-fifth of an inch; breadth, three-twentieths of an inch. (Gould.)

TYPE in United States National Museum. Type locality, Puget Sound, Washington.

RANGE. Puget Sound to Magdalena Bay, and Socorro Islands, Lower California.

Described as *Chiton dentiens.*

## Lepidochitona flectens Carpenter, 1864

*Supplementary Report,* British Association for the Advancement of Science, 649. *Proceedings,* Philadelphia Academy of Natural Sciences, **60**:1865. Tryon and Pilsbry, *Manual of Conchology,* **14**:75 and **15**:64; Pl. 15, figs. 34–37.

Mantle-margin scarcely granular. Rosy, very small, scarcely sculptured: valves beaked and waved as in *M. simpsonii:* eaves and incisors normal. (Carpenter.)

I. t. parva, subelongata, rosea, elevata; jugo acuto; areis lateralibus vix definitis; marginibus valvarum excurvatis, suturis incurvatis, apicibus valde prominentibus; valvis granulis minutis, haud crebris, subradiatim sparsis, ominoque minutissime punctulatis; mucrone conspicuo, antico; intus, sinu suturali lato, planato; subgrundis haud porrectis; laminis lateralibus uno-terminalibus quod xi.–fissatis: limbo palii vix minutissime granulato. Long., .35; lat., .24; div., 110°. (Carpenter.)

Shell small, subelongate, roseate; jugum acute; lateral areas scarcely defined. Valve margins excurved, suture incurved, apices very prominent; valves having minute, not very close granules sparsely subradiating, all over very minutely punctulate. Mucro scarcely projecting. Terminal valves with 11, median 1 slit. Girdles very minutely granulate. **Length, 8¾**; breadth, 6 mm.; div., 110°. (Carpenter.) (Tryon and Pilsbry, *Manual of Conchology.*)

TYPE in United States National Museum. Type locality, Puget Sound.

RANGE. Vancouver Island, B.C., to San Diego, California.

Described as *Trachydermon flectens* Carpenter.

## SECTION CYANOPLAX Pilsbry, 1892

## Lepidochitona hartwegii Carpenter, 1855

*Proceedings,* Zoölogical Society of London, 231. Tryon and Pilsbry, *Manual of Conchology,* **14**:45; Pl. 14, figs. 81–85.

C. t. ovata, subelevata (ad angulum 125°), sublaevi; olivacea, macula migra in utroque jugi latere picta; valvis sine lineis diagonalibus, partim detritis, lineis incrementi conspicuis, tota superficie minutissime irregulariter subgranulosis; marginibus rotundatis, suturis magnis; limbo tenui, granulis minimis, confertissimis, irregulariter munito; intus valvarum marginibus arcuatis, lobis prominentibus, sinu late, haud alto; ad jugum vix impressus. (Carpenter.)

Shell oval, rather low, the dorsal ridge obtusely rounded; dull olive-green, generally having a pair of lighter stripes on the ridge of each valve with a black blotch outside of the light dashes. Girdle rather narrow, dense, microscopically closely granulated. The tail valve is convex as a whole, but the subcentral umbo is not conspicuous. The entire surface is very closely microscopically granulated and bears larger wart-like granules irregularly scattered over the minute sculpture, these warts being much more numerous upon the lateral areas (which are otherwise rather ill-defined) and the terminal valves. The interior is of an intense blue-green color. Sutural plates rounded, leaving a wide, angular sinus. Insertion plates shorter than the eaves, blunt, the anterior valve having the teeth bi- or tri-lobed, the posterior valve having them crenulated. Slits of anterior valve 10–11; median valves 1; posterior valves 9–12. Eaves spongy. Gills extending forward to the front end of the foot. Length, 30; breadth, 17 mm. Length, 27; breadth, 18 mm. (Tryon and Pilsbry, *Manual of Conchology.*)

TYPE in Museum Cuming. Type locality, Monterey, California.

RANGE. Forrester Island, Alaska, to Gulf of California.

Described as *Chiton hartwegii* Carpenter.

## Lepidochitona hartwegii nuttalli Carpenter, 1855

*Proceedings,* Zoölogical Society of London, 231. Tryon and Pilsbry, *Manual of Conchology,* **14**:46.

C. t. *Chitoni hartwegii* simili, sed latiore, depressa (ad angulum 130°), superficie granulis majoribus, maxime marginem versus; marginibus quadratis, suturis nullis; intus valvarum marginibus valde arcuatis, sinu lato, alto. Long., 1.05; lat., .8; alt., .24 poll. (Carpenter.)

Umbo flatter; valves broad, non-swelling, squared at the sides, and not beaked or waved. Posterior valve having 11, central 1, anterior 8 slits. (Tryon and Pilsbry, *Manual of Conchology.*)

TYPE in Museum Nuttall. Type locality, Monterey, California.

RANGE. Straits of Juan de Fuca to Turtle Bay, Lower California.

Described as *Chiton nuttalli* Carpenter.

## Lepidochitona lowei Pilsbry, 1918

*Nautilus,* **31**:127.

Oval, rather depressed, dirty buff, a little darker toward the beaks, which project somewhat. Surface of the valves finely, closely and evenly granose throughout, the granules oblong. The anterior valve and posterior areas of the posterior valve have a few very weak, low, radial impressions.

The mucro of the posterior valve is slightly post-median, but little raised, the slope behind it being convex toward the edge. The interior is white, stained buff or pinkish near the sinus, where it is conspicuously porous. The eaves are wide and closely porous throughout. Teeth smooth, those of the posterior valve being directed forward. Girdle is densely covered with minute elongate scales. Length, about 17; width, 12 mm. (San Pedro.) Length, about 19; width, 13 mm. (San Pedro.) Length, about 16; width, 11 mm. (White Point.) (Pilsbry.)

TYPE in Academy of Natural Sciences, Philadelphia, No. 117955. Type locality, San Pedro, California.

RANGE. San Pedro, California.

Described as *Trachydermon lowei* Pilsbry.

## Lepidochitona fackenthallae Berry, 1919

*Lorquinia,* January 1919, 5.

Chiton of moderate size, best described in few words by comparison with *C. lowei* Pilsbry, which it resembles very closely, but is apparently somewhat larger; light brown in color; has the eaves lower, very much more finely porous, and (especially on *valvei*) more decidedly projecting; has less pronounced sculpture, though of similar nature to that of *C. lowei*, and the head valve less crenulate; interior pure white instead of tinted: tail valve with fewer (10 instead of about 13) blunter and coarser teeth; all valves have a much less porous sinus and lines leading into slits. (Berry.)

TYPE in Berry Collection. Type locality, Pacific Grove, California.

RANGE. Known only from type locality.

## Lepidochitona raymondi Pilsbry, 1894

*Nautilus,* 8 :46.

Shell longer and narrower than *T. Hartwegii*. Back somewhat keeled, varying in elevation. Color (1) olivaceous green mottled with white, sometimes with dark lateral streaks as in *Hartwegii,* sometimes ruddy at the ridge, of (2) uniform blackish, or (3) dark brown, uniform or with whitish flecks. Valves rather strong, slightly beaked when unworn, the posterior (sutural) margins straight or slightly concave. Intermediate valves rather rounded where they join the girdle, scalloping the inner border of the latter; not distinctly divided into areas. Lateral area hardly or not raised (the diagonal being indistinct) evenly sculptured with minute, equal granules. Central areas also evenly sculptured throughout with similar granules, slightly finer on the ridge. End valves with the same equal sculpture, the tail valve with the mucro central and a little projecting. Interior light blue, with darker stains at bases of the sutural laminae and behind the rather strong blue-white callus. Sinus and sutural laminae as in *Hartwegii*. Slits in valve i, 8; valves ii–vii, bilobed. All teeth longer than the narrow, porous eaves. Girdle narrow, black or with small whitish spots, leathery, very minutely papillose. Length, 23; breadth, 11 mm. (type). (Pilsbry.)

TYPE in Philadelphia Academy of Sciences. Type locality, San Francisco, California.

RANGE. Seward, Alaska, to San Pedro, California.

Described as *Trachydermon (Cyanoplax) raymondi* Pilsbry.

### Subgenus SPONGIORADSIA Pilsbry, 1893

#### Lepidochitona aleutica Dall, 1878

*Proceedings of the United States National Museum,* 1:1. Tryon and Pilsbry, *Manual of Conchology,* 14:85 and 15:65; Pl. 15, figs. 30–33.

T. t. parva, rufo-cinera, oblonga, fornicata; jugo acutissimo; mucrone submediano, apicibus prominentibus; areis lateralis inconspicuis; tota superficie quincuncialiter minute reticulata; intus, valv. ant. 16–, post. 11–, centr. 2–fissatis; dent. parvis, perspongiosis, late separatis; subgrundis spongiosis, curtis; sinu parvo; zona squamulis minutis obsita. Long., 6; lat., 3 mm. (Dall.)

Shell small, rufous-ashen, oblong, vaulted, the jugum very acute; mucro submedian; apices prominent; lateral areas inconspicuous, entire surface minutely reticulated in quincuncial pattern. Interior: anterior valve with 16, central with 2, posterior with 11 slits; teeth small, very spongy, widely separated; eaves spongy, sinus small. Girdle beset with minute scales. (Dall.) (Tryon and Pilsbry, *Manual of Conchology.*)

TYPE in United States National Museum. Type locality, Western Aleutians.

RANGE. Aleutian Islands.

Described as *Trachyradsia aleutica* Dall.

### Genus NUTTALLINA Carpenter, 1873

Shell elongated, valves projecting anteriorly; mucro posterior, elevated; laminae acute, smooth, elongate; central valves bifissate; sinus not laminated, planate; girdle spinose. Differs from *Acanthopleura* in the smoothness of the sharp teeth, in their great length and Radsioid slitting; in the throwing forward of the rest of the shell, as in *Katherina,* and in the deep spongy flat sinus which interrupts the sutural laminae. (Tryon, *Structural and Systematic Conchology.*)

TYPE. *C. scabra* Reeve.

DISTRIBUTION. Mediterranean, Japanese seas, coast of California.

## Nuttallina californica Reeve, 1847

*Conchologia Iconica,* Chiton; Pl. 16, fig. 89. Tryon and Pilsbry, *Manual of Conchology,* 14:279; Pl. 54, figs. 23, 24, and Pl. 56, figs. 12–18.

Chit. testa oblongo-ovata, valvis medio subrostratis, undique subtilissime corrugato-crenulatis, terminalibus caeterarum areisque lateralibus laticostatis; fusco-viridi, ligamento corneo, pilis setisque dense obsito.

Shell oblong, ovate, valves somewhat beaked in the middle, very finely corrugately granulated throughout, terminal valves and lateral areas of the rest broad-ribbed; brownish green, ligament horny, thickly beset with hairs and bristles. (Reeve.)

TYPE in Museum Cuming. Type locality, California.

RANGE. Straits of Juan de Fuca to San Diego, California.

Described as *Chiton californicus* Reeve.

## Nuttallina fluxa Carpenter, 1864

### (New name for *N. scabra* Reeve)

*Supplementary Report,* British Association for the Advancement of Science, 649. Tryon and Pilsbry, *Manual of Conchology,* 14:280; Pl. 54, figs. 21, 22, and Pl. 56, figs. 19, 20.

Green, mottled with orange-red; not beaked; with only marginal and diagonal ribs. (Carpenter.)

Shell similar to *N. californica,* but having the individual valves very much shorter in proportion to their width; the outer layer of the median valves produced at the sides anteriorly, curving broadly forward and laterally upon the sutural plates; the median ridge and sulci more or less obsolete. Tail valve shorter, with less posterior mucro. Color of valves lighter, more variegated. Girdle rusty brown or alternately blackish and white; bearing rather sparsely scattered, white, spike-like spines, sometimes having one or two at each suture. Length, 29; breadth, 13 mm. (This is Pilsbry's description of *N. scabra* Reeve, in *Manual of Conchology,* 14.)

This species has hitherto been confused with the preceding (*N. californica* Reeve) by most collectors. Carpenter, however, distinguished it and gave the name *fluxa.* Although the individual valves are much shorter in this species than in *N. californica,* they overlap much less, so that the total length of the animal is about the same in the two species. The elevation is about equal also, but in *californica* the girdle is generally wider at the sides, especially in old specimens. In size, individuals of the northern species considerably exceed any of the southern I have seen, although the majority of specimens do not differ much. (Tryon and Pilsbry, *Manual of Conchology.*)

TYPE in United States National Museum? Type locality ?

RANGE. Point Conception, California, to Gulf of California.

## Nuttallina thomasi Pilsbry, 1898

*Proceedings,* Philadelphia Academy of Natural Sciences, 289:1898.

General form oblong, rather depressed, not keeled dorsally; surface granulose when not eroded. Color blackish or dark brown, with a whitish band on each side of the median line of back or irregular whitish maculations; girdle dark. Intermediate valves short and wide, with a slight or hardly noticeable depression on each side of the jugum, and others in front of and behind the scarcely defined, obsolete, diagonal convexity, the anterior and posterior margins subparallel, slightly arcuate. Anterior valve granulate, without radial ribs, the posterior margin excavated mesially. Posterior valve with tegmentum slightly wider than the anterior, the obtuse mucro somewhat behind the middle. Interior blue-green, with the area behind the valve-callus dark brown, or livid purplish, with the light sutural laminae and blue-green area behind the sinus. Length, 15; breadth, 8 mm. or smaller. (Pilsbry.)

TYPE in Philadelphia Academy of Natural Sciences. Type locality, Monterey, California.

RANGE. Known only from type locality.

## Genus SCHIZOPLAX Dall, 1878

Shell and zone like *Tonicella;* central valves with a median slit. Branchiae subambient. (Tryon, *Structural and Systematic Conchology.*)

TYPE. *Chiton Brandtii* Middendorff.

DISTRIBUTION. Arctic to Sitka, Alaska.

## Schizoplax multicolor Dall, 1920

*Nautilus,* 34:22.

Chiton depressed, broad, wider behind than in front, maroon varied with white streaks, with a rather wide girdle, the surface of which is covered with soft bristles like those of *Mopalia muscosa,* among which are sparsely scattered, irregularly disposed, longer translucent spicules; surface of the valves minutely uniformly reticulate under the lens, appearing smooth to the unaided eye; the mesial suture evident, the fifth valve widest, the posterior valve very small with a subcentral inconspicuous vertex at the anterior third; anterior valve with nine, middle valves with two, posterior valves with two slits, the interior lines of which are marked by a row of minute pores; the middle of the valves on each side of the median suture conspicuously porous internally. Length in alcohol, 8; maximum breadth, 6 mm. (Dall.)

TYPE in United States National Museum, No. 383018. Type locality, St. Paul Island, Bering Sea.

RANGE. Bering Sea.

### Schizoplax brandtii Middendorff, 1846

*Bulletin,* Imperial Academy of Sciences, St. Petersburg, **6:** 117. Tryon and Pilsbry, *Manual of Conchology,* **14:** 47; Pl. 11, figs. 32–37.

Chiton minutus testa externa ovata, nitidula fusco olivacea lituris undulatis longitudinalibus et maculis laete aeruginosis circumscriptis elegantissime notata. Valvae laeviusculae, sub lente incrementi strias distinctas, et superficem minutissime ac aequaliter granulosam exhibent; areis lateralibus carina tumida vix distinctis. . . . . In summa linea mediana valvarum intermediarum, superne sulcus decurrit linearis longitudinalis, cui inferne opponitur canalis profunde et abrupte incisus, commissura antica valvarum singularium interruptis. . . . . Incisura lateralis *Valvarum* intermediarum supra nondum, infra vix conspicua. . . . . Valva antica dentibus marginalibus 14. . . . . Valva ultima margine postico convexo dentibus 13. . . . . Pallium marginale viride glabrum setis microscopicis fimbriatum. Branchiarum series ab initio secundae tertiae ad finem tertiae quartae partis totius corporis longitudinis porrigitur. Lamellularum Branchialium numerus circiter 22. . . . . Adulti longitudo 0.16 decim. (Middendorff.)

Shell oval, rather elevated, the jugum rounded; olive-brown, streaked, maculated or clouded with blue, sometimes chestnut. Umbo central, irregularly planate. Lateral areas scarcely defined. Entire surface smooth, seen under a lens to be finely granulated in quincunx. Girdle narrow, olive-ashen, maculated, closely beset with minute spines, but appearing almost smooth to the naked eye. The median valves have a sharply cut longitudinal median sulcus. The jugal sinus is moderate, deep, scarcely laminate, conspicuously spongy. Eaves small, very spongy. Anterior valve with 11, posterior 11, and median valves with one slit. Length, 16; width, 5 mm. (Tryon and Pilsbry, *Manual of Conchology.*)

This very remarkable species is very prettily marbled with olive, chestnut, and blue; the girdle generally dark olive, dashed with ashy spots and in fine specimens having a pubescent appearance. The slit is occupied by a cartilaginous substance of a dark brown color, most visible from within. (Tryon and Pilsbry, *Manual of Conchology*.)

TYPE in Academy, St. Petersburg. Type locality, Insulae Scantar et Sinus Tuguricus maris Ochotici, . . . . Ins. Sitka, Alaska.

RANGE. Shantar Bay, Okhotsk Sea; Bering, Pribilof, and Aleutian Islands, and south to Cape Fox, Alaska.

Described as *Chiton brandtii*.

## Family ISCHNOCHITONIDAE

### Genus ISCHNOCHITON Gray, 1847

Valves external, having sharp, slit, insertion plates, the teeth not buttressed. Eaves solid (rarely somewhat porous in s. g. *Trachydermon*); girdle covered with imbricating scales, either flat or convex, smooth or striated. Gills typically extending the entire length of the foot, but in some species they are short in front or at both ends. (Tryon and Pilsbry, *Manual of Conchology*.)

TYPE. *C. longicymba* Blainville.

DISTRIBUTION. New Zealand, Australia, Japan, West Indies, Cape of Good Hope, west coast of North and South America.

### Subgenus STENOPLAX

### Ischnochiton fallax Carpenter, 1892

Tryon and Pilsbry, *Manual of Conchology*, 14:59; Pl. 16, figs. 17–18.

Almost exactly like *Ischnochiton magdalenensis* in form and sculpture, but more roseate; the central areas pitted; lateral areas having close radiating wrinkles interrupted by lines of growth. Interior: Posterior valves having 9, central 1, anterior 10 slits; teeth acute; eaves conspicuous; sinus moderate, scarcely laminate, but the jugal part of the valves produced forward. Girdle having very minute granules. Length, 27½; breadth, 12½ mm.; divergence 120°. (Tryon and Pilsbry, *Manual of Conchology*.)

TYPE in ? Type locality, Monterey, California.

RANGE. Vancouver Island to Todos Santos Bay, Lower California.

### Ischnochiton biarcuata Dall, 1903

*Proceedings of the Biological Society of Washington*, 16:176.

Animal about 18 mm. long and 7 mm. wide (in the dry state); girdle narrow, with very small, close-set, more or less imbricating, brownish scales; valves rounded evenly above, only the lateral areas distinct; anterior valve with 7 or 8, median with 1, posterior valve with 11 slits; interior of valves rose-pink; exterior ashy, marbled with lilac and brown, an obscure lilac median line on the medial valves; sculpture of undivided central areas formed by two sets of arcuate radiations crossing each other obliquely and with the inter-reticulations impressed or punctate, so that an irregularly zigzag effect is produced by the arrangement of the punctations; lateral areas irregularly concentrically vermiculate, the spaces between the elevated ridges deeply minutely punctate, with somewhat of a zigzag effect here also; the sculpture of the anterior valve resembles that of the lateral areas; of the posterior valve the mucro is low, subcentral and inconspicuous, the central area sculptured like that of the medial valves, the posterior area like the anterior valve; the sutural plates are quite short and the sinus smooth and wide. There is no noticeable mucro to the medial valves. The peculiar sculpture of this species separates it from those already described from this region. In a general way it recalls the very young of *I. magdalenensis* Hinds. (Dall.)

TYPE in United States National Museum, No. 109308. Type locality, Santa Barbara Channel near Avalon, Catalina, California.

RANGE. Santa Barbara to San Diego, California.

## SECTION STENORADSIA Carpenter, 1878

### Ischnochiton acrior (Carpenter) Pilsbry, 1892

Tryon and Pilsbry, *Manual of Conchology,* 14:61; Pl. 14, figs. 86–89.

Shell much like *I. magdalensis,* but broader and flatter, with much sharper sculpture. Olive-green, pink where worn, or sometimes light flesh-colored, with the jugum or central areas often green. Girdle dark ashen or olive. The anterior valve is notably concave, as in *I. conspicuus.* The lateral areas are much raised, and sculptured with acute radiating riblets, sometimes splitting; central areas having acute longitudinal riblets. Interior pink, with a blue spot at the jugal sinus. Sutural plates wide, pink; sinus broad, deep, angular. Anterior valves having 13–15, central valves 2–4, posterior valves 10–13 slits. Teeth long, sharp, smooth. Eaves wide, dark-blue. Girdle wide, tough, covered with very small, solid, short scales. Length, 110; breadth, 55 mm. Length, 75; breadth, 40 mm. (Tryon and Pilsbry, *Manual of Conchology.*)

TYPE in Academy of Natural Sciences, Philadelphia, No. 35702. Type locality, Point Abreojos, Lower California.

RANGE. San Miguel Island, California, to Gulf of California.

## Ischnochiton magdalenensis Hinds, 1843

*Zoölogy of the Voyage of H. M. S. tSulphur, Molluscae,* **54**; Pl. 19, fig. 1.

Testa elongate-ovali, subelevata, medio pallida seu subcarnosa; valvarum area laterali alaeformi radiatim corrugato-sulcata, transversim grosse lineata, madiata et antica longitudinaliter corrugato-sulcatis; ligamento scabriusculo.

In shape elongated, and the middle valves being scarcely of greater breadth than the others, with the margins somewhat linear. The lateral spaces, and which with much propriety and convenience might be called alae or wings, are crossed by rude lines which appear to indicate periods of cessation of growth. The character of the coloring of the dorsal ridge is disposed to vary. In an individual of this species the phenomenon occurred of the existence of nine valves; and though Chitons are not indisposed to diminish the number by one or more, it seems far less frequent that they should increase them, as this is the only instance within my experience. The number of species of Chitons on the west coast of America is very great, and they extend throughout its vast extent from Chile to a high northern latitude. They most usually abound in numbers, but are limited in their geographic range. Those of the coasts are nearly always distinct from those of the deep water. So very prone are they to restrict themselves to narrow portions of the coast, that it would be exceedingly easy to convey a close idea of any particular locality by pointing out the species inhabiting it. (Hinds.)

TYPE in ? Type locality, Magdalena Bay, Lower California.

RANGE. Coos Bay, Oregon, to Magdalena Bay, Lower California.

## Ischnochiton sarcosus Dall, 1902

*Proceedings of the United States National Museum,* **24**:558.

Chiton rather elongate, marbled with scarlet and white, paler mesially, rather low and not carinate; the girdle densely set with small, curved, smooth bristles of different sizes, dark red and white mixed; under side of the girdle chocolate color, and the interior of the valves flesh pink; dorsal angle somewhat more than 110°, the jugal region being rounded off; intermediate valves with prominent lateral areas hardly concentrically or radially sculptured, but finely punctate all over and more or less serrate near the girdle on the posterior margin; jugum not defined, central area of the valves longitudinally sculptured with low inosculating wrinkles which sometimes form diamond-shaped interstitial excavations; the whole sculpture has an obsolete appearance; anterior valve finely punctate, feebly concentrically sculptured; posterior valve the same, with a low subcentral mucro as in *I. magdalenensis;* anterior valve with about 8, posterior about 10 slits, intermediate valves with 2 slits; sinus wide, entire; pleural laminae wide, short; long., in the dry state, 36; lat., 15; alt., 5 mm. (Dall.)

TYPE in United States National Museum, No. 109043. Type locality, San Martin Island, Lower California.

RANGE. San Pedro, California, to Gulf of California.

## Ischnochiton conspicuus (Carpenter) Pilsbry, 1892

Tryon and Pilsbry, *Manual of Conchology,* **14**:63; Pl. 15, figs. 91–96.

Shell large, elongated, moderately elevated; green (or rarely earthy-brown), but where eroded at the beaks it is pink. Interior pink, with a blue spot at the jugal sinus. Lateral area much raised, having acute radiating riblets or striae; central area apparently smooth, but microscopically granulated, sometimes having some faint longitudinal striae at the jugum. Front slope of the anterior valve concave; girdle densely beset with short bristles, giving it a velvety aspect. The entire surface is very densely, microscopically granulated, where not eroded. The lateral areas have fine acute radii and often the black edge of each valve is crenulated by oblique, sharp, little folds. The color is often varied by darker little flames on the central areas. The posterior valve is large and depressed. The sutural plates are large, the sinus deep and angular. Insertion plates having in anterior valve 9, central valves 2 or 3, posterior valve 10 slits. Teeth sharp and thin in posterior and central valves, but blunt and bi- or tri-lobed in anterior valve. Eaves wide, solid, light blue-green colored. Length, 82; breadth, 36 mm. Length, 89; breadth, 41 mm.

TYPE in Academy of Natural Sciences, Philadelphia, No. 35704. Type locality, San Diego, California.

RANGE. Monterey, California, to Gulf of California.

## Ischnochiton marmoratus Dall, 1919

*Proceedings of the United States National Museum,* **55**:503.

Chiton small, elegantly marbled with white, gray, and brown, with a velvety girdle covered with mottled gray, white, and brown densely set spinules; anterior valve with about 12 radial rows of prominent pustules, the rows sometimes bifid, and the ground microscopically decussate; slits 8–10; posterior valves smaller, with 8–9 slits, the mucro slightly behind the center with a peripheral irregularly double row of pustules while the central area is coarsely axially reticulate; intermediate valves with one slit, hardly mucronate, the jugal and pleural tracts axially sculptured with beaded threads finer on the jugum; lateral areas with two or three radial rows of prominent pustules which near the girdle project like little cylinders in an irregular manner; interior whitish, with hardly perceptible jugal sinuses and very narrow sutural laminae. Length, 7; width, 4; height, 1.5 mm. (Dall.)

TYPE in United States National Museum, No. 218735. Type locality, Pacific Grove, Monterey Bay, California.

RANGE. Known only from type locality.

### Ischnochiton brunneus Dall, 1919

*Proceedings of the United States National Museum,* **55**:504.

Chiton of moderate size, flattish, of a yellowish brown with a few black flecks on the firgle, which is armed with smooth oval slightly imbricating scales; anterior valve with 11, posterior with 14, intermediate with single slits; anterior valve with about 16 beaded radial ribs sometimes bifurcate, the posterior margin denticulate, the interspaces microscopically reticulate; posterior valve flattish, the mucro slightly behind the center, with about 14 sometimes bifid rows of pustules with the same reticulate ground; the central area is sculptured like the pleural tracts of the intermediate valves; the latter valves have a nearly smooth mucro and feeble dorsal carination; the pleural tracts have 8 or 10 axial undulated threads on each side, beneath which are more closely set, coarser, transverse threads with deeply punctate interstices; the lateral areas have three or four radial ribs, the posterior annulate, denticulating the posterior margin, the others beaded, the substratum everywhere with the minute reticulation. Length, 11; width, 12; height, 3 mm. (Dall.)

TYPE in United States National Museum, No. 58734a. Type locality, San Diego, California.

RANGE. Known only from type locality.

### Ischnochiton listrum Dall, 1919

*Proceedings of the United States National Museum,* **55**:504.

Chiton of moderate size, straw color, with occasional black blotches on the girdle, which is densely minutely scaled; though the shell is flattish, the dorsum is nearly keeled, but the valves are hardly mucronate; the gill row lacks one-fourth of being ambient; the anterior valve has about 16 thread-like, beaded, radial riblets, the two marinal ones are large with half a dozen sharp denticles on the edge; the whole shell is covered under its stronger sculpture with very minute reticulation; the posterior valve is smaller and has 14 riblets like those on the anterior valve; the central area is smaller, the mucro being low and behind the center; it is sculptured like the pleural tracts; intermediate valves with a small nearly smooth jugum, the pleural tracts with about a dozen axially straight slender threads on each tract, with subequal and closer transverse threads in the interspaces with deep interstices; the lateral areas have two or three slender beaded radii on a flat substratum anteriorly, but on the posterior side there is a wider rib with five or six widely spaced annulations, which are projected backward like the teeth of a comb, beyond the margin; internally the valves are whitish. Length, 12; width, 8; height, 2.3 mm. (Dall.)

TYPE in United States National Museum, No. 58734b. Type locality, San Diego, California.

RANGE. Known only from type locality.

### Ischnochiton ritteri Dall, 1919

*Proceedings of the United States National Museum,* **55**:505.

Chiton of moderate size, brownish, darker on the prominences, with a rather wide girdle set with smooth oval convex whitish imbricating scales, the valves nearly mucronate, almost keeled; anterior valve with about 20 flattish radii with narrower interspaces, the whole covered with a minute oblique decussation; internally whitish, with 16 slits, the insertion plates smooth, the eaves very narrow; posterior valve with a nearly central low mucro, the posterior tract sculptured like the anterior valve but more faintly, the central area punctate-reticulate; internally whitish, with 12 slits, the sinus defined by small notches at each end, straight-edged, the sutural laminae narrow, elongate; intermediate valves (the first larger than the others) with no defined jugal area, but deeply punctate-reticulate, the punctations coarser toward the margins of the pleural area, lateral areas with three flattish radii with narrower interspaces, all covered with the minute reticulation; internally white with two slits at each end, one sometimes obsolete; the sinus wide, concave, shallow, between two faint notches, the sutural laminae narrow; gill rows ambient. Length, 22; width, 15; height, 5 mm. (Dall.)

TYPE in United States National Museum, No. 218759. Type locality, channel at Juneau, Alaska.

RANGE. Known only from type locality.

### Ischnochiton retiporosus Carpenter, 1864

*Supplementary Report,* British Association for the Advancement of Science, 649. Tryon and Pilsbry, *Manual of Conchology,* **15**:75; Pl. 16, figs. 47, 50–53.

Mantle-scales very small, close, smooth. Central pattern in net-work, 3–6 side ribs. (Carpenter.)

Shell small, subelongate, ashen-colored; much elevated, the jugum arcuate. Valves subquadrate, apices concealed, lateral areas little defined, having 3–6 rounded, obsolete riblets, here and there with acute projecting granules. Central areas pitted all over, interstices small, punctate. Terminal valves with more acute, close, narrow riblets. Mucro little raised, anterior. Inside with broad sutural sinus. Insertion plates of intermediate valves with one slit on each side, terminal valves with about 12 slits. Girdle bearing close, minute scales, which are little elongated. Length, 11; breadth, 7 mm. (Pilsbry.)

Type in United States National Museum. Type locality, Puget Sound.

Range. Victoria, British Columbia, to San Pedro, California.

### Ischnochiton retiporosus punctatus Whiteaves, 1886

*Transactions of the Royal Society of Canada,* 4: Section 4, 1887.

Shell small, elongated, rather strongly elevated, back distinctly angulated; color pale-cream or nearly white, but with a few small patches of reddish-brown on the girdle and a narrow and noncontinuous series of variously interrupted and broken-up yellowish-brown spots upon the median line of all the valves but the anterior one. Girdle squamose, the scales closely imbricating much broader than high, and distinctly striated when viewed under an achromatic microscope with half-inch objective. Mucro central and tolerably distinct where not worn off; anterior valve marked with faint but rather numerous radiating striae and concentric lines of growth. Central areas of all the valves but the anterior one regularly pitted as viewed by an ordinary simple lens of moderate power, the pitting being most distinct near the suture. These pits are the interstices between longitudinal lightly curved and convergent raised lines, and the curved raised lines of growth. Lateral areas of the valves (exclusive of the anterior one) not so distinctly pitted, but more or less marked with faint radiating striae, especially those of the anterior portion of the shell; those of the third to the seventh, both inclusive, each bearing from four to six distant, isolated, prominent, and rounded tubercles on each side, three being usually placed on each of the lines which separate the lateral from the central areas. Length, of the only specimen collected, about 14; maximum breadth of the same, 6½ mm. (Whiteaves.)

Type in Ottawa Museum? Type locality, Discovery Passage, at Station No. 7.

Range. Duncan Bay, British Columbia.

## Ischnochiton venezius Dall, 1919

*Proceedings of the United States National Museum,* **55**:509.

Chiton small, warm yellow-brown with a whitish "bloom," the girdle mottled with brown and white, covered with a pavement-like coating of minute rounded scales; anterior valve with 14, posterior with 12, and the intermediate valves with one slit at each end; anterior valve nearly semi-circular, with over 20 radial rows of small pustules; for the rest covered with a very minute oblique decussation; posterior valve smaller, with mucro low and slightly behind the center, the posterior tract decussated minutely and with a few scattered pustules near the margin, the central area reticulate; intermediate valves mucronate, minutely decussate; pleural tracts with fine low axial threading and punctate interspaces; lateral areas with 2 or 3 somewhat irregular radial rows of pustules on a minutely decussated ground. Length, 7; width, 3.5; height, 2 mm. (Dall.)

TYPE in United States National Museum, No. 216792. Type locality, near Venice, California.

RANGE. Known only from type locality.

## Ischnochiton lividus Middendorff, 1847

*Beitrage zu einer Malacologia Rossica,* 1:124; Pl. 13, figs. 3, 4. Tryon and Pilsbry, *Manual of Conchology,* **14**:76; Pl. 6, figs. 22–24.

Chiton testa externa ovali carinata livida. Valva antica, valvarum inter-mediarum areis lateralibus et valvae ultimae area postica laeviusculis obsolete radiatim striatis. . . . . Valvarum intermediarum area centrali et valvae ultimae area antica longitudinaliter costulata. . . . . Costulae rarae subtiles albidae, numero 18 ad 20 quarum imprimis medianae et praecipue ubi postica versus valva in mucronem brevem protracta, sub lente tuberculiferae vel quasi e tuberculorum contiguorum serie confluente, ex-ortae. . . . . Mucro valvarum intermediarum et valvae ultimae antice in partem glenoidalem prominulam ulrinque incisura parvula praecise limitatam, exit. . . . . Valva antica dentibus marginalibus 9. . . . . Striis radiatis obsoletis circiter 14. . . . . Valva ultima, margine postico convexo, dentibus 14. Pallium marginale nudum epidermide glaberrima. Branchiarum series ante initium secundae tertiae partis totius corporis longitudinis inserta, ad vicinitatem ani usque pergit. . . . . Lamellulae branchiales numero 26. Longitudo adulti, 0.23 decim. (Middendorff.)

Shell elevated, tegmentum smooth, shining, livid. Anterior valve, lateral areas, and posterior valve obsoletely radiately rib-striate. Central areas longitudinally costulate, the riblets separated, sharply cut, whitish, 18–20 in number. Lateral areas depressed. Gills median. Length, 23 mm. (Tryon and Pilsbry, *Manual of Conchology.*)

TYPE in Academy of Sciences, St. Petersburg. Type locality, Patria, Ins Sitcha.

RANGE. Known only from type locality.

## Ischnochiton interstinctus Gould, 1846

*Proceedings of the Boston Society of Natural History*, 2:145. *Report of the United States Exploring Expedition, Mollusca*, 315; Pl. 23, fig. 423.

Testa solida, oblongo-ovalis, valde convexa, cinereo et nigro variegata, concentrice undulato-striata, granulis sparsis nigris aspersa, ad dorsum subcarinata et longitudinaliter nigro bifasciata; areis lateralibus vix distinctis; margo aculeis inaequalibus curtis curvatis nigris et canescentibus indutus. Long., 1$\frac{3}{10}$; lat., $\frac{3}{4}$ poll. (Gould.)

Shell solid, elongated oval, slightly narrowed anteriorly, convexly elevated and slightly carinated along the back. General color a hoary white, here and there clouded with black, and with an interrupted black stripe each side of the summit. The surface is very minutely and irregularly granulated, so as to appear rugose under the lens; occasionally, and more especially on the lateral areas, these granules are shining jet black, so that the surface, when closely examined, appears sprinkled with isolated black dots. The valves are concentrically striated, the striae a little undulating, deep, distinct, and somewhat imbricated at the anterior margin, becoming fainter at the apex and sides. The lateral areas are very slightly raised, not distinctly defined, and with faint indications of a few radiating lines; a few minute, radiating lines may also be seen on the central areas, near the margin. Margin rather broad, clothed with short, unequal, slightly curved, black, and hoary spines, which are generally disposed in alternate clusters of black and white. Interior pale red and violet, somewhat iridescent. (Gould.)

TYPE in United States Museum. Type locality ?

RANGE. Aleutian Islands to Sitka, Alaska, and Catalina Island, California.

Described as *Chiton incanus* Gould.

## Ischnochiton radians Carpenter, 1892

Tryon and Pilsbry, *Manual of Conchology*, 14:121 and 15:75; Pl. 16, figs. 48, 49.

Shell rather large, wide, elevated, the jugum acute; olivaceous, elegantly radially streaked with brown. Interior blue-green with two brown rays. Valves delicate, flat, produced in the sinus in front; mucro in front of the middle; scarcely elevated; entire surface quincuncially granulated; lateral areas scarcely defined, and with the end valves obsoletely subrirulate. Interior: posterior valve having 9–10, central valves 1, anterior valve 10–11 slits; teeth acute; eaves small; sinus wide, short, scarcely laminate. Girdle regularly covered with solid, coarsely striated, moderate-sized scales. Length, 21; breadth, 15 mm.; divergence, 110°. (Carpenter.) (Tryon and Pilsbry, *Manual of Conchology.*)

TYPE in ? Type locality, Monterey, California.

RANGE. Prince of Wales Island, Alaska, to San Pedro, California.

## Ischnochiton scabricostatus Carpenter, 1864

*Supplementary Report,* British Association for the Advancement of Science, 649. Tryon and Pilsbry, *Manual of Conchology,* 14:121, and 15:76; Pl. 16, figs. 55, 56.

Small, arched, orange; rows of prominent granules over shagreened surface. Lobes blunt, slightly rugulose, close to eaves. 8–20 fm. (Carpenter.)

Shell small, orange-colored, elevated; valves gothic arched, jugum acute. Entire surface very closely granulated; lateral areas well defined, with three subobsolete, radiating series of large granules; beaks scarcely apparent; umbonal margin slightly turned inward, and having dark spots giving a false appearance of teeth. Central areas having separated longitudinal series of narrow, subobsolete riblets, furnished with large granules. End valves having similar radiating granulose riblets. Umbo of posterior valve scarcely central, little projecting. Interior; sutural plates separated by a large flat sinus; insertion plates subobtuse, the end valves having 10–12, central valves 1 slit; eaves moderate, subconspicuous. Girdle wide, closely clothed with imbricating, elongated, transversely striated scales. Length, 7½; breadth, 4½ mm.; divergence, 100°. (Carpenter.) (Tryon and Pilsbry, *Manual of Conchology.*)

TYPE in United States National Museum. Type locality, Catalina Island, California.

RANGE. San Pedro and Catalina Island, California, to Cerros Island, Lower California.

Described as *Lepidopleurus scabricostatus* Carpenter.

## Ischnochiton aureotinctus Carpenter, 1892

Tryon and Pilsbry, *Manual of Conchology,* 14:123.

Shell resembling *I. scabricostatus,* but the lateral areas are scarcely defined, radial and longitudinal series of lirae none. Entire surface equally granulated. Orange spotted with red. Interior: posterior valve having 11, central valve 1, anterior valve 13 slits. Length, 6¼; breadth, 4 mm.; divergence, 100°. (Carpenter.) (Tryon and Pilsbry, *Manual of Conchology.*)

TYPE in ? Type locality, Catalina Island, California.

RANGE. Catalina Island to Cerros Island, Lower California.

### Ischnochiton veredentiens Carpenter, 1864

*Supplementary Report,* British Association for the Advancement of Science, 649. Tryon and Pilsbry, *Manual of Conchology,* **14**:122. *Proceedings of the California Academy of Sciences,* **3**:211.

Margin similar. Small, arched, sculptured like *Mertensii,* but with two rows of bosses, one of which dentates the sutures. 10–20 fm. (Carpenter.)

I. t. parva, albida, rosaceo tincta; valvis gothice arcuatis, jugo subacuto; tota superficie minute granulosa; areis lateralibus conspicue definitis, minoribus, costis diagonali et suturali validis instructis, bullis valde expressis munitis; valv. term. costulis subobsoletis radiantibus; areis centralibus clathris longitudinalibus utroque latere circ. viii. distantibus, expressis, subgranulosis, supra jugum obsoletis; interstitiis a costulis subradiantibus decussatis; umbonibus conspicuis; marginibus umbonalibus a costis bulliferis valde indentatis, dentibus viii.–x. jugum versus obsoletis, marginibus haud intortis; mucrone submediano, vix extante; marginibus lobatis eleganter a clathris pectinatis: intus, sinu maximo, planato, interdum serrato; laminis insertionis acutis, late unifissatis, valv. term. circ. viii-fissatis, subgrundis conspicuis; limbo pallii squamis majoribus, planatis, tenuibus, vix striatis. Long., 0.25; lat., 0.10; div., 90°. (Carpenter.)

Shell small, whitish, tinged with roseate; valves gothic-arched, the jugum subacute; entire surface minutely granulose; lateral areas conspicuously defined, having strong diagonal and sutural ribs, provided with well-projecting, rounded grains; end valves having subobsolete radiating riblets; central areas having about 8 distant, raised, subgranulose, longitudinal ribs on each side, the ribs obsolete at the jugum; interstices latticed with subradiating riblets; umbones conspicuous, the umbonal margin deeply toothed by a wart-bearing rib, the teeth 8 to 10, obsolete toward the jugum; margin scarcely intorted. Umbo of posterior valve submedian, slightly projecting; margin elegantly pectinated. Interior: sinus large, flat, valves having about 8 slits; eaves conspicuous. Girdle having large flat, striated scales. Length, 6¼; breadth, 2½ mm.; divergence, 90°. (Carpenter.) (Tryon and Pilsbry, *Manual of Conchology.*)

TYPE in ? Type locality, Catalina Island.

RANGE. Point New Year to Catalina Island, California.

## Ischnochiton newcombi Carpenter, 1892

Tryon and Pilsbry, *Manual of Conchology,* **14**:120.

Shell small, wide, tumid, with obtuse, rounded jugum. Mucro median, rather elevated. Ashy or olivaceous, elegantly clouded; central areas and entire surface granulated, the granules being quincuncially arranged, close and regular; lateral areas scarcely defined, having about 4 lines of tubercles which are hardly elevated. Posterior valve having about 30, anterior valve 40 such lines. Interior: posterior valve having 12, median 1, anterior valve 11 slits; teeth acute; eaves apparent; sinus wide, short, scarcely laminate. Girdle imbricated with large striated and rather regular scales. Length, 8¾; breadth, 6¼ mm.; divergence, 100°. (Carpenter.) (Tryon and Pilsbry, *Manual of Conchology.*)

Type in ? Type locality, Catalina Island, California.

Range. Known only from the type locality.

## Ischnochiton serratus Carpenter, 1864

*Annals and Magazine of Natural History,* **13**:315. Tryon and Pilsbry, *Manual of Conchology,* **14**:122; **15**:78; Pl. 16, figs. 42–46.

I. testa parva, cinerea, olivaceo hic et illic, praecipue ad suturas, punctata interdum sanguineo maculata; ovali, subdepressa, suturas indistinctis; tota superficie minutissime granulata; ar. diag. valde distinctis, costis latissimis obtusis ii.–v. munitis, interstitiis nullis; marginibus posticis eleganter serratis; ar. centr. costis acutis, parallelis, utroque latere circ. xii.; jugo obtuso, haud umbonato; costis transversis, subradiantibus, fenestratibus, interstitiis impressis: mucrone mediano, obtuso; valv. term. costis obtusis ut in ar. diag., circ. xx.; intus valvarum mediarum lobis bifissis, terminalium circ. ix.-fissis; lobis suturalibus magnis; limbo pallii squamis majoribus, imbricatis, vix striatulis. Long., 34; lat., 2 poll. (Carpenter.)

Shell small, ashy, dotted here and there and especially at the sutures with olive, sometimes spotted with red; oval subdepressed, the sutures indistinct; entire surface most minutely granulated; lateral areas strongly defined, provided with two to five very wide, obtuse ribs, no interstices; posterior margins elegantly serrated. Central areas having about 12 acute parallel ribs on each side; jugum obtuse, scarcely umbonate; with subradiating latticing ribs, interstices impressed. Mucro median, obtuse. End valves having about 20 obtuse ribs, like the lateral areas. Interior: median valves bifissate, end valves with about 9 slits. Sutural plates large. Girdle imbricated with large, scarcely striated scales. (Tryon and Pilsbry, *Manual of Conchology.*)

Type in ? Type locality, Cape San Lucas, Lower California.

Range. San Diego, California, to Gulf of California.

## Ischnochiton corrugatus Carpenter, 1892

Tryon and Pilsbry, *Manual of Conchology,* 14:123.

Shell similar to *C. sanguineus* in form and varied coloring; entire surface granulose; central areas having impressed punctate wrinkles, hardly lirate; lateral areas strongly longitudinally corrugated. Interior: posterior valve with 8–10, anterior valve 10–9, central valves 1 slit. Sinus wide, flat. Girdle thin, covered with very close, very small striatulate imbricating scales. Length, 13¾; breadth, 7 mm.; divergence, 90°–100°. (Pilsbry.)

TYPE in California State Collection. Type locality, Catalina Island, California.

RANGE. Catalina Island, California, to Todos Santos Bay, Lower California.

## Ischnochiton berryi Dall, 1919

*Proceedings of the United States National Museum,* 55:507.

Chiton of moderate size, deep crimson with irregular blotches of brown and white on its lateral slopes, and a crimson girdle of rounded, densely imbricated small scales; gill rows ambient; anterior valve with a nearly smooth mucro, from which radiate about 20 more or less corrugated, sometimes bifurcating threads, the whole overspread like the rest of the shell by an almost microscopic decussation; anterior slits about 12; posterior valve much smaller with a subcentral mucro, from which radiate threads, like those on the anterior valve, in all directions; posterior slits 11; intermediate valves with one slit, jugal area minutely decussated, almost carinate; lateral areas with usually three corrugated ribs, the nodulation of the posterior rib undulating the margin of the valve, but there may be one or two minor intercalary ribs; the pleural areas are axially ribbed with deep decussated interspaces, the ribs growing stronger as they recede from the jugal region; interior pink, the jugal sinus hardly marked, axially striated, merging laterally into narrow sutural laminae. Length of shell, 11; width, 7; height, 2.5 mm. (Dall.)

TYPE in United States National Museum, No. 193375. Type locality, Pacific Grove, California.

RANGE. Monterey to San Pedro, California.

## SECTION LEPIDOZONA Pilsbry, 1892

### Ischnochiton willetti Berry, 1917

*Proceedings of the California Academy of Sciences,* series 4, 7:236, figs. 1, 2.

Shell rather large, regularly elliptical; elevated, with an angular dorsal ridge and arculate side slopes. General surface indistinctly granulose.

Anterior valve with 35–45 low, occasionally bifurcating, radiating ribs, separated by shallow distinct grooves and bearing a variable number (normally 8 to 10) of small, well-separated, distinct, rounded pustules. Median valves: lateral areas distinctly raised, sculptured like the anterior valve, but the 5 to 7 ribs relatively wider and more flattened, the defining grooves sometimes, but not always, sharply chiseled; central areas on each side sharply sculptured with 20–25 narrow, faintly beaded, longitudinal ridges, their interstices traversed by low, rather irregular cross-ridges, becoming nearly or quite obsolete at the jugum. Posterior valve with mucro sculptured like the anterior valve, but the 28–30 ribs rather less distinct and the grooves less conspicuous; region in front sculptured like the central areas of the intermediate valves. Interior of central valves thickened across the middle. Teeth with a distinct fossa separating them at base from body of shell, their edges roughened, those of the posterior valve almost cremate. Anterior valve with 11, second valve with 2–1, third to seventh valves with 1–1, posterior valve with 12 slits. Girdle wide, regular; covered dorsally with a closely imbricating armature of large, smooth or barely striated, convex scales, each normally bearing a short, striated, nipple-shaped process projecting upward from the dorsal end. Color of outer surface of shell, reddish-brown of varying intensity in different specimens, sometimes nearly black; girdle lighter. Interior of shell, light salmon. Maximum length of the type and largest specimen, 29.5 mm.; width, 16 mm. (Berry.)

TYPE in Berry Collection, No. 3700. Type locality, Forrester Island, Alaska, 15–20 fathoms.

RANGE. Known only from type locality.

## Ischnochiton mertensii Middendorff

*Bulletin,* Imperial Academy of Sciences, St. Petersburg, **6**:118. Tryon and Pilsbry, *Manual of Conchology,* **14**:125; Pl. 26, figs. 20–26.

Chiton testa externa ovali elevata opaca, aspera, fusco-cinerea; . . . . valva valvae ultimae area postica, valvarum denique intermediarum areis lateralibus radiatim expresse granulosocabris. . . . . Valvarum intermediatum areis centrilibus et valvae ultimae area antica longitudinaliter esculpte-costatis, costis medianis postica versus dichotomia; costarum interstitia lamellulis erectis transversis in loculamenta dissepta. . . . . Valva antica dentibus marginalibus 13 et radiis granulosocabris 26. . . . . Valva ultima margine postico convexo dentibus 13 et radiis granulosocabris 20. . . . . Valvarum intermediarum radiis granulosocabris quinque. Pallium marginale epidermide fuscocinerea, squamis aspera, obtectum. . . . .

Squamae hae in series oblique decurrentes ordinate. Long., 0.23 decim. (Middendorff.)

Shell oval, elevated, with angular dorsal ridge, and straight sideslopes. Varying in color from orange-red to claret-red, or even dark red-brown, and either unicolored or speckled and blotched with white. The lateral areas are elevated, and sculptured with radiating rows of elevated pustules standing upon a smooth, almost flat ground; the pustules of the sutural row often irregular. Central areas having acute, narrow, parallel, raised riblets, the intervals between them regularly latticed across, except at the dorsal ridge, where the riblets have a tendency to diverge, and cross-hatching is obsolete. End valves radially pustulose. Mucro central, low, flat, and inconspicuous. Interior white or blue-white, the median valves when detached showing broad red-brown rays posteriorly, the end valves with crescents of the same color. Sutural plates low, sinus flat, angular, finely toothed. Anterior valve having 10–11, central valve 1, posterior valve 10 slits; teeth rather short and obtuse, and usually distinctly roughened; eaves rather wide. Girdle firm, compactly covered with regular, solid, oval, shining scales, which are usually smooth but frequently are superficially or obsoletely striated. Length, 35; breadth, 21 mm. (Tryon and Pilsbry, *Manual of Conchology.*)

TYPE in Academy, St. Petersburg. Type locality, California.

RANGE. Sitka, Alaska, to San Pedro, California.

Described as *Chiton mertensii.*

## Ischnochiton cooperi (Carpenter) Pilsbry, 1892

Tryon and Pilsbry, *Manual of Conchology,* 14:127; Pl. 26, figs. 27–30.

Shell oval and elevated, with angular dorsal ridge and straight sideslopes. Sculpture like *I. mertensii.* Color olivaceous, or dull earthy brown, indistinctly clouded more or less with light blue, especially upon the side areas. The lateral areas are raised, and bear irregular rows of rounded pustules, the young having four rows, the adult 6 to 8. A strong lens reveals a fine, subobsolete granulation of the nearly flat surface between the pustules. The central areas have a fine but distinct and even radial striation, over which run acute narrow raised threads parallel to the dorsal ridge; upon the ridge these threads are seen to be more or less diverging, especially upon the second valve. The end valves are radially ridged, the ridges bearing elongated pustules, or showing scars where such pustules have been. Mucro low, flat. Interior bluish, the valves marked under their umbones with dark teeth roughened but rather sharp; eaves wide, dark, minutely punctulate, but solid, not spongy. Girdle compactly covered

with small imbricating, deeply striated scales. Length, 40; breadth, 24 mm.
(Pilsbry.)

Type in Academy of Natural Sciences, Philadelphia, No. 118659. Type
locality, Bolinas Bay, California.

RANGE. Mendocino County to Catalina Island, California.


## Ischnochiton cooperi acutior (Carpenter MS) Dall, 1919

*Proceedings of the United States National Museum,* **55**:508.

On comparing specimens named by Carpenter in the collection of the
United States National Museum, the only differences I could perceive
were that the specimens of the variety *acutior* were lighter and brighter
in color, more emphatic in sculpture, and apparently younger shells. In a
group where color is often without systematic value, these differences
seem hardly worthy of a name, analogous mutations being most common
among Chitons. (Dall.)

Type in United States National Museum. Type locality, Lower California.

RANGE. San Diego, California, to Las Animas Bay, Lower California.


## Ischnochiton clathratus Reeve, 1847

*Conchologia Iconica,* Chiton, Pl. 18, fig. 113. Tryon and Pilsbry, *Manual of Conchology,* **14**:128; Pl. 26, figs. 31–34.

Chit. testa ovata, elevatiuscula, valvis terminalibus caeterarum areisque
lateralibus radiatim rugoso-granatis, centralibus peculiariter crebriclathra-
tis; nigricante-viridi; ligamento subtiliter granoso-coriaceo, obscure tessellato.

Shell ovate, rather elevated, terminal valves and lateral areas of the
rest radiately roughly grained, central areas peculiarly closely latticed;
blackish-green; ligament finely granosely coriaceous, obscurely tessellated.
(Reeve.)

Type in Museum Cuming. Type locality, not given.

RANGE. Bolinas Bay, California, to Gulf of California.

Described as *Chiton clathratus.*


## Ischnochiton decipiens Carpenter, 1892

Tryon and Pilsbry, *Manual of Conchology,* **14**:123.

Shell exactly like *I. pectinulatus,* but reddish, elevated, the jugum
acute; scales of the girdle small, striated. Length, 15; breadth, 7½ mm.;

divergence, 105°. (Carpenter.) (Tryon and Pilsbry, *Manual of Conchology*.)

TYPE in ? Type locality, Monterey, California.

RANGE. Catalina Island to Monterey, California.

## Ischnochiton sinudentatus Carpenter, 1892

Tryon and Pilsbry, *Manual of Conchology*, 14:128.

Shell oval, much elevated, reddish; jugum acute; mucro median, scarcely elevated; entire surface minutely granulose. Central areas having about 12 subparallel bars, which pectinate the front margin, their interstices decussated; lateral areas having 3–4 granulose radiating lirae, the anterior valve with 26, posterior valve 24 such lirae. Interior: anterior valve with 10, central 1, posterior valve 9 slits; teeth acute; eaves conspicuous, subspongy, crenulated by the sculpture; sinus wide, short, laminate with about 6 teeth. Girdle having moderate-sized, wide, conspicuously striated, rather regularly imbricating scales. Length, 15; width, 7½ mm.; divergence, 100°. (Carpenter.) (Tryon and Pilsbry, *Manual of Conchology*.)

TYPE in ? Type locality, Monterey, California.

RANGE. Known only from type locality.

## Ischnochiton asthenes Berry, 1919

*Proceedings of the California Academy of Sciences,* series 4, **9**: No. 1, Pl. 8. *Lorquinia,* January 1919.

Shell small, carinate, with convex side slopes. Surface granulose. Head valve with 11–12 low, rounded flutings, bearing 2–4 minute pustules each. Median valves with lateral areas distinctly raised, bicostate, each rib bearing 2–5 small pustules; central areas with 15–18 narrow, granulose, longitudinal ribs, finely interlatticed between, but becoming obsolete on jugum. Slit formula 9, 1–1, 12. Girdle scales small, convex, delicately ribbed. Color of shell yellowish-brown, mottled deeper. (Berry.)

TYPE in Berry Collection. Type locality, White's Point, Los Angeles County, California.

RANGE. Known only from type locality.

## Ischnochiton stearnsii Dall, 1902

*Proceedings of the United States National Museum,* 24:557.

Chiton of moderate size, yellowish or buff color; the girdle yellowish-white, covered with subcylindric, blunt, smooth, close-set, large spines, the ends of which have a pebbly appearance, mixed with a smaller proportion of small but rather similar spinules; the ends of the large spines, when worn flat, have a pavement-like aspect; back not keeled, but rather steeply rounded; gills ambient; intermediate valves with a dorsal angle of about 90°, the lateral areas prominent, with about five radial riblets in each, divaricating to seven or ten at the girdle margin, and cut into beads by numerous fine concentric furrows; pleural areas and jugum hardly differentiated, sculptured with fine, slightly irregular, longitudinal wrinkles, finer mesially, crossed by inconspicuous, less elevated, transverse lines; anterior valve with fine, beaded, divaricate radial riblets, the insertion plates and eaves very short, smooth, not spongy, with about 17 slits; the posterior valve with a small, low, subcentral mucro, from which two elevated lines extend to the margin, one on either side, forming two areas, and from which the wrinkled sculpture, less prominent on the anterior area, diverges; posterior slits about 15, lateral slits 2; sinus rather wide, entire; pleural laminae rather wide and short. Long. of animal, about 25; lat., 15; alt., 6 mm. (Dall.)

TYPE in United States National Museum, No. 109024. Type locality, U.S.S. "Albatross" Station 3104, off Farallon Islands, in 391 fathoms.

RANGE. Known only from type locality.

### Ischnochiton golischi Berry, 1919

*Lorquinia,* January 1919.

Shell small, strongly carinate, side-slopes faintly arcuate. Surface minutely granulose. Head valve with about 45 nearly obsolete ribs, weakly grooved between, each bearing 5–7 extremely minute distant granules. Median valves with lateral areas distinctly raised, each with 4–5 almost obsolete, minutely, distantly, granose ribs; central areas on each side with about 20 granulose longitudinal riblets, finer and closer toward the jugum, their interstices granulose but scarcely latticed. Tail valve with perhaps 30 indistinct radiating ribs on the posterior tract. Dorsal girdle scales finely, distinctly striate. Color of the shell warm brownish rosy. (Berry.)

TYPE in Berry Collection. Type locality, off Santa Monica, California, in 100 fathoms.

RANGE. Known only from type locality.

### SECTION RHOMBOCHITON Berry, 1919

### Ischnochiton regularis Carpenter, 1855

*Proceedings,* Zoölogical Society of London, 232. Tryon and Pilsbry, *Manual of Conchology,* 14:142; Pl. 18, figs. 41–46.

C. t. elongata, elevata (ad angulum 110°), fusco-olivacea, jugo acuto; valvis intermediis lineis diagonalibus haud conspicuis; areis lateralibus et valvis ultimis strigis radiantibus, areis centralibus strigis longitudinalibus, parallelis; mucrone parvo; marginibus valvarum subrotundatis, suturis modicis; limbo squamoso, squamis oblongis, irregulariter tessellatis; intus valvarum marginibus haud valve arcuatis, sinu lato, haud alto; ad jugum linea impressa. Long., 1.1; lat., .58; alt., 25 poll. (Carpenter.)

Shell oblong, elevated, dorsally carinated, the side-slopes straight. Surface appearing almost smooth to the naked eye. Color a uniform olive or slaty-blue, the girdle having more or less of a blue or purple ".bloom." The lateral areas are very little raised, and are sculptured with numerous delicate radiating threads, occasionally branching toward the lower margin, and freely branching along the posterior edge of the valve. The terminal valves have similar delicate radii. The central areas have numerous longitudinal, somewhat beaded threads, separated by flat intervals. The posterior valve is elevated, with anterior umbo. The interior is light blue. Sutural plates low, connected across the sinus by a narrow plate which is sulcate above and cut into about 10 teeth by delicate slits. Anterior valve having 14–16, central 2–3, posterior 22 slits. Teeth even and sharp, slightly striated or grooved outside. Eaves solid. Girdle wide, flat, covered with solid, regularly and closely imbricating striated scales. Length, 35; breadth, 19 mm. (Tryon and Pilsbry, *Manual of Conchology.*)

TYPE in Museum Cuming. Type locality, Monterey, California.

RANGE. Mendocino County to Monterey, California.

Described as *Chiton regularis* Carpenter.

## SECTION TRIPOPLAX Berry, 1919

### Ischnochiton trifidus Carpenter, 1864

*Supplementary Report,* British Association for the Advancement of Science, 649. Tryon and Pilsbry, *Manual of Conchology,* 14:141; Pl. 18, fig. 40.

Center-punctures few, deep: 2–4 blunt ribs: side plates with 2 slits. (Carpenter.)

Shell rather large, rather elevated, regularly oval; red-chestnut, maculated with lighter and darker; jugum acute, gothic. Mucro median, flat; entire surface very minutely granulated; central areas having about 8 strongly punctate lines perpendicular to the jugum; lateral areas strongly defined, having 2 to 4 obsolete ribs, sometimes punctate in the interstices. Interior whitish flesh-colored, with two reddish-purple rays diverging from the flat umbones; posterior valve having 13, anterior 13, central valves 2 slits; teeth acute, sometimes serrated at the edge, sometimes striated outside, sometimes smooth. Eaves sub-spongy; sinus small, laminate, the laminae slit at the sides and sometimes in the middle. Girdle having very small, solid, smooth scales. Gills almost ambient. (Carpenter.) Length, 40; breadth, 26 mm.; divergence, 135°. (Tryon and Pilsbry, *Manual of Conchology.*)

TYPE in United States National Museum. Type locality ?

RANGE. Shumagin Islands to Puget Sound.

Described as *Trachydermon trifidus* Carpenter.

## Genus CHAETOPLEURA Shuttleworth, 1853

Valves as in *Ischnochiton;* eaves solid; girdle leathery, more or less hairy; gills extending almost or entirely to the front end of the foot. (Tryon and Pilsbry, *Manual of Conchology.*)

TYPE. *C. peruvianus* Lam.

DISTRIBUTION. West Coast of North America.

### Chaetopleura gemmea Carpenter, 1892

Tryon and Pilsbry, *Manual of Conchology,* **14**:31; Pl. 13, figs. 69–74.

Shell oblong, elevated, red, olive-ashen or yellow; umbo of tail valve depressed, situated behind the middle. Lateral areas decidedly raised, coarsely radiately tuberculate; central areas having longitudinal beaded lirae. Girdle narrow, leathery, sparsely clothed with short hyaline hairs, which are readily rubbed off. The valves are elevated. The central areas are sculptured with elevated, distinctly beaded longitudinal cords, about 15 in number on each side; they become very small or subobsolete upon the jugum. The lateral areas are strongly differentiated; they have 5 to 7 radiating rows of distinct, clear-cut tubercles. The mucro does not rise above the general level of the posterior valve. Inside often tinted with red, often having a red or black spot at the jugal sinus. Sutural plates well-rounded, the sinus rather deep and angular. Anterior valve having 9–12, median 1, posterior valve 7–8 slits. Teeth rather blunt and stout; eaves not spongy, rather wide. Length, 16½; breadth, 8 mm. (Tryon and Pilsbry, *Manual of Conchology.*)

TYPE in ? Type locality, Monterey, California.

RANGE. Monterey, California, to Magdalena Bay, Lower California.

### Chaetopleura beanii Carpenter, 1857

*Mazatlan Catalogue,* 197. Tryon and Pilsbry, *Manual of Conchology,* **14**:32.

L. t. ovata, elevata (ad angulam 110°) fusco-olivacea, seu albido caeruleoque maculata; valvis intermediis valde mucronatis, interstitiis curvatis, marginibus subrotundatis; valva postica depressa, excavata, mucrone minimo, superiore; jugo et areis lateralibus indistinctis; superficie tota granulis instructa, curvatis, sinu planato; marginibus acutis, fissuris circiter x. valvis terminalibus, una in utroque latere, intermediis; marginibus externis prominentibus; limbo palii piluloso, sinpulis parvis, erectix, planatis. (Carpenter.)

Shell ovate, elevated, brown-olive or maculated with whitish and bluish. Intermediate valves strongly mucronate, interstices curved, margins somewhat rounded. Posterior valve depressed, excavated, the mucro small, superior; jugal and lateral areas indistinct; surface covered with granules and ornamented with small close points. Sutural plates large, curved; sinus flat; insertion plates acute, the intermediate valves with one, terminal valves 10 slits. Girdle hairy, spines small, erect, flat. Length, 5¾; breadth, 3¼; alt., 1 mm.; divergence, 110°. (Tryon and Pilsbry, *Manual of Conchology*.)

TYPE in British Museum. Type locality, Mazatlan, Mexico.

RANGE. Unalaska, Alaska, to Mazatlan, Mexico.

Described as *Ledidopleurus beanii* Carpenter.

## Chaetopleura prasinata Carpenter, 1864

*Annals and Magazine of Natural History,* series 3, **13**:314. Tryon and Pilsbry, *Manual of Conchology,* **14**:34.

I. testa I. parallelo forma et indole simili, sed vivide viridi; ar. diag. seriebus bullularum irregulariter ornatis; ar. centr. clathris valde extantibus, acutis, jugo obtuso parallelis, utroque latere circ. xvi.; valv. term. seriebus bullularum circ. xviii.; mucrone submediano, inconspicuo; umbonibus haud prominentibus; tota superficie minutissime granulosa; intus valvarum lobis mediarum i.-term circiter x.-fissis; sinu lato, planato; suturis planatis; limbo pallii anguto, minutissime squamulis furvicaceis creberrime instructo; interdum pilulis intercalatis. Long., 8; lat., 4 poll. (Carpenter.)

Similar to *parallela,* but vivid green; lateral areas irregularly ornamented with series of tubercles. (Tryon and Pilsbry, *Manual of Conchology*.)

TYPE in United States National Museum. Type locality, Cape San Lucas, Lower California.

RANGE. San Diego, California, to Cape San Lucas, Lower California.

## Chaetopleura parallela Carpenter, 1864

*Annals and Magazine of Natural History,* **13**:314. Tryon and Pilsbry, *Manual of Conchology,* **14**:34; Pl. 12, figs. 50.

I. testa ovata, subelevata (ad angulum 120°); rufo-fusca, olivaceo tincta; valvis latis, marginibus parum rotundatis, interstitiis parvis; valvis intermediis valde, insculptis; arcis lateralibus seriebus granulorum a jugo radiantibus circiter vi.; interdum irregularibus, granis rotundatis, separatis, extantibus; areis centralibus clathris creberrimis, jugo parallelis, horridis, extantibus, interdum granulosis, ornatis; valvis terminalibus seriebus granulorum, circ. xx., interdum bifurcantibus, ut in areis lateralibus, ornatis; mucrone vix conspicuo; limbo pallii angusto, pilulis furvicaceis creberrimis minutis conferto; lobis valvarum bifidis, terminalibus fissuris circ. xi., a parte externa simplici disjectis. Long., 7; lat., 48; alt., 16 poll. (Carpenter.)

Shell rather acutely carinated, the side slopes nearly straight. There are 17 beaded longitudinal threads on each side of the central areas, and from 6–7 rows of separated, rounded, erect tubercles on each lateral area. The lateral areas are decidedly elevated; the mucro is slightly prominent. The girdle has sparsely scattered hyaline short hairs. The gills continue as far as the front end of the foot. The outside is reddish, marbled with darker and white, the girdle dark ashen. The interior bluish-white. (Tryon and Pilsbry, *Manual of Conchology*.)

TYPE in United States National Museum. Type locality, Cape San Lucas, Lower California.

RANGE. San Diego, California, to Cape San Lucas, Lower California, and West Columbia.

## Chaetopleura lactica Dall, 1919

*Proceedings of the United States National Museum, 55:509.*

Chiton small, cream-colored, often maculated with pale green, with high, almost carinate back, the girdle spongy with a few small spinules and sparse hairs on a leathery basis; anterior valve with 8 or 9 slits, posterior with 2 slits, intermediate valves with single slits; anterior valve with about 10 feeble radii, otherwise minutely punctate-reticulate; posterior valve small, with subcentral mucro and similar sculpture; intermediate valves with the jugal and pleural tracts not separated, mucronate, sculptured with obliquely axial very fine threads with punctate interspaces; the lateral areas bounded by a slender, sometimes beaded rib in front and a marginal thickening behind, otherwise sculptured like the rest of the valve; internally whitish with a wide and shallow jugal sinus and prominent sutural laminae. Length, 8; width, 4.5; height, 2 mm. (Dall.)

TYPE in United States National Museum, No. 172900. Type locality, Catalina Harbor, California.

RANGE. Known only from type locality.

SECTION DENDROCHITON Berry, 1911

## Chaetopleura thamnopora Berry, 1911

*Proceedings,* Philadelphia Academy of Natural Sciences, 487; text figures 1–7; Pl. 40.

Shell small, oblong, rather narrow for *Mopalia,* much elevated and strongly carinated, the side slopes nearly straight. Valves sharply beaked in front; the lateral areas fairly well-defined, not raised, having a few very faint radial grooves, but without well-marked sutural or diagonal ribs. Central areas ornamented with a series of about nine or ten very strong low, broad, longitudinal riblets, curved and converging toward the median ridge; their intervals of nearly equal breadth, not latticed or otherwise sculptured. On the jugal tract these ribs are obsolete or wanting. Entire surface minutely granular-porous, but not so distinctly as in *M. heathi,* and due largely to the numerous sense organs which under a high power appear with great clearness. Anterior valve everywhere very finely granulous, otherwise without distinct sculpture. Central area of posterior valve reduced, but similar in ornamentation to those of the median valves; mucro anterior to the middle of the tegmentum; posterior slope steep, slightly concave; posterior margin of the tegmentum in general semicircular, but rather abruptly (though not extensively) squarish truncate or emarginate at the middle. Anterior valve with 7 slits; intermediate valves with 1–1 slits; tail valve "ischnoid," with a regular crescentic insertion plate cut by 6 slits. Sinus broad, rather shallow; in last valve narrower and minutely crenulate. No median sinus behind, and no indications of an approach to this condition other than the above-noted squaring of the tegmentum. Girdle narrow, apparently nude even under a hand lens, but shown by high power to be well-clothed above with numerous exceedingly minute, but not very crowded, ovid spicules developed into stout elongate spines at the margin. The spinelets of the lower surface are longer, flatter, and more pointed than those of the upper, being somewhat intermediate in character between these and the marginal ones. Opposite each suture is a pore from which springs a group of about 6 recurved bristle-like hairs surrounding a single much larger and longer bristle, which branches freely, and in living or alcoholic material is a prominent feature even to the unaided eye. Being very brittle, these structures are frequently broken off, but their stumps or pores are always evident and show a very regular arrangement. Apparently homologous with the sutural tufts are two similar ones on each side of the head valve, one on each side of the tail valve, and one in the median line in front and behind. In addition there is a second series of much smaller but equally distinct tufts lying just outside of the first and in a more or less regular alternation with them. The long central bristles have a thickened sheath-like base from which are given off slender, more or less recurved, hair-like processes. All together there are 22 of these tufts besides an equal or slightly larger number of minor (alternating) ones. The typical form seems to be pink with sutural spots of brown and more or less green suffusion. The girdle shows alternating bands of burnt sienna and pale buff. The latter are sutural in position and there are also small intersutural spots of the same color. The interior is rose, paler toward the edges, but not so vivid as in *M. heathi*. Length, 9; width, 5 mm. (Berry.)

TYPE in Berry Collection. Type locality, off Monterey, California.

RANGE. Resurrection Bay, Alaska, to San Martin Island, Lower California.

## Chaetopleura gothica Carpenter, 1864

*Supplementary Report,* British Association for the Advancement of Science, 649.
Tryon and Pilsbry, *Manual of Conchology,* 14:74 and 15:65; Pl. 15, figs. 28, 29.

Blunt parallel riblets along very arched back. Sutural lobes united at sinus; eaves not spongy. 8–20 fm. (Carpenter.)

Shell small, much elevated, green, elegantly tinted with rose and olive; valves gothic-arched, the dorsal ridge acute; lateral areas small, arcuately distinctly defined, granulose; umbones prominent; umbonal margin having a tessellated color pattern and incurved. Central areas longitudinally ribbed, ribs rounded, close, not much elevated, the interstices small, sometimes slightly decussated. End valves sculptured like the lateral areas, the tail valve having the umbo median and somewhat elevated. Interior having the sutural plates scarcely separated, the jugal sinus very broad and shallow; insertion plates of median valves having a single slit, terminal plates with 8–10 slits; the plates are obtuse; eaves hardly elevated. Girdle most minutely scaly, the granules very close, rounded, smooth; edge with small suberect hairs. (Carpenter.) Length, 5; breadth, 2½ mm.; div., 80°. (Tryon and Pilsbry, *Manual of Conchology.*)

TYPE in United States National Museum. Type locality, Catalina Island, California.

RANGE. Monterey to Catalina Island, California.

Described as *Trachydermon Gothicus* Carpenter.

## Genus **PALLOCHITON** Dall, 1878

Valves exposed, solid, the anterior and median having sharp insertion teeth slightly roughened outside; slits of anterior valve not corresponding to anything in the external sculpture; tail valve having the mucro at the posterior end, the teeth sharp and all strongly directed forward; eaves solid; sinus notched at sides. Girdle leathery, with a few deciduous hairs, but no pores.

This genus closely resembles in sculpture as well as structural characters *Chaetopleura Shuttlew.*, differing only in the posterior position of the mucro and the consequent throwing forward of the tail-valve insertion teeth—characters of not great importance. The slits correspond in position to nothing in the external sculpture. The girdle is decidedly like *Chaetopleura,* being very sparsely hairy (or smooth) and lacking all appearance of pores. (Tryon and Pilsbry, *Manual of Conchology.*)

TYPE. *P. lanuginosus* Carpenter.

DISTRIBUTION. San Diego, California, to Point Abreojos, Lower California.

## Pallochiton lanuginosus (Carpenter MS) Dall, 1879

*Proceedings of the United States National Museum,* 1:297; Pl. 3, fig. 21 and 4:287. Tryon and Pilsbry, *Manual of Conchology,* 14:257; Pl. 56, figs. 1–11.

Minor lateral normal, bi-alate; major lateral tridentate, shaft normal. (Dall.)

Shell oblong or ovate, rather elevated, carinated, the side-slopes nearly straight. Surface lusterless, color very variable; sometimes dull brown or purplish-brown; sometimes green along the ridge and purple or lilac dotted with black or olive at the sides; or having the sides of some valves scarlet to even snow-white. The median valves are rather acutely beaked in the young, beaks eroded in old specimens. Lateral areas but little raised, sculptured with gem-like pustules scattered irregularly on a flat (microscopically puncticulate) ground; the pustules often few, rarely wanting on some valves; central areas sculptured with many closely beaded longitudinal threads, which converge slightly at the ridge, and diverge toward the outer sides of the valves; on the second valve they diverge at the ridge. Head valve pustulose like the lateral areas. Tail valve much depressed, the mucro at the posterior end; anterior area wide, sculptured like the central areas; posterior area very narrow, vertical, sometimes pustulose. Interior bluish-white, darker at jugum and posteriorly; rarely flesh or pink tinted. Sutural plates very broad; sinus narrow, deep, angular and notched at the sides. Anterior valve having 8–9, central 1 slit, the teeth long, sharp, a little rugose outside; posterior valve having 10–11 slits, the teeth chisel-shaped, strongly directed forward, smooth and rather sharp. Eaves of anterior and median valves narrow, of posterior valve wider, solid. Girdle rather fleshy, leathery when dried, somewhat encroaching at the sutures, and smooth or clothed with sparse delicate hairlets. (Tryon and Pilsbry, *Manual of Conchology.*)

Type in United States National Museum. Type locality ?

Range. San Diego, California, to San Ignacio Lagoon, Lower California.

## Genus CALLISTOCHITON Carpenter, 1882

Valves conspicuously sculptured; the insertion-plates short, smooth or nearly so, festooned, being curved outward at the ribs and slit there, thickened outside at the edges of the slits, the latter corresponding in position to the ribs of the outer surface. Sinus squared. Mucro median or post median, generally depressed. Girdle poreless, densely clothed with minute striated or smooth scales. (Tryon and Pilsbry, *Manual of Conchology.*)

Type. *Chiton pulchellus* Gray.

Distribution. Australia, Red Sea, Japan, Gulf of Mexico and western Americas from southern California to northern Chile.

### Callistochiton palmulatus Carpenter, 1892

Tryon and Pilsbry, *Manual of Conchology,* 14:262; Pl. 58, figs. 12–16.

Shell similar to *C. pulchellus,* but more flattened, the dorsal ridge acute; mucro subcentral, depressed, the posterior area strongly swollen; sculpture stronger; central areas having about 10 sub-parallel acute lirae on each side, pectinating the sutures, interstices deeply punctate. Anterior valve having 11 ribs, of which the outer two are joined; posterior valve 7 very strong ribs bifurcating behind. Interior: anterior valve having 11 slits, central 1, the teeth normal; posterior valve having 26 slits, the teeth crowded, minute, palmate. Eaves very strong; sinus small, strongly laminate, the laminae deeply slit on each side. Girdle imbricated with striated scales. Length, 11¼; breadth, 7½ mm.; divergence, 135°. (Carpenter.) (Tryon and Pilsbry, *Manual of Conchology.*)

Type in California State Collection, No. 1077. Type locality, Santa Barbara, California.

Range. Monterey, California, to San Marco Island, Lower California.

### Callistochiton palmulatus mirabilis Pilsbry, 1892

Tryon and Pilsbry, *Manual of Conchology,* 14:263; Pl. 58, figs. 7–11.

Shell oblong, elevated, the back angular, side-slopes convex. Surface lusterless, dull brownish, the lateral areas and end valves blackish. Valves not beaked; the lateral areas widely separated by the eroded beaks, greatly elevated, each split by a deep median sulcus, the two ribs thus formed bearing coarse transverse grains. Central areas sculptured with about 15 narrow longitudinal cords, parallel at the dorsal ridge except on the second valve in which they diverge forward; the interstices wider than the cords, and finely latticed across. Anterior valve having 9 stout radiating ribs, strongly granose, and with the exception of the two outer ones, they are generally not bifid. Posterior valve much higher than the anterior, the mucro somewhat in front of the middle, the area behind it enormously developed, elevated and convex, sculptured with 4, 5, or 6 primary stout ribs, each of which splits into two; the two outer ribs are broader, and split into several riblets. Interior bluish-white; sutural-laminae slightly connected across the squarish sinus. Anterior valve having 9, central 1, posterior valve 22 slits; teeth short, somewhat roughened. Eaves broad, calloused. Girdle narrow, thin, covered with very densely imbricating, minute, deeply striated scales. Length, 16; breadth, 7 mm. (Pilsbry.)

TYPE in Philadelphia Academy of Sciences. Type locality, San Diego, California.

RANGE. Known only from the type locality.

## Callistochiton crassicostatus Pilsbry, 1892

Tryon and Pilsbry, *Manual of Conchology*, 14:264; Pl. 58, figs. 1–6.

Shell oblong, elevated, the dorsal ridge very obsoletely angular, side-slopes arched. Surface lusterless, green or brown. Valves not beaked, the lateral areas extremely prominent, unevenly granulated, the concentric riblets being cut by one or several radiating grooves. Central areas having strong longitudinal bars, converging Λ-like on the ridge (even on the second valve), the intervals very closely and finely latticed across. Anterior valve having seven very strong ribs, each divided by a shallow median groove. Posterior valve elevated, having the mucro directly over the posterior edge, the hinder area not higher than the area in front of it; posterior slope vertical, convex, sculptured with five very strong, deeply separated ribs, which are granose above, and subdivide into several riblets each toward the lower margin. Interior bluish-white; sutural-plates slightly connected across the rounded sinus. Anterior valve having 9, central valves 1, posterior valve 13–20 slits; teeth short, rather sharp and smooth, hardly projecting below the eaves, thickened along the slits outside; eaves broad, solid. Girdle narrow, thin, covered with excessively minute, closely imbricating, striated scales. Length, 22; breadth, 9 mm. (Tryon and Pilsbry, *Manual of Conchology*.)

TYPE in Philadelphia Academy of Sciences. Type locality, Monterey, California.

RANGE. Forrester Island, Alaska, to San Diego, California.

## Callistochiton decoratus (Carpenter) Pilsbry, 1892

Tryon and Pilsbry, *Manual of Conchology,* **14**:269; Pl. 58, figs. 17–20.

Shell oblong, moderately convex, obtusely subangular along the dorsal ridge, the side-slopes slightly convex. Surface rather shining, varying in color from olive-buff to dark green, uniform or having darker flammules and spots. Apices of the median valves eroded. Lateral areas raised, each rendered bicostate by a central sulcus, and having a few uneven longitudinal impressed grooves, giving it a terraced aspect. Central areas having numerous parallel longitudinal ribs, the intervals closely latticed across, this sculpture obsolete on the ridge, where there is a broad V-shaped smooth area, which rarely shows a few diverging subobsolete ribs. Anterior valve having 11 rounded ribs. Posterior valve less elevated than the anterior, highest at the front margin, the mucro depressed, post-median; posterior area having 9–10 rounded ribs. Interior bluish, generally marked with olive or green at jugum, bases of the sutural-plates, and slit-rays. Sutural plates wide, separated by a square sinus notched at each side. Anterior valve having 9 or 11, central 1, posterior valve 9–12 slits; teeth rather sharp, smooth; slit-rays distinct and porous. Eaves narrow. Girdle narrow, ashy-brown, covered with minute, striated, closely imbricating scales. Length, 20; breadth, 9–10 mm.; divergence, about 110°. (Tryon and Pilsbry, *Manual of Conchology.*)

TYPE in Academy of Natural Sciences, Philadelphia, No. 118687. Type locality, Todos Santos Bay, near San Tomas River, Lower California.

RANGE. Santa Barbara, California, to San Tomas River, Lower California.

## Callistochiton decoratus infortunatus Pilsbry, 1892

Tryon and Pilsbry, *Manual of Conchology,* **14**:266; Pl. 59, figs. 37–42.

Shell large, regularly ovate, the dorsal ridge obtuse; valves arcuate; mucro obtuse, median. Olivaceous, sometimes spotted with paler on the ribs and jugum. Central areas having about 12 parallel lirae on each side, decussated, the interstices having square depressions. Lateral areas having two very strong rounded tuberculose ribs. Anterior valve with 9, posterior with 7–8 elegantly spreading ribs. Interior: posterior valve having 7–8, central 1, anterior 9 slits; teeth concave outwardly, obtuse, slit at the apices of the ribs, sometimes with an intercalated slit or abnormally serrate; teeth of posterior valve very obtuse, hardly sloping; eaves small, delicate. Sinus wide, flat, but little angular, sometimes crenulated by the riblets of the exterior. Girdle irregularly imbricated with flattened scales, each one about six-striated. Length, 17½ ; breadth, 8¾ mm. (Tryon and Pilsbry, *Manual of Conchology*.)

TYPE in Philadelphia Academy of Natural Sciences. Type locality, Ecuador.

RANGE. Monterey, California, to Ecuador.

Described as *Callistochiton infortunatus* Pilsbry.

### Callistochiton decoratus punctocostatus Pilsbry, 1896

*Nautilus*, 10:50.

Similar to *C. decoratus* in sculpture of end valves and lateral areas ; but the central areas have no wide, smooth triangle at the ridge, such as types of *decoratus* have, being somewhat irregularly pitted toward the beaks, and with rows of pits on each side of a small oblong smooth tract at the ridge ; most valves pitted also on the ridge anteriorly. (Pilsbry.)

TYPE in Philadelphia Academy of Natural Sciences. Type locality, Long Beach, California.

RANGE. San Pedro, California, to San Tomas River, Lower California.

### Callistochiton diegensis Theile, 1910

*Zoologica*, 22: Heft. 56, Pl. 2, 86 ; Pl. 9, figs. 4–10

Pilsbry hat (*Man. Conch.*, v. 15, p. 87) zu *C. decoratus* eine Form von San Diego gezogen, die beträchtliche Verschiedenheiten aufweist, diese scheinen mir derartig zu sein, das kein Grund ersichtlich ist, diese Form mit *C. decoratus* zu vereinigen. Pilsbry hat zwar das ganze Tier abgebildet (t. 16, f. 54), doch gehen die Unterschiede daraus nicht genügend hervor, daher habe ich die Schalenteile abgebildet. Der Rücken ist stumpfgekielt, die Seiten fast gerade ; die Farbe gelblich. Das vorderste Stück (figs. 4, 5) hat (bei dem mir vorliegenden Exemplar) 12 Rippen, die etwas knotig und durch niedrige Falten verbund sind, so das zwischen eine Reihe von Maschen sichtbar ist ; das Articulamentum hat 9 Einschnitte. Die Mittelstücke (fig. 6) sind auf einem schmalen Mittelstreifen glatt, seitlich mit mehreren, durch quere Fäden verbundenen Längsrippchen besetzt ; die hintere der beinen Rippen auf den Seitenfeldern ist, ähnlich wie es auch auf dem vordersten Stück der Fall ist, geteilt und am Hinterrande mit Körnchen besetzt. Die Apophysen sind durch eine schmale, von Einschnitten begrenzte verbunden, die auch sonst kleine Einschnitte aufweist. Das hinterste Stück (figs. 7–9) hat hinter dem zentralen Apex einen schwachen Eindruck und fällt erst in der Nähe des Randes ziemlich steil ab ; es sind 10 Rippen und 13 Einschnitte vorhanden (abnorm?), so das sie sich nur teilweise entsprechen. Die Randschuppen sind etwas 180″ bret, meist mit 6 flachen Rippeb besetzt. (Theile.)

TYPE in Berlin? Type locality, San Diego, California.

RANGE. Known only from type locality.

## Callistochiton acinatus Dall, 1919

*Proceedings of the United States National Museum, **55**:510.*

Chiton small, yellowish-white, strongly sculptured; the anterior valve with nine nodulous ribs with narrower interspaces, a minutely granulose surface and about 12 inconspicuous slits; posterior valve with 18 slits and 6 nodular ribs; this valve is much smaller, the central tract is narrow and coarsely reticulate; intermediate valves somewhat mucronate with one slit at each end; the sculpture on the lateral areas consists distally of two very prominently pustulate ribs, toward the dorsal ridge there are only crowded minor postules; the jugal and pleural tracts are coarsely reticulate, the interspaces deep; the girdle yellowish with densely crowded microscopic imbricating scales; length, 6; breadth, 3; height, 1.5 mm. (Dall.)

TYPE in United States National Museum, No. 218773. Type locality, San Pedro, California.

RANGE. Known only from the type locality.

## Callistochiton celetus Dall, 1919

*Proceedings of the United States National Museum, **55**:510.*

Chiton of moderate size, pale brownish, with a mottled white and brownish girdle scaled like the preceding species (*C. acinatus*); anterior valve with 11 nodulous ribs, which, except the closer set posterior pairs, have about equal interspaces; this valve has 9 slits, each corresponding to a rib, the posterior valve has 24, but only 6 ribs which are mostly split for a short distance distally; this valve is conspicuously convex, with about 6 large smooth nodules to each rib, the central tract is narrow and coarsely axially threaded; in the intermediate valves with 1 slit at each end, the lateral areas comprise two strong ribs with vertically compressed nodules, 8 or 10 to a rib, there is a very narrow smooth spot on the hardly mucronate jugum, the pleural tracts have near the jugal area 6 or 7 close-set axial threads and beyond them about the same number of sharp straight axial threads with wider interspaces crossed by much finer, closer, transverse threadlets; internally the valves are whitish and the sutural laminae appear to join each other in front of the obsolete sinus. Length, 10; width, 6; height, 2 mm. (Dall.)

TYPE in United States National Museum, No. 218770. Type locality, San Pedro, California.

RANGE. Known only from type locality.

## Callistochiton aepynotus Dall, 1919

*Proceedings of the United States National Museum,* 55:511.

Chiton yellowish-white, with a keeled back, the girdle as usual in the group; anterior valve with 10 annulate ribs and a slit for each rib; posterior valve small, low, with only about 5 feeble ribs, and 5 slits, the central area reticulate; intermediate valves with a narrow smooth line at the jugum, and single slits; the pleural tracts sharply obliquely reticulate with deep interspaces; the lateral areas bounded by 2 strong annulate ribs, the posterior rib wider, the interspaces regularly punctate-reticulate; interior white, the jugal sinus almost obsolete. Length 15; width, 7; height, 5 mm. (Dall.)

TYPE in United States National Museum, No. 225448. Type locality, Puget Sound, in 37 fathoms.

RANGE. Known only from the type locality.

## Callistochiton fisheri Dall, 1919

*Proceedings of the United States National Museum,* 55:512.

Chiton light greenish-gray, small, strongly sculptured, the girdle covered with minute closely crowded gray scales giving a velvety aspect; anterior valve with 12 strong annulate radial ribs with narrower interspaces, internally with 11 slits; posterior valve with the mucro behind the center, the central area coarsely irregularly reticulate, the posterior area elevated, with six strong annulate ribs, a slit to each rib, a roundly excavated jugal sinus and narrow rectangular sutural laminae; the interior is dark green; intermediate valves with one feeble slit at each end, excavate rather than mucronate at the jugal area which is obliquely recticulate, passing into the pleural tracts which are small, axially threaded, the interstices with minute transverse threads; lateral areas with two annulate ribs, the posterior stronger, denticulating the posterior edge of the valve; internally pale greenish with a wide shallow straight-edged jugal sinus and narrow sutural laminae. Length, 10; width, 5; height, 3 mm. (Dall.)

TYPE in United States National Museum, No. 110353. Type locality, Glory of Russia Bay, Tanaga Island, Aleutians.

RANGE. Known only from the type locality.

## Callistochiton chthonius Dall, 1919

*Proceedings of the United States National Museum,* **55**:511.

Chiton of moderate size and dark reddish-brown color, including the girdle which exhibits small lozenge-shaped imbricating scales uniform over the surface; anterior valve with 11 subcarinate radial ribs, the two posterior tending to bifurcate, and nine slits; the ribs are hardly nodulous, and in the type the interspaces are reticulated only on the left side of the axis; posterior valve with 13 slits, nine nodulous riblets, the mucro subcentral, the central area axially threaded with a median smooth keel; intermediate valves with a smooth narrow keel at the jugum, the pleural tracts with a strong rectangular reticulation; the lateral areas with two strong keeled ribs, the posterior rib bifurcate for most of its length and the distal ends sometimes divided; these ribs are more or less nodulous, denticulating the posterior edge of the valve; the interspace between the main ribs is deep and angular; the insertion plates have a single slit; the interior of the posterior valve dark green, of the other valves greenish-white; the jugal sinus is very shallow, straight-edged and wide, the sutural laminae rather narrow. Length, 21.5; width, 10; height, 5 mm. (Dall.)

TYPE in United States National Museum, No. 109488. Type locality, San Pedro, California.

RANGE. Known only from the type locality.

## Callistochiton cyanosus Dall, 1919

*Proceedings of the United States National Museum,* **55**:511.

Chiton of moderate size, rounded back, and rather low dorsum, the color pale blue with interrupted bands of straw-color, the girdle armature as usual, but rather finer and pale in color; anterior valve with 11 keeled ribs, a slit to 9 of them; these ribs are crossed by fine concentric threads with about equal, not punctate, interspaces; posterior valve low, with subcentral mucro similarly sculptured, the central area rather larger than the ribbed portion with 10 or 11 prominent axial threads on each side, the interspaces crossed by similar threads, making a conspicuous reticulation; in the posterior portion are 8 carinate ribs each with a slit; the intermediate valves with one slit on each end, the lateral areas bounded by 2 strong keeled ribs crossed by small equal threads, not punctate in the interspaces, denticulating the posterior edge of the area; jugal areas showing a narrow triangular smooth space, on each side of which are 15 or more strong straight axial threads, the subequal interspaces reticulated by smaller threads and the interstices deep; interior bluish-white with a wide straight-edged jugal sinus and rather broad sutural laminae. Length, 13; width, 8; height, 3 mm. (Dall.)

TYPE in United States National Museum, No. 109317. Type locality, Long Beach, California.

RANGE. Long Beach and San Pedro, California.

## Family CHITONIDAE

### Genus **TONICIA** Gray, 1847

Anterior and posterior valves with many slits, middle with one; teeth sharp, serrate; eaves short, spongy; sinus squared, denticulate; girdle smooth or downy; gills encircling. (Tryon, *Structural and Systematic Conchology.*)

TYPE. *C. elegans* Fremb.

DISTRIBUTION. Western America, West Indies, New Zealand, Australia, Red Sea, Philippines.

### **Tonicia pustulifera** Dall, 1919

*Proceedings of the United States National Museum,* **55**:516.

Chiton of moderate size, polished, yellowish mottled with dark green, dark brown, and brown dots, the back rounded, the girdle yellowish, densely covered with minute short spinules giving a velvety aspect and with no fringe of longer spines at the border; anterior valve with 8, posterior with 11, intermediate valves with single slits, the insertion plates minutely radially striated; anterior valve more than semicircular with a profusion of minute pustules corresponding to sense organs and leaving a puncture when worn off; otherwise smooth; posterior valve smaller with the mucro much in front of the center, the central area narrow, sculpture as in the anterior valve, internally with a pink spot, the jugal sinus small and shallow, the sutural laminae long and narrow; intermediate valves with the first axially longer than the others, the valves slightly mucronate, the jugal area narrow and smooth except on the first valve which has a few axial grooves, the pleural tracts smooth except for a few very feeble wavy subaxial threads often obsolete on one side of the same valve, stronger near the jugum and on the posterior valves, and a few scattered very minute pustules; there are faint traces in spots, of a microscopic decussation; lateral areas without ribs, feebly indicated, bearing oblique rows of minute pustules; internally white, pinkish under the jugum, the jugal sinus narrow, deep, straight-edged, the sutural laminae broad, arcuate: the gill rows ambient. Length about 19; width, 10; height, 5 mm. (Dall.)

TYPE in United States National Museum, No. 218736. Type locality, San Pedro, California.

RANGE. Known only from type locality.

## Family MOPALIIDAE

## Genus **MOPALIA** Gray, 1847

Shell regular; laminae lengthened; anterior valve with six or more slits, the others with a single slit; last valve sinuate behind; sinus narrow; mucro median, depressed; sutures indented; girdle wide, bristly, sometimes fissured behind, sometimes projecting anteriorly. (Tryon, *Structural and Systematic Conchology.*)

TYPE. *C. blainvillei,* Brod.

DISTRIBUTION. Northern Pacific, Alaska, Japan, Lower California.

### Mopalia ciliata Sowerby, 1840

*Annals and Magazine of Natural History,* new series, 4:289. *Conchological Illustrations,* fig. 79. Tryon and Pilsbry, *Manual of Conchology,* 14:303; Pl. 64, figs. 64–73.

Ch. testa depressa, dorso subrotundato, valvis subreniformibus, ad latera disjunctis; areis lateralibus oblique granoso-sulcatis; costa granulosa utrinque marginata; areis centralibus granoso-sulcatis, valva prima radiatim costata; margine ciliato. Long., ⅞; lat., ½ poll.

The valves are flattish, with the edges arched and slightly beaked, and not united at the lateral extremities; central areas longitudinally grooved; a slightly raised granular rib separates the lateral from the central areas; these are obliquely grooved so as to meet the grooves of the central areas at acute angles on the rib; margin brown, covered with light brownish hairs; the colors are variegated, green, yellow, and black. (Sowerby.)

Shell oblong, rather depressed, the dorsal ridge carinated (sometimes rounded), side-slopes straight or somewhat convex. Surface lusterless, finely sculptured, variously colored, usually either (1) typical coloring, verdigris green maculated with black or black-brown, the girdle yellow or (2) maculated with maroon and sometimes touched with rich chestnut on the ridge, or having some valves or parts of valves vivid scarlet, or scarlet mixed with olive and snow-white, or entirely white; or (3) light olive-buff with brownish girdle. Valves somewhat beaked, the lateral areas bounded by a riblet, rather coarsely granulated, with larger granules along the posterior margin. Central areas sculptured with longitudinal, curving riblets somewhat granulated. Posterior valve small, with posterior mucro, broadly emarginate or waved at the hinder margin. Interior bluish-white or light blue-green. Sinus broad and rather rounded, spongy or roughened. Sutural plates arcuate. Anterior valve having 8 slits, median valves 1 slit. Posterior valve having a broad, deep, rounded caudal sinus, and a single slit on each side. Girdle wide, yellow or brown, generally notched behind, more or less sparsely clothed with curling, strap-like, brown hairs, which bear near their bases a bunch of minute, white, acute spines. Length, 46–50; breadth, 25 mm.; divergence, 125°. (Pilsbry.)

TYPE in Museum of G. B. Sowerby. Type locality unknown.

RANGE. Vancouver Island, British Columbia, to Lower California.

Described as *Chiton ciliatus.*

## Mopalia ciliata wosnessenskii Middendorff, 1847

*Bulletin,* Imperial Academy of Sciences, St. Petersburg, 6:119. Tryon and Pilsbry, *Manual of Conchology,* 14:305; Pl. 64, figs. 69–73.

Chiton testa externa ovata opaca e fusco-viridescenti, straminea vel latericia undique seriatim tuberculosa. Valva antica et valvarum intermediarum areae laterales, radiatim tuberculosae, tuberculis quincunce dispositis. Radii octo tuberculorum duplo majorum prominentium, in incisurae marginale sexcurrunt, dentes novem disjungentes. . . . . Valvae ultimae area postica tuberculorum obsoletorum seriebus concentricis, margine postico bidentato. . . . . Valvarum intermediarum areis centralibus et valv. ultimae area antica longitudinaliter undulato costatis, costarum interstitiis scrobiculatis. . . . . Costae medianae, postica versus divergentes in summa caruna conniventes, angulorum acutorum seriem hic effingunt. . . . . Areae laterales a centralibus tuberculorum majorum prominentium serie sejunctae. . . . . Valva antica semiorbicularis dentibus marginalibus 9. . . . . Valva ultima margine postico emarginato dentibus 2. Pallium marginale latum, postice bilobum, epidermide e fusco-rubicunda, setis rubicundis confertis crinita, obtectum. Cutis pallii, epidermide liberata, porosa. Long. adulti, 0.53 decim. (Middendorff.)

Shell elongated, the back roundly arched, not carinated; dull-colored, varying from light olive or green to drab, generally with blackish patches on each side of the middle, and more or less mottled throughout with dusky. Sculpture much fainter than in typical *ciliata.* Girdle apparently lacking the white spicules described above. (Tryon and Pilsbry, *Manual of Conchology.*)

TYPE in Academy of Sciences, St. Petersburg. Type locality ?

RANGE. Unalaska, Alaska, to Santa Rosa Island, Pacific Grove.

## Mopalia ciliata elevata Pilsbry, 1892

Tryon and Pilsbry, *Manual of Conchology,* **14**:300; Pl. 64, figs. 82, 83.

More elevated, acutely carinated, more or less pitted superficially, and either painted with the pattern of typical lignos on a white, creamy, or green ground, or having concentric zigzag blackish streaks. Divergence about 90°. The interior is bright green or blue. (Tryon and Pilsbry, *Manual of Conchology.*)

TYPE in Philadelphia Academy of Natural Sciences. Type locality, Puget Sound?

RANGE. Forrester Island, Alaska, to Puget Sound.

Described as *Mopalia lignosa* form *elevata* Pilsbry.

## Basiliochiton lobium Berry, 1925

*Proceedings,* Philadelphia Academy of Natural Sciences, **77**:28–9; Pl. 11, figs. 1–2.

Shell rather small, with a slight luster; elongate-oval, well-elevated, the side-slopes nearly straight; valves distinctly beaked behind. Entire surface unsculptured and nearly smooth except for the numerous very fine lines of growth and traces of an extremely minute granulation barely discernible under even high magnifications. Texture of shell somewhat streaky radially.

Anterior valve short, crescentic, its slope very steep in front and slightly convex; faint traces of low radial ribs leading to the insertion slits possibly discernible. Intermediate valves with lateral areas scarcely definable and not raised. Posterior valve small, with well-elevated median mucro, and an initially sharply falling, then more gradual slope behind. Articulamentum subtranslucent, porcellaneous, whitish. Sutural laminae thin, rather broadly arcuate, not continuous across the fairly wide, distinctly spongy sinus. Specimen not disarticulated, but anterior valve evidently with about 8 intermediate valves with 1–1, posterior valve with 9 slits, all slits having porous rays leading into them, which are especially conspicuous in the head valve. Girdle narrow, rather delicate and inclined to curl in dried material; covered dorsally with very minute granular scales, giving an effect like fine sandpaper, and an abundant, but easily broken, marginal armature of long, dagger-like spinelets; a few similar spinelets noted here and there on the upper surface may be adventitious from the margin; sutural tufts present, consisting of a group of 3–12 somewhat falciform spinules situated on whitish spots a little way out from each suture, 7 rather smaller tufts around the head valve, and 5 similar ones around the tail valve; best preserved tufts indicative that these are remnants of coarsely spinulose, recurved, plume-like setae, bearing spinules on the upper or posterior margin only. Ctenidia extending to valve iii. Color of outer surface of shell deep olive buff, shading to vinaceous fawn near the girdle and clouded with citrine drab along the ridge; sutural margin of head valve, all the tail valve, and vicinity of a sharply marked "jugal triangle" on valve iii, finely mottled dark olive. Girdle darker, with conspicuous whitish spots around the setae. Maximum length of type, 13.2 mm.; maximum width, 6.1 mm.; altitude, 2.8 mm.; length of shell only, 13.0 mm.; width of tegmentum of valve iv, 5.1 mm., of valve v, 5.2 mm. (Berry.)

TYPE in Berry Collection, No. 4908. Type locality, tide pool in the "Devil's Slide," La Jolla, California.

RANGE. Known only from the type locality.

## Mopalia muscosa Gould, 1846

*Proceedings of the Boston Society of Natural History,* **2**:145. *Report of the United States Exploring Expedition, Mollusca.*

Testa ovalis, depressa, scabra, ad dorsum obtusa, cinereo bifasciata et plumose striata; valvis magnis, lateraliter disjunctis; areis lateralibus parvis, granulis subquadratis radiantibus arcuatim tessellatis; areis centralibus sulcis acutis confertis flexuosis subparallelis longitudinaliter aratis; valva antica magna, semicirculari, decemcostata et granulis subquadratis insculpta; valva posteriori parva, costa transversali inconspicua subterminali: margo latus, filis corneis inaequalibus muscosis indutis. (Gould.)

Shell depressed, rather broad-oval, obtusely ridged along the back, everywhere sculptured. The color is pale olive, ashy about beaks, and with an ashy-white band along each side of the dorsal ridge. Valves large, narrowing toward the sides, so that their edges do not come in contact; the lateral valves are small, slightly raised, coarsely granulated with square granules arranged in radiating lines; central areas deeply incised with sharp, flexuous, but nearly parallel ridges, the presenting edges of which are slightly denticulated; these are arranged along the back in a plumosely diverging series, and over the rest of the spaces they are nearly longitudinal; the spaces intervening between the ridges are indented with little alveoli. The anterior valve is large and semicircular, having about ten radiating, raised lines, and granulated like the lateral areas; posterior valve very small, partially umbonated near its extremity, where it is crossed by a faint, transverse ridge, nearly parallel with the margin. Margin broad, covered with coarse, moss-like, horny fibers or threads, of unequal size and length, flexible when wet, very brittle when dry. Length, 2; breadth, 1⅛ in. (Gould.)

TYPE in United States National Museum? Type locality, Puget Sound, Washington.

RANGE. Shumagin Islands, Alaska, to Rosario, Lower California.

## Mopalia muscosa hindsii Reeve, 1847

*Conchologia Iconica,* Chiton; Pl. 12, fig. 67 *a–b.* Tryon and Pilsbry, *Manual of Conchology,* 14:296; Pl. 62, figs. 99, 100.

Chit. testa oblonga-ovata, antice subattenuata, valva antica terminali nonafariam carinata, conspicue granato-clathratis, retusa, postica parva, umbonata, retusa, caeteris undique diversimodo granato-clathratis, areis lateralibus non elevatis, margine antico tenuicarinato; olivaceo-viridi, interdum luteo-albido variegata; ligamento corneo, setis brevibus sparsim obsito.

Shell oblong-ovate, a little attenuated anteriorly, anterior terminal valve nine-keeled, conspicuously granosely latticed, the posterior small, umbonated, retuse, the rest diversely granulosely latticed throughout, lateral areas not raised, distinguished by a fine keel along the front margin; olive-green, sometimes variegated with yellowish-white; ligament horny, sparingly beset with short bristles. (Reeve.)

TYPE in Mr. Cuming's Cabinet. Type locality, not known.

RANGE. Sitka, Alaska, to Gulf of California.

Described as *Chiton hindsii* Reeve.

## Mopalia muscosa porifera Pilsbry, 1892

Tryon and Pilsbry, *Manual of Conchology,* 14:297; Pl. 62, figs. 93, 94.

The shell is rather small, thin and high, but not to as great an extent as *M. acuta.* It is sculptured as in *M. hindsii,* but the diagonal and sutural ribs are more prominent, and sculptured with strong, transverse beads exactly as in *M. imporcata,* the suture being dentated by them. Color of valves olivaceous, clouded obscurely with smoky, and having a black stripe each side of the dorsal lighter stripe. The posterior valve has a depressed mucro near the posterior margin, which is very slightly waved inward. Girdle firm and leathery, minutely and evenly papillose all over, blackish with spots of orange; having a small pit or pore at each suture, with a series of pores alternating with these a little outside of the middle of the girdle. Sometimes some of the sutural pores bear large hairs, curling outward; and sometimes some or many of the pores are completely absent. Length, 23; breadth, 13½ mm.; divergence, 105–115°. The interior is a deep blue-green, or gray-blue with a purple-pink blush. (Tryon and Pilsbry, *Manual of Conchology*.)

TYPE in Philadelphia Academy of Natural Sciences. Type locality, Bolinas Bay, California.

RANGE. Known only from the type locality.

## Mopalia muscosa laevior Pilsbry, 1918

*Nautilus,* **31**:126. Tryon and Pilsbry, *Manual of Conchology,* 14:300; Pl. 63, figs. 60–61.

This name has long been used in the collection for the form figured and described in *Manual of Conchology*, Vol. 14, p. 300, pl. 63, figs. 60–61. (Pilsbry, *Nautilus.*)

One of the principal mutations of *Mopalia muscosa* is toward still smoother forms entirely lacking pitted or reticulated sculpture, having only a few subobsolete longitudinal wrinkles on the ridge, the pointing in concentric streaks (following growth marks) on each valve. (Tryon and Pilsbry, *Manual of Conchology*.)

TYPE in Philadelphia Academy of Natural Sciences. Type locality, Olympia, Washington.

RANGE. Known only from the type locality.

## Mopalia muscosa lignosa Gould, 1846

*Proceedings of the Boston Society of Natural History,* **2**:142. Tryon and Pilsbry, *Manual of Conchology,* 14:299; Pl. 63, figs. 58–59.

Testa solidula, ovata, tectiformis, caesia, lineolis. fuscis inaequalibus subradiantibus marmorata, systemate duplici punctorum majorum et punctorum minorum impressa; valvis planulatis, angulatis, sine rostris; areis lateralibus haud elevatis, vix distinctis; valva anteriori parva; valva posteriori vix umbonata: intus aeruginosa. Long., 1½; lat., ⅞ poll. (Gould.)

Shell oval, elevated, carinated or angular at the dorsal ridge, the side-slopes straight. Surface lusterless, apparently smooth; grayish, greenish, or bluish with radiating streaks, lines and flammules of brown or purple-brown. Under a lens the lateral areas appear nearly smooth; the central areas being closely and finely pitted all over. Girdle narrow, sparsely hairy. Interior white and light blue. Length, 60; breadth, 35 mm.; divergence, 120°. (Tryon and Pilsbry, *Manual of Conchology*.)

TYPE in ? Type locality, Puget Sound.

RANGE. Sitka, Alaska, to Magdalena Bay, Lower California.

## Mopalia muscosa swanii Carpenter, 1864

*Supplementary Report,* British Association for the Advancement of Science, 648.
Red, ridge arched; less sculptured. (Carpenter.)
TYPE in United States National Museum. Type locality, Tatooche
Island.
RANGE. Shumagin Islands, Alaska to Monterey, California.
Described as *Mopalia kennerleyi* var. *swanii* Carpenter.

## Mopalia muscosa kennerlyi Carpenter, 1864

*Supplementary Report,* British Association for the Advancement of Science, 648.
*Proceedings,* Philadelphia Academy of Natural Sciences, **59**:1865.

Grayi, antea, p. 603, nom. preoc. Sculpture fainter: olive with red:
ridge angular; post. valve waved. (Carpenter.)

M. t. *M. muscosa* forma, indole, sculpturaque simili; sed multo magis
elevata; plus minusve rubente, plus minusve olivaceo variegata, intus pallida; granis lateralibus fere aequalibus; liris centralibus haud acutis, interstitiis rarius cancellatis; suturis undatis, apicibus valvarum prominentibus;
valva antica octoradiata, radiis granulosis, margine octies inciso; valvis intermediis utraque semel incisis; valva postica mucrone obsoleto, sinu postico
alto, angustiore, marginibus anticis valde alatis, lateribus posticis semel
incisis. (Carpenter.)

Shell like *M. muscosa* in form, character, and sculpture; but much
more strongly elevated; more or less reddish; more or less mottled with
olive, within pale; with lateral grains nearly equal; with central lirae not
acute; with interstices more rarely cancellate; with wavy sutures, with
apices of the leaves prominent; with anterior shell octo-radiate, with granular radii, with the margin eight times incised; with the intermediate leaves
incised once on both sides; posterior leaf with obsolete edge; with deep
narrow posterior sinus, with anterior margins strongly winged, with hinder
sides incised once. (Translation.)
TYPE in ? Type locality, Puget Sound?
RANGE. Shumagin Islands, Alaska, to Monterey, California.

## Mopalia muscosa acuta Carpenter, 1855

*Proceedings,* Zoölogical Society of London, 232. Tryon and Pilsbry, *Manual of Conchology,* **14**:297; Pl. 64, figs. 75–81.

C. t. ovata, valde elevata (ad angulum 105°), tenui; olivacea, interdum
maculis tenebrosioibus; lineis diagonalibus vix monstrantibus; areis latera-

libus et valvis ultimis tenuissime granulosis, granulis longis, irregulariter radiantibus; areis centralibus iisdem lineis longitudinalibus undatis instructis; marginibus valvarum subquadratis, suturis parvis; jugo acuto, mucrone inconspicuo; limbo angusto, sublaevi, tenui; intus virescente, valvarum marginibus et jugo impresso albidis; valvarum marginibus vix arcuatis, sinu parvo, inciso. Long., .9; lat., .5; alt., .2 poll. (Carpenter.)

Shell rather small, strongly elevated, the dorsal ridge acute; valves thin; sculpture minute and delicate, varying between a minute granulation and a fine pitting; the sutures delicately denticulate; diagonal riblets obsolete or delicately raised. Color olivaceous obscurely mottled with dusky, the tail valve having a light ray behind. Interior blue-green. Tail valve slightly waved upward in the middle behind, and having either a simple narrow caudal slit, or a slit with a small tooth set in its apex. Lateral slits of the tail valve generally double on one side. Girdle narrow, sparsely hairy. Length, 22½; breadth, 12½ mm.; divergence, 105°. (Carpenter's type.) (Tryon and Pilsbry, *Manual of Conchology.*)

TYPE in Academy of Natural Sciences, Philadelphia, No. 35729. Type locality, Santa Barbara, California.

RANGE. Santa Barbara to San Diego, California.

## Mopalia cirrata Berry, 1919

*Lorquinia,* January 1919, 5.

Shell of moderate size, oval, with slightly arcuate side-slopes. Head valve with eight strong, somewhat tuberculose radial ribs and weaker sutural ribs, the interstices checkered by pits and squarish grains like basket work. Median valves with strong diagonal ribs; sutural ribs more or less obsolete; the areas between with even, basket-like sculpture; central areas on each side with some twenty arcuate, granulose riblets, strongly converging toward the jugum, and conspicuously latticed by a somewhat weaker system of curved transverse riblets; a narrow tract along the jugum smooth. Tail-valve small, with a slightly elevated mucro and narrow, abrupt, caudal sinus. Color of shell a soiled buff with brown suffusion and mottling. Girdle moderately wide, minutely scaled above, and bearing an inner sutural series of extremely long, slender setae, furnished with slender branching aciculae; series of minor setae near the margin. (Berry.)

TYPE in Berry Collection. Type locality, "Albatross" Station 4263, Dundas Bay, Alaska.

RANGE. Known only from type locality.

## Mopalia imporcata Carpenter, 1864

*Supplementary Report,* British Association for the Advancement of Science, 648.
Tryon and Pilsbry, *Manual of Conchology,* 14:301; Pl. 62, fig. 98.

Pale: central areas ribbed: post. valve slightly notched. Indications of
sutural pores in these two species, if confirmed, will require a new genus.
(Carpenter.)

Shell small, oblong, strongly elevated and acutely carinated, the side-
slopes straight. Buff-white, slightly stained with rust-brown. The median
valves are acute at the umbo, but there is no projecting beak; lateral areas
bounded by a strongly elevated, narrow, crenulated, diagonal rib, and
having a wider rib at the sutural margin, also crenulated, thus denticulating
the suture; the space between the two lateral ribs being finely corrugated-
granose. Central areas sculptured with strong, curved, longitudinal ribs,
which converge forward somewhat, toward the median keel; the intervals
between these ribs being closely and finely latticed across by threads radiating from the beaks. Anterior valve having 8 strong, narrow, raised ribs,
with one wider rib at each sutural edge. Posterior valve depressed, the
mucro situated at the posterior end. Interior white; sinus very small and
shallow. Anterior valve having 8, median 1 slit; teeth but little thickened
along the slits. Posterior valve having a rather wide, moderately deep,
rounded, tail sinus, and a single slit on each side. Girdle leathery, dusty,
with a hair-pore at each suture, and some scattered or alternating hairs.
Length, 10; breadth, 6 mm.; divergence, 95°. (Tryon and Pilsbry, *Manual of Conchology.*)

TYPE in United States National Museum. Type locality, Puget Sound.
RANGE. Forrester Island, Alaska, to San Pedro, California.

## Mopalia imporcata lionotus Pilsbry, 1918

*Nautilus,* 31:126.

This chiton agrees closely with *M. imporcata* except that there is a
narrow, smooth jugal tract. The lateral areas are granose between the
coarsely tubercular diagonal and sutural ribs. The anterior valve has 10
ribs. Posterior valve is depressed behind the mucro, which is at the
posterior fourth. Interior light Niagara-green, darker posteriorly on each
valve. The girdle bears branching processes, often like the branches of
spines on a cactus. These are scattered, sometimes sutural. Length, 15.5;
width, 9 mm.; divergence, 95°. (Pilsbry.)

TYPE in Philadelphia Academy of Natural Sciences. Type locality,
White Point, California.
RANGE. San Pedro and White Point, California.

## Mopalia egretta Berry, 1919

*Lorquinia*, January 1919, 5. *Proceedings of the California Academy of Sciences*, series 4, 9:13; Pls. 6, 7.

Closely allied to *M. sinuata*, but with the side slopes arcuate, lateral areas distinctly raised, sutures strongly dentate, and with the major girdle setae sutural in position, long, erect, and branching into numerous, nearly straight, slender, needle-like aciculae. Color brownish red with maroon and buff mottlings. (Berry.)

TYPE in Berry Collection. Type locality, Forrester Island, Alaska, in 20 fathoms.

RANGE. Forrester Island, Alaska, to Puget Sound, Washington.

## Mopalia sinuata Carpenter, 1864

*Supplementary Report,* British Association for the Advancement of Science, 648. Tryon and Pilsbry, *Manual of Conchology,* 14:303; Pl. 62, figs. 95–97.

Small, raised sharp back, red and blue, engine-turned; posterior valve deeply notched. (Carpenter.)

Shell oblong, elevated, and strongly carinated, the side-slopes straight. Color whitish, clouded with delicate blue-green and maculated with rich tawny brown. Median valves hardly beaked, the lateral areas not raised, but strongly defined by an elevated diagonal rib; sculptured with two oblique series of fine riblets forming a latticed pattern. Central areas having a series of longitudinal curved riblets converging toward the median keel, crossed by curved radiating threads a little finer and less prominent. Anterior valve having 8 (not counting the posterior sutural borders) strong, radiating ribs narrower than the latticed intervals. Posterior valve depressed, the mucro being at the posterior third. Interior bluish-white. Sinus very small and narrow. Anterior valve having 8 slits, median one slit; teeth thickened outside at the edges of the slits. Posterior valve having a deep rounded median sinus behind (which is continued upward in a superficial excavation to the mucro), and a single slit on each side. Girdle rather narrow, leathery, "dusty," bearing a few hairs, with a rounded pore at each suture. Length, 11½; breadth, 7 mm.; divergence, 105°. (Tryon and Pilsbry, *Manual of Conchology.*)

TYPE in United States National Museum. Type locality, Puget Sound.

RANGE. Forrester Island, Alaska, to San Francisco, California.

## Mopalia phorminx Berry, 1919

*Proceedings of the California Academy of Sciences,* series 4, 9: No. 1, 10; Pls. 3, 4, 5. *Lorquinia,* January 1919.

Shell small, with straight side-slopes. Surface finely granulose. Head valve with ten major radiating series of coarse pointed pustules, one or two similar but smaller series interposed in alternation. Median valves with lateral areas bounded by rib-like diagonal and sutural series of coarse pustules, and with some 2–5 minor series of pustules interpolated between. Central areas on each side with 10–14 sharp, flexuose, longitudinal ribs, the outer ribs more or less broken posteriorly, the breaks forming radial series of granules in front of the diagonal rib. Tail valve with mucro strongly posterior, little elevated; caudal sinus strong. Slit formula 8, 1–1, 1–1. Girdle narrow, armed above with a microscopic spinulation, and both major sutural and several minor series of beard-like setae. Color brownish-cream, with yellowish-brown mottlings. (Berry.)

TYPE in Berry Collection. Type locality, off Point Pinos, California.

RANGE. Known only from type locality.

## Mopalia lowei Pilsbry, 1918

*Nautilus,* **31**:125.

The Chiton is rather small, oblong, moderately elevated, carinate, the lateral slopes straight. The valves are irregularly mottled with ferruginous, sea-green, and olive. The anterior valve has ten radial ribs, those at the suture bearing compressed tubercles, the others rounded tubercles. The intervals are also tuberculose, with some interstitial granulation. Valves ii to vii have low, tuberculose sutural and diagonal ribs, the lateral areas tuberculose and granular. Central areas with the jugal tract closely striate longitudinally, the striae converging forward near the beaks, elsewhere subparallel, but slightly irregular in places. Pleural tracts having longitudinal ribs, near the ridges converging forward somewhat, becoming divergent toward the lateral borders. These are intersected by a system of much weaker curved ribs radiating forward and laterally, forming oblong tubercles on the longitudinal ribs. The posterior valve is short, nearly flat, with a broad, shallow posterior sinus, the scarcely raised mucro being at the posterior third. The interior is nearly white, strongly striate across the central part, where some valves may show a green or pink stain. The posterior valve has a rather deep posterior sinus and a single slit on each side. The girdle is rather narrow in dry specimens, and bears coarse processes covered with sharp spines. Length, 23; width, 12.2 mm. (Pilsbry.)

TYPE in Philadelphia Academy of Natural Sciences, No. 117951. Type locality, San Pedro, California.

RANGE. Known only from the type locality.

## Mopalia heathi Pilsbry, 1898

*Proceedings,* Philadelphia Academy of Natural Sciences, 288.

Oblong, rather elevated, carinated, with nearly straight side-slopes; surface smoothish to the naked eye, lusterless, and in color (1) olive-green with some lighter spots, or purplish maculation, or slight roseate suffusion, or (2) vivid red, with scattered blue spots. Valves shaped as in *M. lignosa,* but without a median anterior projection of the tegmentum; the intermediate valves very faintly radially triculcate at the sides, the anterior two grooves defining the low, slight, and inconspicuous diagonal rib, the lateral areas not raised; entire surface very finely and evenly granulate, the granules small, rather pointed, separated, intervals very minutely, radially wrinkle-granulate. Anterior valve with a few faint, shallow, radial furrows. Posterior valve with semicircular posterior outline, the mucro in front of the middle of tegmentum, profile of the surface in front of it convex, that of the posterior slope decidedly concave. Interior deep rose color or slightly purplish; sutural laminae and sinus about as in *lignosa.* Teeth rather long and somewhat roughened, as in *lignosa.* Valve i with 8 slits; ii–vii with 1–1; valve viii entirely *Ischnoid,* with regular, crescentic insertion plate, cut by 7 or 8 slits, which are somewhat closer posteriorly; no sinus behind. Girdle leathery, nude except for solitary or two or three closely grouped long bristles at all or part of the sutures, one on each side of the head valve, and two behind the tail valve. Gills about 25 on each side, not extending quite to the anterior end of the foot. Length, 25; width, 12 mm. (dried specimen), or smaller. (Pilsbry.)

TYPE in Philadelphia Academy of Natural Sciences. Type locality, Pacific Grove, near Monterey, California.

RANGE. Monterey, California.

## Mopalia chloris Dall, 1919

*Proceedings of the United States National Museum,* 55:513.

Chiton low, wide, of moderate size, of a dark bronze green, with leathery girdle which when fresh had a few slender, sparsely distributed hairs upon it (lost in the dry specimen); anterior valve with eight slits, externally with about ten radiating, conspicuous threads with a few feebler, shorter, intercalary threads, over the whole of which is a faint, very minute, oblique decussation; posterior valve small, with a very low subcentral mucro and a faint concave wave medially behind, over which is a white ray; it has two slits and a narrow rounded jugal sinus with wide sutural laminae, the finer sculpture is like that of the anterior valve with one or two radial threads; intermediate valves with one slit, the jugal tract acute, with the pleural tract sculptured with oval punctures between oblique, almost obsolete, minute threads; the lateral areas defined by a single cord, otherwise similarly sculptured; interior bluish-white, the jugal sinus narrow, rounded, the sutural laminae broad, the insertion plates smooth; gill rows ambient. Length, 19; width, 11; height, 3 mm. (Dall.)

TYPE in United States National Museum, No. 293686. Type locality, San Diego, California.

RANGE. Known only from type locality.

## Mopalia chacei Berry, 1919

*Lorquinia,* January 1919.

Moderately small, oval, with distinctly arcuate side slopes. Head valve with ten clear-cut, heavy, pustulose ribs, their interstices heavily granulose. Median valves with strong diagonal and weaker sutural ribs, both pustulose, the lateral areas coarsely granulose between them; central areas on each side with 18–20 clear-cut, slightly granulose, longitudinal ribs, generally straight but on jugum becoming finer and converging in front, interstices latticed by a system of numerous very fine transverse riblets. Tail-valve small, with depressed mucro and shallow posterior sinus. Color of shell dull greenish-brown, more or less mottled with various tones of brown or green. Girdle narrow for a *Mopalia,* bearing a primary sutural series of short setae, spinose on the upper margin, and 2 or 3 gradually smaller series of similar setae outside of these. (Berry.)

TYPE in Berry Collection. Type locality, La Jolla, California.

RANGE. Known only from type locality.

## Mopalia goniura Dall, 1919

*Proceedings of the United States National Museum,* 55:513.

Chiton of small size with a high arched back, of a yellowish color flecked with scarlet; the girdle red, velvety, with numerous sparsely scattered large brown spines (broken off in the type) chiefly near the inner border, and smaller ones scattered near the outer part; gill rows about two-thirds the length of the foot; anterior valve with 8 slits, posterior with 4, intermediate valves with single slits; anterior valve with 10 radii, the two marginal wider, otherwise the surface is covered with punctate reticulation; posterior valve small, the mucro at the posterior third, sculpture of the posterior tract in radial lines of pustules, the central area has very similar ornamentation; the posterior sinus is narrow Λ-shaped, the apex reaching the mucro, the anterior sinus also narrow and acute, the sutural laminae broad; jugal and pleural areas of the intermediate valves not separated, the sculpture of oblique reticulation with emphatically punctate interstices; lateral areas similarly sculptured, bounded on each side by a slender rib, internally whitish with a narrow notch-like sinus. Length, 12; width, 6; height, 4 mm. (Dall.)

TYPE in the United States National Museum, No. 208703. Type locality, Granite Cove, Port Althorp, Alaska.

RANGE. Granite Cove, Port Althorp, Alaska, to Puget Sound.

## Mopalia celetoides Dall, 1919

*Proceedings of the United States National Museum*, 55:514.

Chiton with sculpture almost exactly like No. 218770, except that the latter has more globular nodules on the lateral ribs and the pleural tracts are less coarsely and evenly reticulated. In the present species the anterior valve has 10 radial ribs with wider channeled interspaces minutely reticulated; the posterior valve, however, entirely different, having a posterior mucro with a small rounded sinus below it, and not ribs, the surface being taken up by the central area which has very straight axial threads minutely annulated, with much finer and closer transverse threads visible in the interspaces; gill-rows extending forward as far as the fourth valve; posterior valve with 2 slits, anterior with 8, intermediate with one at each end; intermediate valves with the lateral areas bounded by two strong ribs, the posterior wider, denticulated; jugum minute with a small smooth mucro, the back slightly keeled in front of it, the pleural tracts with about 11 strong axial threads. Length, 12; width, 5.5; height, 2.5 mm. (Dall.)

TYPE in United States National Museum, No. 218771. Type locality, Forrester Island, Alaska.

RANGE. Known only from the type locality.

## Genus **PLACIPHORELLA** Carpenter, 1878

Valves very broad and short, the middle ones much broader than those toward the ends; head valve narrowly crescentic, tail valve still smaller, with posterior mucro. Sinus small; insertion-plates short and thick, the teeth lobed or rugose. Slits 8 or more in the anterior, 1 in the median, 2 in the posterior valve, which has also a shallow, posterior sinus. Eaves spongy. Girdle widest, often very wide, in front, bearing sparsely scattered, scaled hairs.

This genus differs from *Mopalia* in the rotund contour of the valves taken together, and their extreme shortness individually, as if the shell had been crowded together from the ends. The slits are practically as in *Mopalia*, being normally (or at least originally) 8 in the head valve; but the number is frequently increased by the splitting of some teeth. The mantle-edge is produced in front and fringed with long, fleshy, finger-like processes. The longer hairs of the girdle are extremely peculiar, being covered with imbricating scales like a snake skin, on a corneous core. Nothing of the sort has been found in any other group of Chitons. (Tryon and Pilsbry, *Manual of Conchology.*)

TYPE. *Placiphorella velata* Carpenter. Carpenter's MS name.

DISTRIBUTION. California, Lower California, Japan, Bering Sea, Peru.

## Placiphorella pacifica Berry, 1919

*Lorquinia,* January 1919, 6.

Shell obovate, depressed. Anterior valve with 12–14 low, radiating ribs. Central areas of median valves nearly smooth; lateral areas raised, ornamented with 2 low, weakly nodulose ribs. Tail valve small, with sub-terminal mucro. Head valve with 16 slits, median valves with 1–1 slits, tail valve slitless. Girdle narrow behind; expanding into a broad lobe in front; covered dorsally by minute, microscopic spinelets, and occasional slender spinose setae and tufts of spines. (Berry.)

TYPE in Berry Collection. Type locality, Station 4245, Kasa-an Bay, Alaska, in 95–98 fathoms.

RANGE. Known only from type locality.

## Placiphorella borealis Pilsbry, 1892

Tryon and Pilsbry, *Manual of Conchology,* 14:309; Pl. 66, figs. 14–17.

Shell similar in general characters to *P. velata.* Brown at the sides, light along the middle. Surface dull, showing growth-lines. Valves not beaked, having a slight forward bend in the middle at the sinus, but not "false beaked" there as *P. velata* is. Lateral areas more strongly two-ribbed. Anterior valve sculptured with numerous very low, wide, radiating riblets. Posterior valve having the mucro near the posterior margin, depressed, the slope in front of it rising, convex rather than concave; posterior margin waved inward. Interior light blue-green. Sutural plates separated in all the valves by an angular, spongy sinus. Anterior valve having 11 slits, the teeth being peculiarly curved outward at its edges. Posterior valve much less callous inside than that of *P. velata,* the insertion-plate uneven, roughened, but having a single, well-developed slit on each side; the median tail-notch deep and wide. Eaves very spongy. Girdle unknown to me. Breadth of anterior valve 16, length of front slope, including teeth, 4½ mm. Breadth of median valve, 22; length from sinus to beak, 4½ mm. Breadth of posterior valve, 15; length direct from sinus to sinus, 5 mm.; divergence, 135°. Measurements of breadth include the insertion plates. (Pilsbry.)

TYPE in Philadelphia Academy of Natural Sciences. Type locality, Bering Island, Bering Sea.

RANGE. Bering Sea and Island, also Kuril Islands.

## Placiphorella velata (Carpenter) Pilsbry, 1878

*Proceedings of the United States National Museum*, 1:298, 303, 307; Pl. 2, fig. 36.
Tryon and Pilsbry, *Manual of Conchology*, 14:306; Pl. 66, figs. 6–12.

Shell roundly oval, broad, rather depressed, quite obtusely angled. Surface lusterless. Light colored along the middle, mainly olivaceous on the sides, especially the lateral areas; the central areas variously streaked longitudinally with buff, chestnut, and pink. Median valves not beaked, marked by growth lines; the lateral areas somewhat raised, having a rounded wide diagonal rib and another at the sutural margin, the space between them more or less excavated. Central areas having a "false beak" or narrow projection forward at the dorsal ridge (only visible when the valves are separated). Anterior valve crescent-shaped, sculptured with light concentric growth-lines only. Posterior valve small, slightly waved inward behind, the mucro far backward, recurved and elevated, the slope in front of it concave in profile, unless the mucro is eroded. Interior white, slightly blue-tinted. Sinus in valves i to vi represented by a very shallow wave, the sutural plates continuous, being connected by a plate which fills the sinus like a keystone; in valves vii to viii the sinus is deeper and more distinctly angular. Anterior valve having 8 slits, the teeth obtuse, short, often bilobed or conspicuously rugose. Central valves having one slit, teeth wedge-shaped, thicker at the edges of the slit. Posterior valve having a very heavy callus supporting the short, rugose insertion-plate, which is interrupted posteriorly by a broad, shallow sinus, and has one slit on each side (occasionally two on one side). Girdle very broad in front, reddish, fading to yellow toward the outer edge, irregularly and sparsely beset with scaly hairs, of which one or two are generally to be seen in each suture; a close fringe of short (broken) hairs adorns the girdle edge. Length, 50; breadth, 38 mm.; divergence, 135°. Length, 30; breadth, 24 mm.; divergence, 130°. (Pilsbry.)

TYPE in Academy of Natural Sciences, Philadelphia, No. 35756. Type locality, Todos Santos Bay, Lower California.

RANGE. Trinidad, California, to Todos Santos Bay, Lower California.

## Placiphorella stimpsoni Gould, 1859

*Proceedings of the Boston Society of Natural History,* 7:165. Tryon and Pilsbry, *Manual of Conchology,* 14:307; Pl. 62, figs. 84–87.

T. tenuis, rotundato-ovata, depressa, fastigata, fusca, rufo, rosaceo viridi et flavo marmorata vel lineata, concentrice striata; valvis angustis planatis; areis linea elevata finitis; valva antica parvula, crescentica; v. postica minima, emarginata. Ligamentum coriceum antice valde dilatatum, pilis fimbriatum. Long., 1.5; lat., 1 poll. (Gould.)

Shell broadly oval, depressed, subangular, with straight side-slopes. Color whitish along the middle, the sides mottled and streaked with greenish-yellow, olive and blue, the colors so blended as to give a general effect of dark olivaceous. Valves not beaked, sculptured with unequal growthlines. Lateral areas having a moderately prominent diagonal rib, with an inconspicuous wider, lower, sutural rib, the space between excavated. Anterior valve narrowly crescentic, concentrically striated, but lacking all radiating sculpture. Posterior valve (figs. 84, 85) depressed, the mucro near the posterior margin, which is slightly and rather broadly waved inward. Interior a delicate blue-green tinted. Sinus a rather shallow, rounded wave in the earlier valves, becoming deeper, narrower, and angular, in the eighth valve. Anterior valve having 8 slits, median valve having one slit on each side; the eaves narrow, tenaciously adhering to the girdle. Posterior valve small, its greatest width, including insertion and suturalplates, measuring hardly more than one-half the width of the widest median valves; the short posterior insertion-plate rising from a very heavy callous rim; having one oblique slit on each side, and an almost imperceptible wave where the posterior sinus should be. Girdle rather narrow, brown, leathery, sparsely beset with spinules (the stumps only remaining in the specimen before me), of which there is one at each suture and a marginal fringe. (Tryon and Pilsbry, *Manual of Conchology.*)

TYPE in United States National Museum, No. 1646. Type locality, Hakodadi Bay, Japan.

RANGE. Bering Island, and on the east and south, to Cerros Island, Lower California.

## Placiphorella rufa Berry, 1917

*Proceedings of the California Academy of Sciences,* series 4, 7:241; text figs. 3, 4.

Shell of moderate size, broadly oval in outline, depressed. Dorsal ridge only moderately elevated, the side slopes little convex. Anterior valve crescentic, showing strong, irregular lines of growth and occasional weak traces of radial grooves. Median valves with the lateral areas raised into diagonal and sutural ribs having a groove-like depression between; otherwise unsculptured except for the strong, uneven lines of growth, especially prominent where they traverse the ribs; not beaked. Posterior valve small, only about half as wide as the anterior valve; sculptured by two strong, oblique ribs converging to the depressed, distinctly posterior, but not marginal, mucro. Interior of valves calloused. Anterior valve with 9 strong, distinctly cut insertion teeth, the 3 anterior concave, and all coarsely grooved on their outer surfaces; slits 8 in number, their porous apices connected with the apex of the valve by lines of minute pores. Central valves with wide, gently arcuate sutural laminae, scarcely if at all connected across the rather deep sinus; slits 1–1. Posterior valve with a barely indented flattening instead of a posterior sinus; slits 1–1. Eaves distinctly spongy. Girdle of moderate width posteriorly, broadening in front to the wide lobe characteristic of the genus, so that the maximum diameter of the animal is in the region of the head. To the naked eye or under lenses of moderate power the girdle appears smooth and nude over its entire area dorsally (that it is actually thus devoid of spines I am not prepared to state) save for a marginal palisade of minute hyaline spinelets, and three or four submarginal series of armored chaetae, the members of the inner series of which are conspicuously the largest and extend clear around the girdle, about 12 chaetae in this series encircling the anterior lobe, the remainder placed one opposite each suture and two directly behind the tail valve. The ventral surface of the girdle is clothed with very minute scattered spinelets, visible only under quite a high power. In dried specimens the girdle becomes excessively thin and leathery. The ventral surface of the anterior lobe shows no evident radial striation even under moderate magnification. Color of shell in alcohol, a warm brownish-red with more or less lighter painting, except the anterior portion of the head valve which is uniformly paler. Girdle brownish-buff, the anterior lobe sharply paler. Interior of shell tinted a soft salmon-flesh. (Berry.)

TYPE in Berry Collection, No. 1411. Type locality, Forrester Island, Alaska, 15–25 fathoms.

RANGE. Known only from type locality.

## Family ACANTHOCHITONIDAE
## Genus **ACANTHOCHITONA** Gray, 1821

Insertion-plates thrown forward, laminated; anterior valve with five slits, middle and posterior valves each one; teeth long, sharp, smooth; eaves small; gills median; sinus deep, broad, spongy; girdle hairy, with long, fasciculated spiculae. (Tryon, *Structural and Systematic Conchology.*)

TYPE. *C. fascicularis* Auct.

DISTRIBUTION. Europe, Africa, New Zealand, Australia, Japan, China, West Indies, Western America.

## Acanthochitona diegoënsis Pilsbry, 1893

Tryon and Pilsbry, *Manual of Conchology,* 15:25; Pl. 12, figs. 52–54.

Shell oblong, rather elevated, carinated, the side-slopes straight. Color buff or light gray, mottled on the sides with olive or olivaceous black; girdle light green with whiter sutural tufts. The intermediate valves are rather minutely and acutely beaked when not eroded; are wide posteriorly, tapering anteriorly, but the girdle does not encroach much at the sutures. Dorsal areas narrowly triangular, having about a dozen, flattened, longitudinal striae separated by narrower grooves. Latero-pleural areas covered with a rather coarse but regular scale-like granulation, the granules flat, oblong. Posterior valve having the tegmentum covering the greater part of the articulamentum, somewhat diamond shaped, wider than long, with the prominent mucro at the posterior third. Interior blue-green; sinus rather wide, angular; sutural laminae moderate sized, rounded. Posterior valve obscurely bilobed behind, gently curved upward in the middle of the posterior insertion plate. Girdle densely clothed with rather long, light green spicules, and having 18 or 20 tufts of longer whiter spicules, the tufts usually not very conspicuous, and sometimes a few of them are lacking. Length, 19, breadth, 9 mm.; divergence, 110°. Length, 18½, breadth, 8 mm.; divergence, 110°. Length, 11, breadth, 6 mm.; divergence, 110°. (Tryon and Pilsbry, *Manual of Conchology.*)

TYPE in Philadelphia Academy of Sciences. Type locality, San Diego, California.

RANGE. Catalina Island and Newport to San Diego, California.

Described as *A. arragonites* var. *diegoensis* Pilsbry.

## Acanthochitona avicula Carpenter, 1866

*Proceedings of the California Academy of Sciences,* 3:211. Tryon and Pilsbry, *Manual of Conchology,* 15:24.

A. t. *A. arragonitei* forma magnitudine, pallio, et indole simillima; sed sculptura et laminis terminalibus diversa; jugo longitudinaliter sulculis circ. vi. instructis, interstitiis quasi planato-squamosis; umbonibus latis; areis diaganalibus haud definitis; lateribus, squamis (quoad magnitudinem) maximis, planatis, ovalibus ornatis, seriebus indistinctis divergentibus instructis; mucrone parvo, antice sito; colore livido et olivaceo-fusco varie tincto; laminis insertionis valv. lat. ut in *A. arragonite;* antica, fissuris v. Long., 0.16; lat. 0.10. (Carpenter.)

Shell very similar to *A. arragonites* in form, size, girdle, and general habit; but the sculpture and terminal laminae are different. Dorsal ridge having about 6 longitudinal grooves, the intervals appearing flatly scaled;

umbones wide; diagonal areas hardly defined; sides ornamented with oval, flattened scales, large for the size of the shell, and in indistinct diverging series. Mucro small, situated in front. Color livid and olivaceous-brown variously stained. Plates of insertion at sides as in *A. arragonites;* anterior plate with five slits. (Carpenter.) (Tryon and Pilsbry, *Manual of Conchology.*)

Type in California State Collection, No. 1072. Type locality, Catalina Island, 10–20 fathoms.

Range. Catalina Island, California, to Gulf of California.

Described as *Acanthochites avicula* Carpenter.

## Genus **KATHERINA** Gray, 1847

Valves two-thirds covered by the expanded girdle, the exposed portion divided into dorsal and side areas, instead of central and lateral. Insertion plates sharp, extremely long, thrown forward; that of the head valve with 7–8 slits; sinus deep, spongy. Tail valve with a wide caudal emargination or sinus, and several slits, often partly obsolete, on each side. Girdle broad, smooth, poreless, leathery. Gills extending the whole length of the foot. (Tryon and Pilsbry, *Manual of Conchology.*)

Type. *C. tunicatus* Wood.

Distribution. Kamchatka, Alaska, Puget Sound, Catalina Island, California.

### Katherina tunicata Wood, 1815

*General Conchology,* 11; Pl. 2, fig. 1. Tryon and Pilsbry, *Manual of Conchology,* 15:41; Pl. 1, figs. 1–11.

Ch. testa octovalvi lævi, valvis albis, supra membranis suborbicularis, limbo coriaceo reflexo.

Shell of eight valves, smooth, valves white, above the membrane roundish, margin coriaceous and turned back. (Wood.)

This singular shell makes a very different appearance from any other species of the genus. The valves instead of being wholly visible, are coated halfway upward with a black membrane, which is reflected from the coriaceous border, and fixed into a groove cut for that purpose round the upper part of each valve. This portion that is seen above the membrane, is roundish, uneven, and emarginate, of a sordid white or flesh color, and collectively resembling a row of dried beans. The remains of a thin cortical substance is to be seen on parts of the naked shell, which, independent of the coriaceous membrane, once covered the valves. The valves within are quite exposed and perfectly white. Length, 4; breadth, 1¼ in. (Wood.)

TYPE in British Museum. Type locality, west coast of North America.
RANGE. Western and southern Bering Sea, to Cook's Inlet, Alaska, and south to Catalina Island, California.
Described as *Chiton tunicata.*

## Family CRYPTOCHITONIDAE

### Genus **CRYPTOCHITON** Middendorff and Gray, 1847

Valves entirely concealed in the leathery girdle, and lacking tegmentum; their posterior margins produced backward in a deep lobe on each side, the lobes united across the median line, causing the apices of all valves to be removed inward from the posterior edge. Slits subobsolete or lacking in the intermediate valves. Girdle covered with minute tufts of short bristles. Gills extending the entire length of the foot. (Tryon and Pilsbry, *Manual of Conchology.*)
TYPE. *Cryptochiton stelleri* Middendorff.
DISTRIBUTION. West coast of America and Japan.

### Cryptochiton stelleri Middendorff, 1846

Chiton amiculatus (Sowerby and Reeve) minime vero Pallasii. Chiton gigas ovatus dorso rotundato, testa interna valvis fragilissimis aditu nullo abditis, incarceratis in pallio totum corpus ambiente coriaceo, obtecto epidermide ex ferrugineo lutescente, pustulis confertis subverrucosa. Epidermidis pustulae apertae (nisi summo apice per se perforatae) sublente seterum rubrarum confertissimarum exhibent fasciculos internos. . . . . Valvae planiusculae omnes incrementi umbone vix perspicuo subcentrali. . . . . Valva prima dentibus quinque latis in margine antico. . . . . Valva secunda et ultimo utrinque incisura laterali. . . . . Valva ultima postice emarginata bidentata. Branchiarum series ab initio secundae quartae partis ad finem septimae octovae partis longitudinis totius corporis porrectae. Lamellarum branchialium numerus circiter 70. . . . . Adulti longitudo, 1.9 decim. (Middendorff.)

Diese Species kann unmöglich mit irgend einer andern verwechselt werden, und begründet sogar durch ihre Eigenthümlichkeiten ein neue Untergeschlecht. Middendorff.

Oblong rather depressed, the bilobed posterior outlines of the valves (in dry specimens) showing through the leathery integument, which completely covers the valves. Color a dull ferruginous or brick-red, very well preserved specimens being rendered much brighter by the closely placed fascicles of brilliant vermilion spines. The valves are wholly concealed, white or flesh-colored, entirely lacking the outer colored layer (tegumentum) of other Chitons; their edges are more or less thinned and crenulated by radial striae. Anterior valve having the apex at the posterior third, and with 4 to 7 slits. Intermediate valves having the apex near the posterior third; formed of two large anterior lobes expanded at the sides, and two smaller, narrow posterior or near the posterior third; deeply sinused in the rear, and usually having a slit on each side of the sinus. Girdle leathery, thick red, densely covered with countless minute fascicles of vermilion spinelets. Length, 15 to over 20 cm. (Tryon and Pilsbry, *Manual of Conchology.*)

TYPE in Academy, St. Petersburg. Type locality, Kamchatka.

RANGE. Bering Island, Kamchatka, Kuril Islands and Okhotsk Sea, Aleutian Islands, Cook's Inlet, to San Miguel and San Nicolas Islands, California.

## Genus **AMICULA** Gray, 1847

Valves almost covered by the extension of the girdle over them, leaving only a small rounded or heart-shaped portion exposed at the apex of each; posterior borders of valves produced backward in rounded lobes at each side, the lobes completely separated by a posterior sinus having the tegmentum at its apex. Posterior valve having a posterior sinus and one slit on each side. Girdle more or less pilose, often having pore rows. The essential features of *Amicula* are its small exposed portion or tegmentum, situated at the posterior edge, and not extending forward to the sinus, its Mopaloid posterior valve, short contour, and short gills. (Tryon and Pilsbry, *Manual of Conchology.*)

TYPE. *Chiton vestitus* Sowerby.

RANGE. North Pacific and Atlantic Oceans. Fossil, Pleistocene Drift, Lower Canada.

## Amicula amiculata Pallus, 1786

*Nova Acta Academiae Scientiarum Imperialis Petropolitana*, **2**: 235; Pl. 7, figs. 26–30.

Maximus est omnium huius generis qui hucusque innotuerunt, quippe qui saepe in longitudinem sex pollicum angulicorum (Stellero obseruante) excrescit, mihique ipsi inter minores plures, quadripollicaris, licet siccus, e Curilis insulis adlatus est. (Pallus.)

Valves covered with cartilage, scabrous and subverrucose outside, the part surrounding the valves being thick, harsh, cartilaginous. The 8 valves are white and very fragile, the first being nearly horse-hoof shaped, crenulated on the front margin; the intermediate valves are shaped as if made of two circular disks, and have a transverse obsolete swelling above. The first seven valves have a pentagonal sharply margined piece (tegmentum), truncated behind, at the angle of the posterior sinus. The eighth valve is angular, as if formed of two pentagons, excavated behind. (Tryon and Pilsbry, *Manual of Conchology.*)

TYPE in Academy, St. Petersburg? Type locality, Kuril Islands.

RANGE. Kuril Islands, and Northern Japan to the Farallon Islands, California.

## Amicula vestita Sowerby, 1829

*Journal of Zoölogy,* **4**:368. *Conchological Illustrations,* Chiton; fig. 128. Tryon and Pilsbry, *Manual of Conchology,* **15**:43; Pl. 8, fig.

Ch. valvus reniformibus, membrana coriacea vestitus, apocibus nudis. Long., 1%10; lat., %10 poll.

The rather elevated points of the valves are alone visible in the living animal: but when the coriaceous membrane which covers them is dry, their edges are easily traced. Little bunches of brownish hairs are scattered over the surface. (Sowerby.)

Oval, rather elevated, the valves nearly covered by a brown (or when young) skin continued upward from the girdle, but their outlines are plainly visible through this integument. The small exposed portion of each median valve is broadly heart-shaped, and situated at the posterior margin of the valve; it is sculptured with strong concentric grooves and a more or less distinct granulation. There is no differentiation into areas. The exposed portion of the posterior valve is heart-shaped, with the mucro inconspicuous, near but slightly behind the middle. Interior pure white. Anterior valve having 6–8 irregularly spaced and unequal slits; posterior valve having a deep sinus behind, and a single small mopaloid slit on each side. Jugal sinus rather small; sutural laminae rather less projecting forward than the posterior rounded lobes on each side. Girdle thin, smooth; adults generally having more or less developed, but always sparsely scattered, small bunches of hairs. Length, 50; breadth, 35 mm. (Tryon and Pilsbry, *Manual of Conchology.*)

TYPE in Zoölogical Society, London. Type locality, Arctic.

RANGE. Arctic Ocean and Bering Sea south to Bering, Pribilof, and Hagemeister Islands. On the Atlantic south to Cape Cod.

## Amicula pallasii Middendorff, 1846

*Bulletin,* Imperial Academy of Sciences, St. Petersburg, **6**: 117. *Sibirische Reise,* 163; Pl. 13, figs. 1–9.

Testa subinterna, occulta, pallio obvoluta, mucronum fossula postica sola externe conspicua; elevata aut subelevata, 98 ad 110 in junioribus, 120 in adultis maximis; ovalis, elongata. Lat., 1; long. 2¾. Valvae albidae, leves, fragiles; tegmentum subnullum, cordiforme, posticum, in imo mucrone; ratio long., 1 ad clivi lat., 1¾–2; articulamentum suturis connatis evanidis; suturae laterales infra sulco obsoleto indicatae, supra sulco profundiori, utroque limite carinuka concomitato; valvae intermediae, transversim si medio dissecares, antica et postico parte congruentes forma; valva ultima anormis, articuli superficie plicatula, apophyses margine tamen integro; apophyses terminal, ⅝.

Limbus luxrians in pallium extenditur, totum animalis dorsum rotundatum obtegens, valvae obcolvens et occultans, solis octo aperturis minutis, rotundatis, in lines medians, quibus aditus ad umbonem valarum patet; color squalido lutescens; epidermis dorsalis undique versum fasciculis pilorum rubicundorum crinita; pili bini, terni xx cet., plerumque octoni et ultra; microscopio spectata epidermis dorsalis inter pilorum fasciculos pubes exhibet, stroma tamen spinulis latentibus nullis ornatum; epidermis ventralis spinulis erectis rarioribus munita, quarum singulae una ab altera spatio, spinulae fere longitudinem aequante, distant. Branchiarum seres postica branchiae parcae, circ. No. 29. Adulti maximi longitudo, 67 mm.; latit., 48; altit., 21. (Middendorff.)

Shell nearly concealed by the girdle, a somewhat heart-shaped tegmentum only being visible at the apex of each valve; elevated at an angle of 98°–110° in the young, 120° in the large adults; oval elongated. Valves white, smooth, fragile, the tegmentum cordiform, posterior. Slits in anterior valve, 6–8; posterior valve, 2. Girdle roundly covering the entire back of the animal, except for 8 small rounded holes along the median line; color dingy buff; dorsal surface bearing all over unequal bunches of reddish hairs, appearing to be sparser in the young. Branchiae extending forward two-thirds the length of the foot. Length, 67; breadth, 48; height, 21 mm. (Tryon and Pilsbry, *Manual of Conchology.*)

TYPE in Academy, St. Petersburg. Type locality, Okhotsk Sea.

RANGE. Arctic Ocean, Bering and Okhotsk Seas, and eastward to the Shumagin Islands, Alaska.

Described as *Chiton pallasii.*

# BIBLIOGRAPHY, VOLUME II, PART III

1767. LINNAEUS, CARL. *Systema Naturae* (Twelfth edition)
1773. PHIPPS, CONSTANTINE JOHN. *Voyage to the North Pole*
1776. MÜLLER, O. F. *Systema Naturae* (Twelfth edition)
1776. FABRICIUS, OTHO. *Fauna Grönlandica,* Hafniae et Lipsiae
1784. MARTYN, THOMAS. *Figures of Nondescript Shells; Universal Conchology,* London
1786. PALLAS, P. S. *Nova Acta Acad. Petropolitana,* Berlin
1788. PALLAS, P. S. *Nova Acta Acad. Petropolitana,* Berlin
1792. GMELIN, J. F. *Systema Naturae,* Leipzig
1800. DONOVAN, EDWARD. *British Shells,* Volume 1
1815. WOOD, WILLIAM. *General Conchology*
1817. FLEMING, ——. *Journal of the Washington Academy of Science*
1817. LEACH, WILLIAM E. *Zoölogical Miscellany*
1822. SAY, THOMAS. *Journal,* Philadelphia Academy of Natural Sciences, Volume 2
1822. SWAINSON, WILLIAM. *Bligh Catalogue,* Appendix
1822. SWAINSON, WILLIAM. *Zoölogical Illustrations*
1824. SOWERBY, JAMES. *Genera of Recent and Fossil Shells,* London
1825. SOWERBY, GEORGE B. *A Catalogue of the Shells in the Collection of the Earl of Tankerville, with an Appendix Describing New Species*
1825. SOWERBY, JAMES. *Genera of Recent and Fossil Shells,* London
1827. BROWN, THOMAS. *Illustrations of the Conchology of Great Britain and Ireland*
1828. GRAY, JOHN EDWARD. Wood, *Index Testaceologicus*
1828. WOOD, WILLIAM. *Index Testaceologicus,* Supplement, London
1829. BRODERIP, W. J., and SOWERBY, G. B. *The Malacological and Conchological Magazine,* Volume 1
1829. BRODERIP, W. J., and SOWERBY, G. B. *Zoölogical Journal,* Volume 4
1829. RANG, A. SANDER. *Annales des Sciences Naturelles,* Paris
1830. MENKE, C. T. *Synopsis Methodica Molluscorum,* Pyrmont
1831. KING, CAPT. PHILLIP P. *Zoölogical Journal,* Volume 5
1833. ESCHSCHOLTZ, J. F. *Zoölogical Atlas,* Berlin
1834. BRODERIP, W. J. *Proceedings,* Zoölogical Society of London
1834. GRAY, JOHN EDWARD. Sowerby, *Genera of Recent and Fossil Shells,* London
1835. SOWERBY, G. B. *Proceedings,* Zoölogical Society of London
1838. COUTHOUY, JOSEPH P. *Boston Journal of Natural History,* Volume 2
1838. SOWERBY, GEORGE B. *Magazine of Natural History,* New Series, Volume 2
1839. SOWERBY, GEORGE B. *Conchologia Iconica,* Chiton
1840. HALDEMAN, S. S. *Monograph of the Freshwater Shells of the United States,* No. 1
1841. DESHAYES, G. P. Guérin, *Magasin de Zoölogie, Mollusca*
1841. GOULD, AUGUSTUS A. *Report on the Invertebrata of Massachusetts*
1842. HINDS, RICHARD B. *Annals and Magazine of Natural History,* Volume 10
1842. MIGHELS, J. W. *Boston Journal of Natural History,* Volume 4

1842. MIGHELS, J. W., and ADAMS, C. B. *Boston Journal of Natural History*, Volume 4

1842. MÖLLER, H. P. C. *Index Molluscorum Groenlandiae*, Hafniae

1842. REEVE, LOVELL A. *Conchologia Systematica*, Volume 2

1842. REEVE, LOVELL A. *Proceedings*, Zoölogical Society of London

1843. HINDS, RICHARD B. *Proceedings*, Zoölogical Society of London

1843. HINDS, RICHARD B. *Zoölogy of the Voyage of H.M.S. "Sulphur," Molluscae*

1843. KOCH, BERGRATH. Philippi, R. A., *Abbildungen und Beschreibungen neue oder wenig gekannte Conchylien*

1844. BEAN, ——. Thorpe, *British Marine Conchology*

1845. JONAS, J. H. *Zeitschrift für Malakozoologie*

1845. PHILIPPI, RUDOLF AMANDUS. *Proceedings*, Zoölogical Society of London

1845. PHILIPPI, RUDOLF AMANDUS. *Zeitschrift für Malakozoologie*

1846. GOULD, AUGUSTUS A. *Proceedings of the Boston Society of Natural History*, Volume 2

1846. HANCOCK, ALBANY. *Annals and Magazine of Natural History*, Volume 18

1846. MIDDENDORFF, A. TH. VON. *Bulletin*, Imperial Academy of Sciences, St. Petersburg, Volume 6

1846. VALENCIENNES, A. *Voyage de la frégate "Venus,"* Atlas

1847. GOULD, AUGUSTUS A. *Proceedings of the Boston Society of Natural History*

1847. MIDDENDORFF, A. TH. VON. *Beitrage zu einer Malakozoologie Rossica*, Volume 1

1847. MIDDENDORFF, A. TH. VON. *Bulletin*, Imperial Academy of Sciences, St. Petersburg, Volume 4

1847. PHILIPPI, RUDOLF AMANDUS. *Abbildungen und Beschreibungen neue oder wenig gekannte Conchylien*, Littorina

1847. REEVE, LOVELL A. *Conchologia Iconica*

1847. SOWERBY, G. B. *Magazine of Natural History*

1848. GOULD, AUGUSTUS A. *Proceedings of the Boston Society of Natural History* Volume 3

1848. JONAS, J. H. *Conchylien Cabinet*

1848. PHILIPPI, RUDOLF AMANDUS. *Zeitschrift für Malakozoologie*, Cassel

1849. GOULD, AUGUSTUS A. *Proceedings of the Boston Society of Natural History*, Volume 3

1849. MIDDENDORFF, A. TH. VON. *Beitrage zu einer Malakozoologie Rossica*, Parts 2, 3

1849. PHILIPPI, RUDOLF AMANDUS. *Zeitschrift für Malakozoologie*, Volume 5

1849. REEVE, LOVELL A. *Conchologia Iconica*

1850. FORBES, EDWARD. *Proceedings*, Zoölogical Society of London

1850. KIENER, L. C. *Species Générale des Coquilles Vivantes*, Trochus

1850. MENKE, C. T. *Zeitschrift für Malakozoologie*

1850. REEVE, LOVELL A. *Conchologia Iconica*, London

1851. ADAMS, A. *Proceedings*, Zoölogical Society, London

1851. GOULD, AUGUSTUS A. *Mexican and California Shells*

1851. MIDDENDORFF, A. TH. VON. *Journal de physique*

1851. MIDDENDORFF, A. TH. VON. *Sibirische Reise*, Volume 2

1852. ADAMS, C. B. *Catalogue of Shells Collected at Panama*

1852. GOULD, AUGUSTUS A. *Boston Journal of Natural History*, Volume 6

1854. ADAMS, A. *Proceedings*, Zoölogical Society of London

1855. CARPENTER, P. P. *Proceedings,* Zoölogical Society of London
1855. REEVE, LOVELL A. *Conchologia Iconica*
1855. SOWERBY, G. B. *Thesaurus Conchyliorum,* Volume 2
1856. CARPENTER, P. P. *Proceedings,* Zoölogical Society of London
1857. CARPENTER, P. P. *Catalogue of the Reigen Collections of Mazatlan Mollusca,* British Museum, London
1857. PFEIFFER, LUDOVICO, M.D. *Proceedings,* Zoölogical Society of London
1859. GOULD, AUGUSTUS A. *Proceedings of the Boston Society of Natural History,* Volume 7
1859. NUTTALL, THOMAS. *Conchologia Iconica*
1859. REEVE, LOVELL A. *Conchologia Iconica,* London
1860. GOULD, A. A. *Proceedings of the Boston Society of Natural History*
1861. MÖRCH, OTTO ANDREAS L. *Proceedings,* Zoölogical Society of London
1864. ADAMS, A. *Annals and Magazine of Natural History,* Series 3, Volume 14
1864. CARPENTER, P. P. *Annals and Magazine of Natural History,* Series 3, Volume 13
1864. CARPENTER, P. P. *Journal de Conchyliologie,* Volume 13
1864. CARPENTER, P. P. *Supplementary Report on the Present State of Our Knowledge with Regard to the Mollusks of the West Coast of North America*
1864. CARPENTER, P. P. *Proceedings of the California Academy of Sciences,* Volume 3
1865. CARPENTER, P. P. *Annals and Magazine of Natural History,* Series 3, Volume 15
1865. CARPENTER, P. P. *Proceedings,* Philadelphia Academy of Natural Sciences
1865. CARPENTER, P. P. *Proceedings of the California Academy of Sciences,* Volume 3
1865. GABB, WILLIAM A. *Proceedings of the California Academy of Sciences,* Volume 3
1865. TRYON, G. W. *American Journal of Conchology,* Volume 1
1866. CARPENTER, P. P. *American Journal of Conchology,* Volume 2
1866. CARPENTER, P. P. *Proceedings of the California Academy of Sciences,* Volume 3
1866. TRYON, G. W. *American Journal of Conchology,* Volume 2
1869. DALL, W. H. *American Journal of Conchology,* Volume 5
1871. DALL, W. H. *American Journal of Conchology,* Volumes 6, 7
1872. DALL, W. H. *Proceedings of the California Academy of Sciences,* Volume 4
1873. DALL, W. H. *Proceedings of the California Academy of Sciences,* Volume 5
1876. HEMPHILL, HENRY. *Proceedings of the California Academy of Sciences,* Volume 7
1878. CARPENTER, P. P. *Proceedings of the United States National Museum,* Volume 1
1878. DALL, W. H. *Proceedings of the United States National Museum,* Volume 1
1884. DALL, W. H. *Proceedings of the United States National Museum,* Volume 7
1885. DALL, W. H. *Proceedings of the United States National Museum,* Volume 8
1885. WEINKAUFF, H. C. VON. *Conchylien Cabinet* (Second edition)
1886. DALL, W. H. *Proceedings of the United States National Museum,* Volume 9
1886. KRAUSE, ARTHUR. *Archiv für Naturgeschichte,* Wiegmann
1886. WHITEAVES, J. F. *Transactions of the Royal Society of Canada,* Volume 4
1887. KEEP, JOSIAH. *West Coast Shells*

1887.  TRYON, GEORGE W. Tryon, *Manual of Conchology,* Volume 9
1888.  PILSBRY, H. A. Tryon and Pilsbry, *Manual of Conchology,* Volume 10
1889.  DALL, W. H. *Bulletin,* Museum of Comparative Zoölogy, Cambridge
1889.  DALL, W. H. *Proceedings of the United States National Museum,* Volume 12
1891.  DALL, W. H. *Proceedings of the United States National Museum,* Volume 14
1892.  CARPENTER, P. P. Tryon and Pilsbry, *Manual of Conchology,* Volume 14
1892.  CARPENTER, P. P. Tryon and Pilsbry, *Manual of Conchology,* Volume 12
1892.  DALL, W. H. *Proceedings of the United States National Museum,* Volume 15
1893.  PILSBRY, H. A. Tryon and Pilsbry, *Manual of Conchology,* Volume 11
1893.  PILSBRY, H. A. Tryon and Pilsbry, *Manual of Conchology,* Volume 15
1893.  STEARNS, R. E. C. *Proceedings of the United States National Museum,* Volume 16
1894.  DALL, W. H. *The Nautilus,* Volume 8
1894.  HEMPHILL, HENRY. *Zoë,* Volume 4
1894.  RAYMOND, WILLIAM J. *The Nautilus,* Volume 8
1895.  DALL, W. H. *Proceedings of the United States National Museum,* Volume 24
1895.  ROCHEBRUNE, A. T. *Bulletin,* Musée d'Histoire Naturelle, Paris
1896.  DALL, W. H. *Proceedings of the United States National Museum,* Volume 18
1896.  PILSBRY, H. A. *The Nautilus,* Volume 10
1897.  DALL, W. H. *Bulletin,* Natural History Society of British Columbia, No. 2
1897.  DALL, W. H. *The Nautilus,* Volume 11
1898.  PILSBRY, H. A. *Proceedings,* Philadelphia Academy of Natural Sciences
1899.  SMITH, E. A. *Proceedings of the Malacological Society of London,* Volume 3
1899.  STEARNS, R. E. C. *Proceedings of the United States National Museum,* Volume 22
1900.  DALL, W. H. *The Nautilus,* Volume 13
1900.  DALL, W. H. *The Nautilus,* Volumes 13, 14
1900.  ORCUTT, CHARLES. *West American Scientist,* Volume 10
1901.  DALL, W. H. *The Nautilus,* Volume 14
1901.  DALL, W. H., and BARTSCH, PAUL. *The Nautilus,* Volume 15
1902.  DALL, W. H. *Proceedings of the United States National Museum,* Volume 24
1902.  DALL, W. H., and BARTSCH, PAUL. *The Nautilus,* Volume 16
1903.  DALL, W. H. *Proceedings of the Biological Society of Washington*
1905.  DALL, W. H. *The Nautilus,* Volume 18
1905.  DALL, W. H. *The Nautilus,* Volume 19
1905.  DALL, W. H. *Proceedings of the United States National Museum,* Volume 34
1905.  WILLIAMSON, BURTON M. *The Nautilus,* Volume 19
1907.  BARTSCH, PAUL. *Proceedings of the United States National Museum,* Volume 32
1907.  BARTSCH, PAUL. *Proceedings of the United States National Museum,* Volume 33
1907.  BARTSCH, PAUL. *Miscellaneous Collections,* Smithsonian Institution
1907.  HEMPHILL, HENRY. *Transactions of the San Diego Society of Natural History,* Volume 1
1908.  DALL, W. H. *Bulletin,* Museum of Comparative Zoölogy, Cambridge, Volume 43
1908.  DALL, W. H. *Proceedings of the United States National Museum,* Volume 34
1909.  DALL, W. H. *Professional Paper 59,* United States Geological Survey
1909.  STEARNS, R. C. S. *Professional Paper 59,* United States Geological Survey

1910. BARTSCH, PAUL. *The Nautilus,* Volume 23
1910. BARTSCH, PAUL. *Proceedings of the United States National Museum,* Volume 39
1910. DALL, W. H. *The Nautilus,* Volume 24
1910. DALL, W. H., and BARTSCH, PAUL. Canada, Department of Mines, Geological Survey Branch, *Memoirs No. 14-N*
1910. THIELE, JOH. *Zoologica,* Volume 22, Heft 56
1911. BARTSCH, PAUL. *Proceedings of the United States National Museum,* Volume 39
1911. BARTSCH, PAUL. *Proceedings of the United States National Museum,* Volume 40
1911. BARTSCH, PAUL. *Proceedings of the United States National Museum,* Volume 41
1911. BERRY, S. S. *Proceedings,* Philadelphia Academy of Natural Sciences
1911. DALL, W. H. *The Nautilus,* Volume 25
1912. BARTSCH, PAUL. *Proceedings of the United States National Museum,* Volume 41
1913. DALL, W. H. *Proceedings of the United States National Museum,* Volume 45
1914. DALL, W. H. *The Nautilus,* Volume 28
1915. BARTSCH, PAUL. *Proceedings of the United States National Museum,* Volume 49
1916. DALL, W. H. *The Nautilus,* Volume 30
1916. DALL, W. H. *Proceedings,* Philadelphia Academy of Natural Sciences
1917. BARTSCH, PAUL. *Proceedings of the United States National Museum,* Volume 52
1917. BERRY, S. S. *Proceedings of the California Academy of Sciences,* Series 4, Volume 7
1917. OLDROYD, I. S. *The Nautilus,* Volume 31
1918. DALL, W. H. *Proceedings of the Biological Society of Washington,* Volume 31
1918. PILSBRY, H. A. Tryon and Pilsbry, *Manual of Conchology,* Volume 14
1918. PILSBRY, H. A. *The Nautilus,* Volume 31
1919. BARTSCH, PAUL. *The Nautilus,* Volume 23
1919. BERRY, S. S. *Lorquinia,* January 1919
1919. BERRY, S. S. *Proceedings,* Philadelphia Academy of Natural Sciences, Series 4, Volume 9
1919. DALL, W. H. *Proceedings of the Biological Society of Washington,* Volume 31
1919. DALL, W. H. *Proceedings of the Biological Society of Washington,* Volume 32
1919. DALL, W. H. *Proceedings of the United States National Museum,* Volume 54
1919. DALL, W. H. *Proceedings of the United States National Museum,* Volume 55
1919. DALL, W. H. *Proceedings of the United States National Museum,* Volume 56
1919. DALL, W. H. *Scientific Results of the Canadian Arctic Expedition,* Volume 8
1920. BARTSCH, PAUL. *Journal of the Washington Academy of Science,* Volume 10
1920. BARTSCH, PAUL. *Proceedings of the United States National Museum,* Volume 58
1920. DALL, W. H. Carpenter, *Proceedings,* Philadelphia Academy of Natural Sciences
1920. DALL, W. H. *The Nautilus,* Volume 34
1921. BARTSCH, PAUL. *Proceedings of the Biological Society of Washington,* Volume 34
1924. OLDROYD, I. S. *Marine Shells of Puget Sound and Vicinity*
1927. BARTSCH, PAUL. *Proceedings of the United States National Museum,* Volume 70

# INDEX TO VOLUME II, PART III

PLATES AND EXPLANATION
OF PLATES

## PLATE 73

PLATE 73

1

2

3

4

5

# PLATE 74

PLATE 74

1

2

3

4

5

6

# PLATE 75

This plate from *Proceedings of the United States National Museum.*

PLATE 75

1

2

3

4

5

6

## PLATE 76

PLATE 76

1

2

3

4

5

6

## Plate 77

This plate from *Proceedings of the United States National Museum.*

| FIGURE | PAGE |
| --- | --- |
| 1. *Bittium nicholsi* Bartsch................................ | Not in text |
| Type. Lower California. | |
| 2. *Bittium nitens* Bartsch.................................... | Not in text |
| Type. | |
| 3. *Bittium eschrichtii icelum* Bartsch.......................... | 19 |
| 4. *Bittium larum* Bartsch...................................... | 34 |
| Type. | |
| 5. *Bittium subplanatum* Bartsch................................ | 23 |
| Type. | |

PLATE 77

1

2

3

4

5

## PLATE 78

This plate from *Proceedings of the United States National Museum.*

PLATE 78

1

2

3

4

5

6

## Plate 79

This plate from *Proceedings of the United States National Museum.*

| FIGURE | PAGE |
|---|---|
| 1. *Bittium mexicanum* Bartsch. . . . . . . . . . . . . . . . . . . . . . . . . . . . | Not in text |
| Type. | |
| 2. *Bittium quadrifilatum* Carpenter. . . . . . . . . . . . . . . . . . . . . . . . . . . . | 24 |
| Type. | |
| 3. *Bittium quadrifilatum* Carpenter. . . . . . . . . . . . . . . . . . . . . . . . . . . . | 24 |
| 4. *Bittium eschrichtii* Middendorff. . . . . . . . . . . . . . . . . . . . . . . . . . . . | 18 |
| 5. *Bittium eschrichtii montereyense* Bartsch. . . . . . . . . . . . . . . . . . . . . | 19 |
| Type. | |

PLATE 79

1

2

3

4

5

## PLATE 80

This plate from *Proceedings of the United States National Museum.*

PLATE 80

## PLATE 81

This plate from *Proceedings of the United States National Museum.*

| FIGURE | | PAGE |
|---|---|---|
| 1. | *Alvania almo* Bartsch<br>Type. | 100 |
| 2. | *Alvania tumida* Carpenter | Not in text |
| 3. | *Alvania oldroydae* Bartsch<br>Type. | 98 |
| 4. | *Alvania clarionensis* Bartsch<br>Type. | Not in text |
| 5. | *Alvania effusa* Carpenter<br>Type. | Not in text |
| 6. | *Alvania lara* Bartsch<br>Type. | Not in text |
| 7. | *Alvania aequisculpta* Keep<br>Type. | 97 |
| 8. | *Alvania ima* Bartsch<br>Type. | Not in text |

PLATE 81

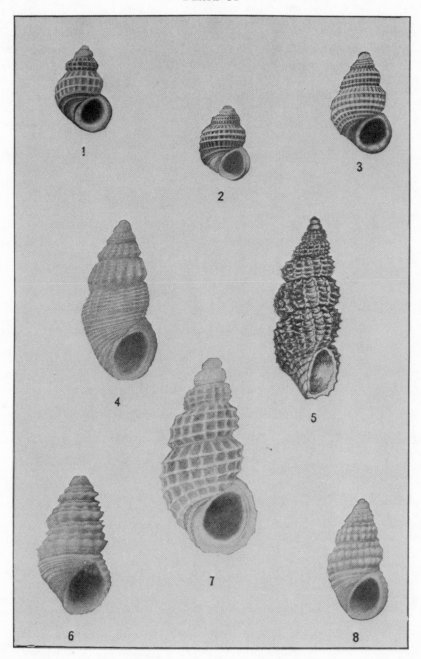

# PLATE 82

This plate from *Proceedings of the United States National Museum.*

PLATE 82

## Plate 83

This plate from *Proceedings of the United States National Museum.*

| FIGURE | | PAGE |
|---|---|---|
| 1. | *Rissoina californica* Bartsch<br>Type. | 106 |
| 2. | *Rissoina dalli* Bartsch<br>Type. | 107 |
| 3. | *Rissoina cleo* Bartsch<br>Type. | 105 |
| 4. | *Rissoina bakeri* Bartsch<br>Type. | 105 |
| 5. | *Rissoina cerrocensis* Bartsch<br>Type. | Not in text |
| 6. | *Rissoina mexicana* Bartsch<br>Type. | Not in text |
| 7. | *Rissoina newcombei* Dall<br>Type. | 108 |
| 8. | *Rissoina coronadoënsis* Bartsch<br>Type. | 107 |
| 9. | *Rissoina histia* Bartsch<br>Type. | Not in text |

PLATE 83

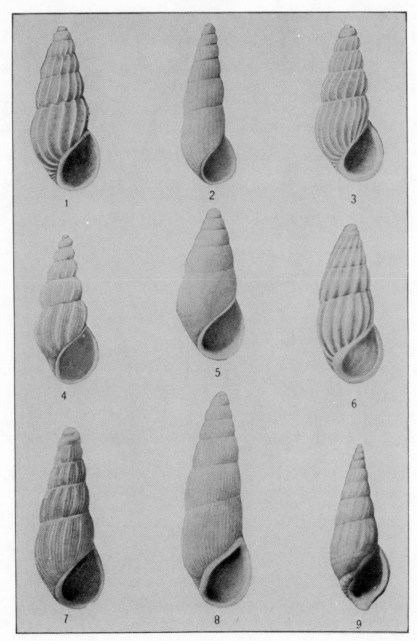

## Plate 84

This plate from *Proceedings of the United States National Museum.*

All figures magnified

| FIGURE | | PAGE |
|---|---|---|
| 1. | *Cingula montereyensis* Bartsch<br>Type. | 81 |
| 2. | *Cingula aleutica* Dall<br>Type. | 80 |
| 3. | *Cingula katherinae* Bartsch<br>Type. | 80 |
| 4. | *Cingula alaskana* Bartsch<br>Type. | 79 |
| 5. | *Cingula martyni* Dall<br>Type. | 79 |
| 6. | *Cingula martyni scipio* Dall<br>Type. | 79 |

PLATE 84

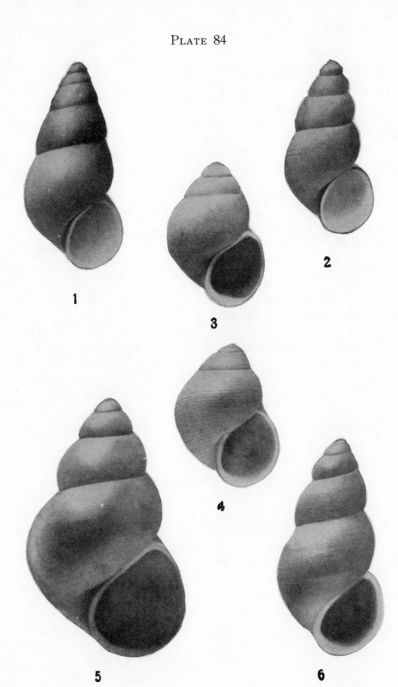

# PLATE 85

PLATE 85

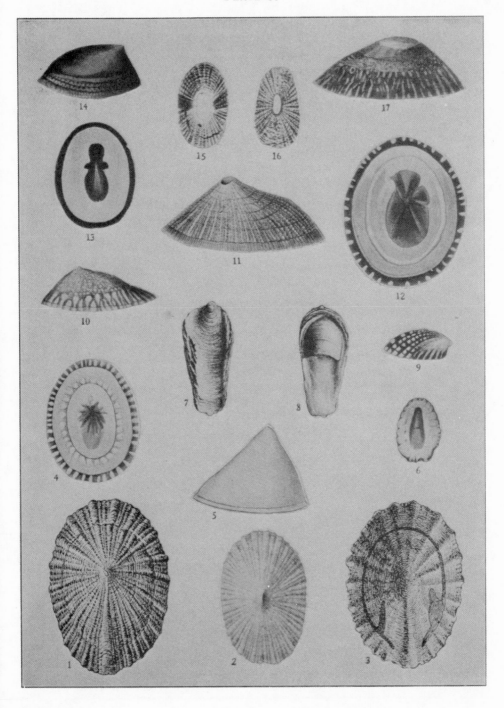

# PLATE 86

PLATE 86

1

2

# PLATE 87

PLATE 87

# PLATE 88

PLATE 88

# PLATE 89

PLATE 89

## PLATE 90

PLATE 90

## PLATE 91

PLATE 91

# PLATE 92

PLATE 92

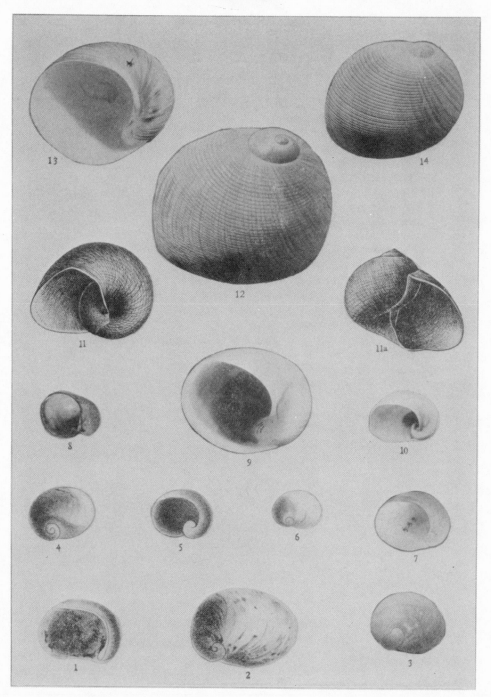

# PLATE 93

PLATE 93

## Plate 94

The figures from *Zoölogical Atlas* are refigured type figures. American Museum of Natural History, New York City.

| FIGURE | | PAGE |
|---|---|---|
| 1, 2. | *Acmaea instabilis* Gould. Tryon, *Manual of Conchology.* | 156 |
| 3, 4. | *Acmaea scutum patina* Eschscholtz. *Zoölogical Atlas.* | 148 |
| 5, 6. | *Acmaea cassis* Eschscholtz. *Zoölogical Atlas.* | 145 |
| 7, 8, 9. | *Acmaea nacelloides* Dall. *American Journal of Conchology.* | 147 |
| 10, 11. | *Acmaea digitalis* Eschscholtz. *Zoölogical Atlas.* | 151 |
| 12, 13. | *Acmaea cassis olympica* Dall. | 147 |
| 14. | *Acmaea scutum pintadina* Gould. Tryon, *Manual of Conchology.* | 149 |
| 15, 16. | *Acmaea asmi* Middendorff. Tryon, *Manual of Conchology.* | 157 |

PLATE 94

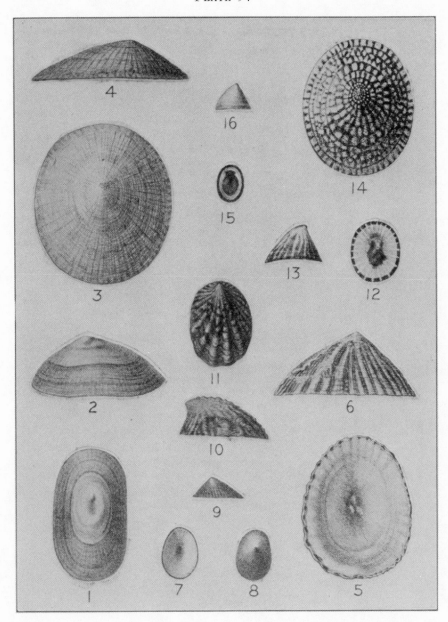

## Plate 95

This plate from *Proceedings of the United States National Museum.*

| FIGURE | PAGE |
|---|---|
| 1. *Alabina diomedeae* Bartsch | Not in text |
| 2. *Alabina occidentalis* Bartsch | 16 |
| 3. *Alabina ignati* Bartsch | Not in text |
| 4. *Alabina turrita* Carpenter | 16 |
| 5. *Alabina monicensis* Bartsch | Not in text |
| 6. *Alabina phanea* Bartsch | 13 |
| 7. *Alabina californica* Dall and Bartsch<br>Fossil. | 12 |

PLATE 95

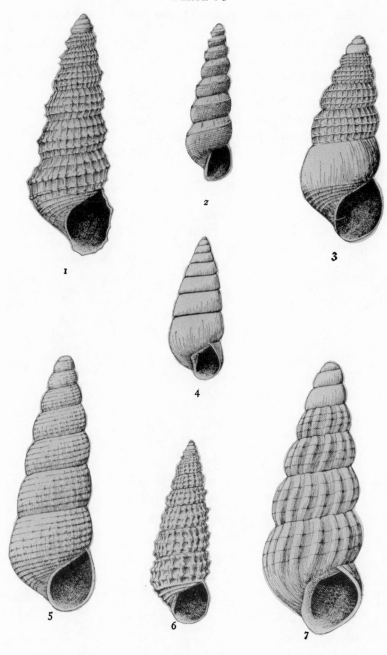

## Plate 96

This plate from *Proceedings of the United States National Museum.*

| FIGURE | | PAGE |
|---|---|---|
| 1. | *Alabina io* Bartsch<br>Fossil. | Not in text |
| 2. | *Alabina hamlini* Bartsch<br>Fossil. | Not in text |
| 3. | *Alabina barbarensis* Bartsch | 13 |
| 4. | *Alabina diegensis* Bartsch | 15 |
| 5. | *Alabina tenuisculpta phalacra* Bartsch | 15 |
| 6. | *Alabina tenuisculpta* Carpenter | 14 |

PLATE 96

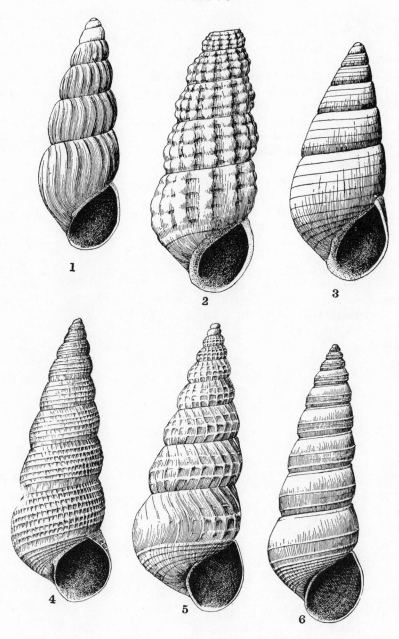

1

2

3

4

5

6

# Plate 97

PLATE 97

## Plate 98

This plate from *Proceedings of the United States National Museum.*

| FIGURE | | PAGE |
|---|---|---|
| 1. | *Calliostoma splendens* Carpenter | 182 |
| 2. | *Tegula brunneum* Philippi | 176 |
| 3. | *Tegula gallina multifilosa* Stearns | 175 |
| 4. | *Simnia barbarensis* Dall | For description, see Part II |
| 5. | *Tegula regina* Stearns | 178 |
| 6. | *Tegula gallina multifilosa* Stearns | 175 |
| 7. | *Tegula regina* Stearns | 178 |
| 8, 9. | *Tegula funebrale* A. Adams | 173 |

PLATE 98

## Plate 99

This plate from *Proceedings of the United States National Museum.*

| FIGURE | | PAGE |
|---|---|---|
| 1, 2. | *Margarites marginatus* Dall | 213 |
| 3. | *Polinices draconis* Dall | 128 |
| 4, 5. | *Margarites albolineatus* Smith | 212 |
| 6. | *Polinices draconis* Dall | 128 |
| 7, 8. | *Margarites beringensis* Smith | 211 |

PLATE 99

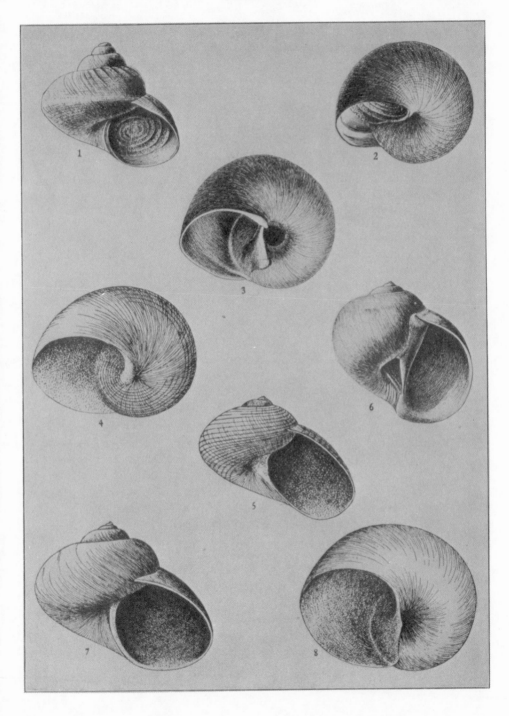

## PLATE 100

This plate from *Proceedings of the United States National Museum.*

PLATE 100

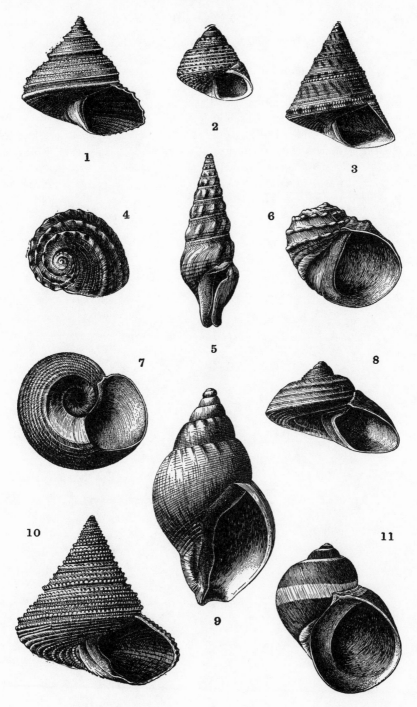

## Plate 101

This plate from *Bulletin 112,* United States National Museum.

| FIGURE | | PAGE |
|---|---|---|
| 1. | *Margarites lirulata* Carpenter | 207 |
| 2. | *Margarites pupilla* Gould | 200 |
| 3. | *Margarites pupilla* Gould | 200 |
| 4. | *Margarites parcipicta* Carpenter | 199 |
| 5. | *Margarites rhodia* Dall | 204 |
| 6. | *Margarites optabilis* Carpenter | Not in text |
| 7. | *Solariella peramabilis* Carpenter | 195 |
| 8. | *Margarites sordida* Hancock | Not in text |
| 9. | *Margarites sordida* Hancock | Not in text |

PLATE 101

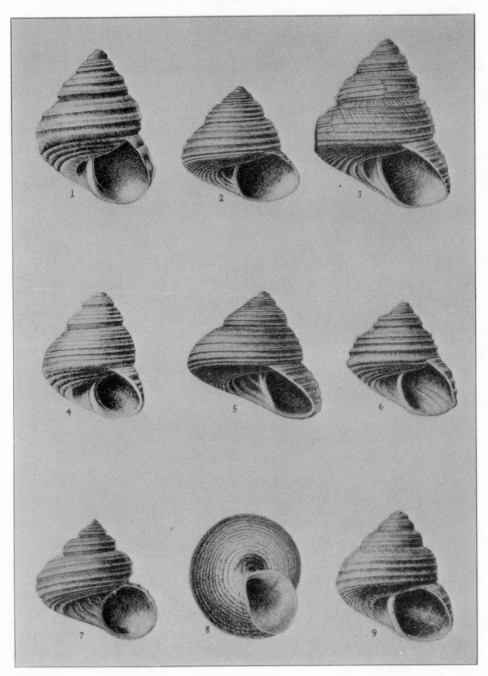

# PLATE 102

PLATE 102

# PLATE 103

PLATE 103

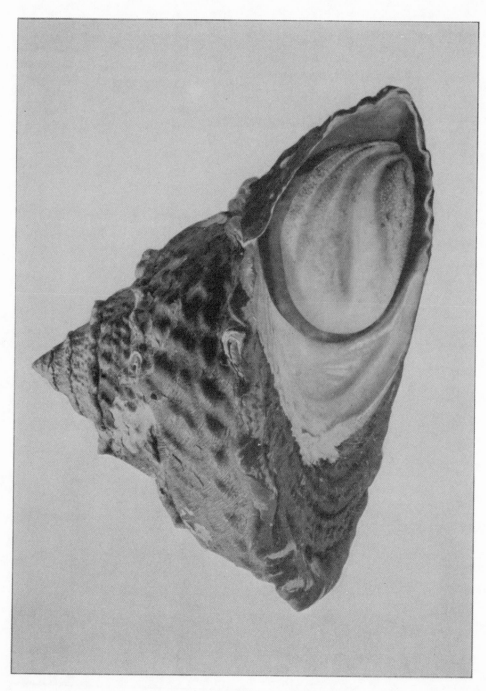

## PLATE 104

This plate from *Proceedings of the United States National Museum.*

| FIGURE | | PAGE |
|---|---|---|
| 1, 2, 3. | *Circulus diomedeae* Bartsch | Not in text |
| 4, 5, 6. | *Leptogyra alaskana* Bartsch | 226 |
| 7, 8, 9. | *Circulus liriope* Bartsch | Not in text |
| 10, 11, 12. | *Scissilabra dalli* Bartsch | 225 |

PLATE 104

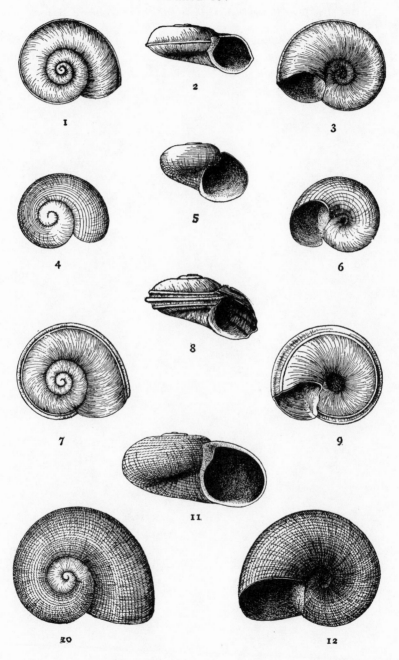

## Plate 105

PLATE 105

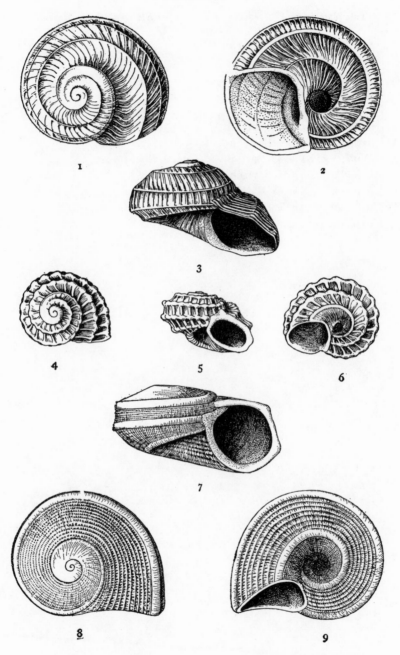

## Plate 106

This plate from *Proceedings of the United States National Museum.*

| FIGURE | | PAGE |
|---|---|---|
| 1, 2, 3. | *Cyclostremella californica* Bartsch | 223 |
| 4. | *Vitrinella williamsoni* Dall | 216 |
| 5. | *Vitrinella eshnauri* Bartsch | 217 |
| 6. | *Vitrinella williamsoni* Dall | 216 |
| 7. | *Vitrinella eshnauri* Bartsch | 217 |
| 8. | *Vitrinella alaskensis* Bartsch | 217 |
| 9. | *Vitrinella eshnauri* Bartsch | 217 |
| 10. | *Vitrinella alaskensis* Bartsch | 217 |
| 11. | *Vitrinella stearnsii* Bartsch | 218 |
| 12. | *Vitrinella alaskensis* Bartsch | 217 |
| 13. | *Vitrinella stearnsii* Bartsch | 218 |
| 14. | *Vitrinella stearnsii* Bartsch | 218 |

PLATE 106

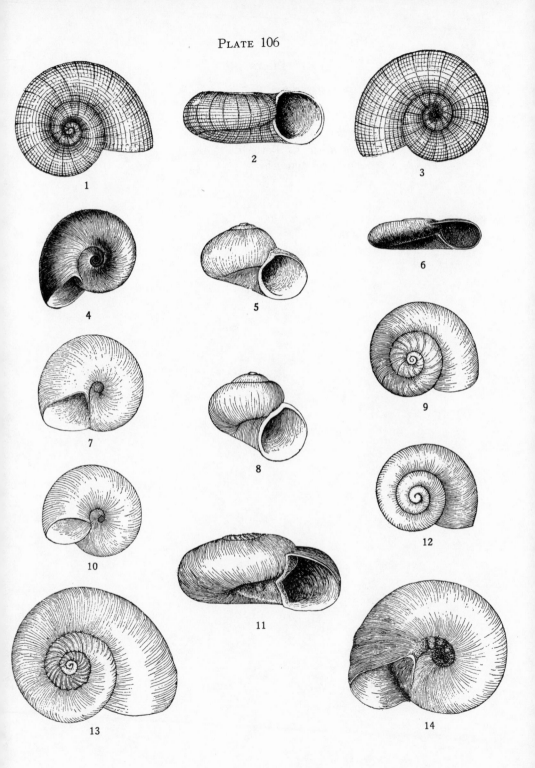

1

2

3

4

5

6

7

8

9

10

11

12

13

14

## PLATE 107

PLATE 107

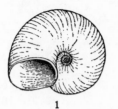

1

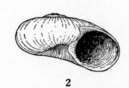

2

3

4

5

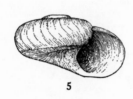

6

7

8

9

10

11

12

# PLATE 108

PLATE 108